Sensory Transduction

Sensory Transduction

Second edition

Gordon L. Fain

Distinguished Professor Emeritus of the Departments of Ophthalmology and of Integrative Biology and Physiology, University of California Los Angeles, USA

OXFORD

UNIVERSITY PRESS

OXFORD
UNIVERSITY PRESS

Great Clarendon Street, Oxford, OX2 6DP,
United Kingdom

Oxford University Press is a department of the University of Oxford.
It furthers the University's objective of excellence in research, scholarship,
and education by publishing worldwide. Oxford is a registered trade mark of
Oxford University Press in the UK and in certain other countries

First Edition published in 2003
Second Edition published in 2020

Impression: 1

Published in the United States of America by Oxford University Press
198 Madison Avenue, New York, NY 10016, United States of America

British Library Cataloguing in Publication Data

Data available

Library of Congress Control Number: 2019941498

ISBN 978–0–19–883502–8 (hbk.)
ISBN 978–0–19–883503–5 (pbk.)

DOI: 10.1093/oso/9780198835028.001.0001

Printed and bound by
CPI Group (UK) Ltd, Croydon, CR0 4YY

To Tim and Nick, who are better.

παῦροι γάρ τοι παῖδες ὁμοῖοι πατρὶ πέλονται,
οἱ πλέονες κακίους, παῦροι δέ τε πατρὸς ἀρείους.
Odyssey. II: 276–277

Acknowledgments (First edition)

This book could not have been written without the help of many friends and colleagues. I am particularly grateful to the Rockefeller Foundation for inviting me to spend 4 weeks at their Study and Conference Center in Bellagio, Italy, where I wrote the first four chapters of the book, and to Shelly Segal and Giana Celli, who made my visit there so enjoyable and productive. Much of the rest of the book was written during two-quarters of leave from teaching granted to me by my department and its then chair, Alan Grinnell. I also express my thanks to the staff of the library at UCLA, who helped me to locate obscure references in books and journals occasionally through interlibrary loan, and to the staff of the Balfour Zoology Library and the library of the Physiological Laboratory in Cambridge, England where I wrote most of Chapter 10. Alan Grinnell read the entire manuscript for me, and individual chapters were read by Michael Bennett, David Corey, Peter Gillespie, Roger Hardie, Sue Kinnamon, John Lisman, Hugh Matthews, John Wood, and Frank Zufall. I am deeply appreciative of all the assistance and encouragement I received from Andy Sinauer and his colleagues, and in particular from Kathleen Emerson, whose many suggestions and corrections greatly improved the book. Finally, my greatest debt is to my wife Margery, who did all the illustrations for the book, and whose patient support has made possible not only this book but almost everything else I have done during our long and happy collaboration.

Acknowledgments (Second edition)

I am extremely grateful to Ellen Lumpkin, David Corey, Jonathan Ashmore, Johannes Reisert, Emily Liman, Roger Hardie, and Michael Do, who took the time to read parts of the text and provided many helpful comments and corrections. My thanks are also due to Julie Musk who edited the text for OUP and made many helpful suggestions, and to the Helen Riaboff Whiteley Center at the Friday Harbor Laboratory, where the final work on the text and figures was done. My greatest debt is again to my wife Margery, who did all of the illustrations and was my ever-present help and counselor. This book could not have been written without her.

Brief contents

1 **The senses**   **1**

2 **Mechanisms of sensation**   **18**

3 **Channels and electrical signals**   **37**

4 **Metabotropic signal transduction**   **57**

5 **Mechanoreceptors and touch**   **76**

6 **Hearing and hair cells**   **99**

7 **Chemoreception and the sense of smell**   **132**

8 **Taste**   **159**

9 **Photoreception**   **178**

10 **Extra sensory receptors**   **217**

Contents

1 The senses   **1**

Early studies of the anatomy of the sense organs   2
The physiology of sensation   2
Cracking the problem: molecular physiology   7
The revolution of molecular biology   9
Piezo proteins: channels mediating touch   13
The code deciphered: sensory transduction   15

2 Mechanisms of sensation   **18**

Sensory membrane   19
Organization of membrane and sensory protein   21
Membrane renewal   22
External specializations   26
Detection of the stimulus   29
Primary and secondary receptor cells   31
Sensitivity of transduction   32
Noise   34
Sex pheromone detection in the male moth   34
Summary   36

3 Channels and electrical signals   **37**

Structure and function of ion channels   37
Structure of the pore   38
Gating   40
Ionotropic receptor molecules   42
Membrane potentials   44
The Nernst equation   45
Ion homeostasis   47
The Goldman voltage equation   48
Driving force and voltage change   48
The voltage response of hair cells   49
The technique of voltage clamping   51
Voltage clamp of the hair cell   52
Voltage-clamped response of the hair cell   54
Ion selectivity   55
Summary   56

4   Metabotropic signal transduction   **57**

G-protein-coupled receptors   60
Heterotrimeric G proteins   61
Effector molecules   62
Second messengers   64
Measuring cell calcium   67
Channels gated by second messengers   68
Mechanism of gating by cyclic nucleotides   70
A metabotropic sensory receptor   73
Summary   73

5   Mechanoreceptors and touch   **76**

Mechanoreception in *Paramecium*   76
Transduction of touch in the round worm *C. elegans*   78
Crayfish stretch receptor   81
Insect mechanoreceptors   85
Mechanoreceptors and touch in mammals   91
Hairy skin   93
Merkel cells   94
Summary   97

6   Hearing and hair cells   **99**

Insect hearing: tympanal organs   99
Insect hearing: Johnston's organ   101
Hair cells   104
Tip links   106
Hair cell transduction proteins   109
The channels   109
Gating and bundle stiffness   112
Adaptation of hair cells   114
Organs of the lateral line   116
The vestibular system   116
The cochlea   121
Endolymph and endocochlear potential   122
Outer hair cells and tuning   124
Electrical resonance   127
Summary   130

7   Chemoreception and the sense of smell   **132**

Chemotaxis   132
Olfaction in insects   135
Insect receptor proteins   138
Coding of olfaction in insects   140
Olfaction in vertebrates: the primary olfactory epithelium   142
Olfactory receptor proteins   143
The mechanism of transduction   144
Desensitization and adaptation   148

Coding in the principal olfactory epithelium   149
The olfactory bulb   151
The accessory olfactory system and vomeronasal organ   154
Summary   156

8   Taste   **159**

Gustation in insects   160
Mammals: taste buds and the tongue   164
Taste transduction: metabotropic receptors   164
    Bitter   165
    Sweet   166
    Umami   166
    Transduction cascade   166
Taste transduction: ionotropic detection   168
    Salty   168
    Sour   172
The coding of taste   174
Summary   175

9   Photoreception   **178**

Photopigment activation   179
Phototransduction   181
The photoreceptors of arthropods   183
Transduction in arthropods   185
Photoreceptor channels in arthropods   189
The role of Ca^{2+} in the regulation of gain and turnoff   192
Vertebrate rods and cones   195
Transduction in vertebrate photoreceptors   197
Ion channels of rods and cones   198
The photocurrent   200
Shutting down the light response   204
Light adaptation   209
Pigment renewal and the recovery of sensitivity after bright light   211
Intrinsically photosensitive retinal ganglion cells   213
Summary   215

10 Extra sensory receptors   **217**

Thermoreception   217
Seeing in the dark: heat receptors as detectors of infrared   221
Electroreception   224
Tuberous receptors and electrolocation   230
Magnetoreception   233
Magnetoreception in migrating birds   235
Summary   238

References   241
Index   273

CHAPTER 1

The senses

Everything we know about the world comes to us through our senses. We experience the world as we do because our organs of sight, hearing, and smell are constructed in a certain way. We could not see color unless we had more than one kind of visual pigment, perceive pitch unless the peak of the traveling wave of the basilar membrane varied with position in the cochlea, or smell different odors unless the nose contained a very large number of olfactory receptor molecules of different selectivity. No biologist would say as Plato did that "the eyes and the ears and the other senses are full of deceit" (*Phaedo*, 83A), that perception is an unreliable pathway to true knowledge (*Theaetetus*, 186C–187A). We could not as a species have survived the hurly-burly of natural selection unless our senses had been and still are fundamentally faithful reporters of the world around us.

Because of the importance of our sense organs in everyday life and the enormous pleasure we derive especially from sight and sound, humans have always been curious how sensation occurs. The ancient Greeks speculated extensively about the nature of the sense organs and were occasionally quite perceptive. Aristotle recognized the five primary senses of sight, hearing, touch, smell, and taste (*de Anima*, Book 3). Plato (*Timaeus* 45B–D) wrongly supposed that the eye emits a kind of fire akin to daylight, which meets a similar fire coming from objects in the world around us. As these fires met, their motion was thought to be communicated to the soul. Aristotle argued against this notion, though he himself gave no clear idea how he thought

vision did occur (Johansen, 1997). On the other hand, he recognized the fundamental importance of moisture in olfaction (see for example *de Sensu*, V; *de Anima*, VII). Moisture must be important, Aristotle reasoned, since fish can smell. How did he know? He doesn't say, and we have to suppose that he or his students had seen fish swimming toward bait. But since the sensation comes to the fish through water rather than through the air, why didn't Aristotle say that fish *taste*? What Aristotle could not have known is that fish, in addition to taste receptors in their oral cavity, have an olfactory organ that has a structure and function very much like our nose. The importance of moisture in olfaction is now absolutely clear: even in terrestrial animals, molecules must pass through the watery mucus of the nose before they can bind to and be detected by olfactory receptor cells.

Some of the most remarkable statements about sensation made by Greek and Roman authors are to be found in the first-century BC *De Rerum Natura* of Lucretius, who based much of his poem on the teachings of the Greek Hellenistic philosopher Epicurus. Lucretius claimed that the distinctiveness of different tastes and odors lies in the shapes of the tiny "seeds" or particles given off to the air or into the mouth by objects tasted or smelled. He thought sweet-tasting substances had smooth round particles, and bitter substances had hooked or barbed particles. He also thought that for both taste and smell the shape of the particles must somehow correspond to apertures within the nose or palate, so that sweet tastes are perceived when smooth particles enter correspondingly

Sensory Transduction. Second Edition. Gordon L. Fain, Oxford University Press (2020). © Gordon L. Fain 2020.
DOI: 10.1093/oso/9780198835028.001.0001

smooth apertures. To account for the variety of taste and odor, he postulated a variety of apertures, some large, some small, some round, and others square or with many angles. This explanation is not too different from our present understanding that scents and many tastes are produced by molecules having different shapes and binding to receptor molecules with appropriately matched binding sites.

Early studies of the anatomy of the sense organs

Although Aristotle and other Greek men of learning certainly performed dissection on animals (Lloyd, 1975), the first systematic anatomical investigations of the human body were undertaken in Alexandria under the reign of the Ptolemies, during the first half of the third century BC (Longrigg, 1988). Herophilus of Chalcedon and Erasistratus of Ceos, taking advantage of a temporary relaxation of religious scruple, first began the dissection of human bodies, and it is to these men that we owe the discovery and first description of the sensory and motor nerves (Solmsen, 1961; Staden, 1989). Much of their work has been preserved—not in their own writings but in the books of Galen written four centuries later. Galen himself also carried out animal dissection (Duckworth et al., 2010), though perhaps not human dissection. Since he lacked even a magnifying glass, his descriptions of the structure of sense organs are rather crude. He understood that hearing is caused by air striking against the ear but seemed not to have noticed the tiny bones of the middle ear and missed altogether the role of the ear drum in transmitting vibrations into the cochlea. He named the principal parts of the eye, probably using terminology borrowed from Herophilus and Erasistratus, and these are the names we still use: sclera, choroid, crystalline lens. As a medical doctor, he knew that if the lens is not perfectly clear and transparent, vision is largely obstructed. He therefore supposed that the lens was the organ of photodetection, containing a "visual spirit" or πνεῦμα that passes down the optic nerve into the brain. The optic nerve as a consequence was supposed to be completely hollow, the only hollow nerve in the body.

Galen's books were probably filled with diagrams, though none has survived. The earliest schema of a sense organ we have is not from Galen himself but rather from a ninth-century AD translation of Galen into the Arab language Syriac. This translation was made by Humain Ibn Is-Hâq, who was born in Mesopotamia, studied medicine, and became an associate of the court physician of the caliph of Baghdad. The drawing in Figure 1.1A is from an English translation of Humain's manuscript (Meyerhoff, 1928). This schema of the eye had an enormous influence, not only on Arab medicine and science but also on the anatomists of the Renaissance, who continued to show the lens in the center of the eye. With a little effort, they should have been able to do the dissection more carefully, preserving the position of the lens in its proper place toward the front. What changed everything was the discovery of the laws of optical refraction and Kepler's solution of the optics of the eye. Kepler explained how images are formed and assigned the primary role in visual detection to the retina instead of to the lens (see Wade, 1998). Once the function of the lens was understood, it became possible for anatomists to do a proper dissection and find the various parts of the eye in their proper places. This is an example of *Lisman's Law*: you have to believe it in order to see it. The cross-section of the eye in Figure 1.1B was made by Descartes (1637/1987), who not only put the lens closer to its actual position but also identified the ciliary muscle and understood its role in changing the shape of the lens during accommodation.

The development of the compound microscope and improved methods for slicing and fixing tissue led to an explosion of information during the nineteenth century about the tissues of the body, including the sense organs (there are useful reviews of older literature in Polyak, 1941; von Békésy, 1960). The most important studies were surely those of the great Spanish neuroanatomist Ramón y Cajal (1911/1998). His clear drawings provided a wealth of information about the shapes of sensory receptors and other cells in sensory organs (Figure 1.2).

The physiology of sensation

These anatomical discoveries helped stimulate the first useful experimentation on the function of the sense organs. The structure of the ear and the role of the ear drum and bones of the middle ear were understood by the middle of the nineteenth century, and Helmholtz (1877/1954) postulated that sound

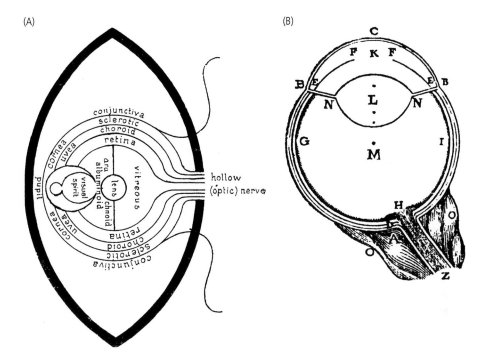

Figure 1.1 Structure of the eye. (A) Diagram of the eye from a ninth-century AD translation of Galen into Syriac by Humain Ibn Is-Hâq, in turn translated into English. (B) More anatomically correct diagram of cross-section of the eye made by René Descartes. ABCB, Cornea and sclera; EF, iris (in actual fact closer to the lens than shown in Descartes' diagram); K, aqueous humor; L, lens; EN, zonule fibers; M, vitreous humor; GHI, retina; H, optic nerve head; O, ocular muscles; and Z, optic nerve. (A from Meyerhoff, 1928; B from Descartes, 1637/1987.)

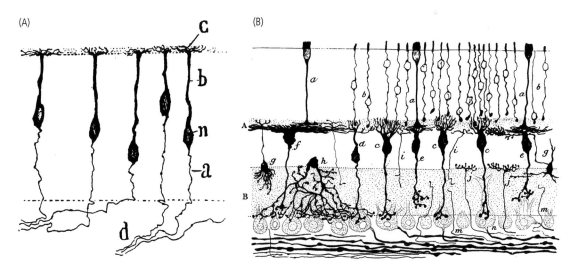

Figure 1.2 Sensory cells from the work of Ramón y Cajal. (A) Bipolar sensory neurons from mammalian olfactory mucosa. a, Axon; b, peripheral process; c, sensory dendrites; d, axon; n, nucleus. (B) Section of retina of an adult dog. A, Outer plexiform (synaptic) layer; B, inner plexiform (synaptic) layer; a, cone fiber; b, rod cell body and fiber; c, rod bipolar cell with vertical dendrites; d, cone bipolar cell with vertical dendrites; e, cone bipolar cell with flattened dendrites; f, giant bipolar cell with flattened dendrites; g, special cells stained very rarely (perhaps inter-plexiform cells); h, diffuse amacrine cell; i, ascendant nerve fibers (probably processes of cell not well stained); j, centrifugal fibers coming from central nervous system; m, nerve fiber (probably again of poorly stained cell); n, ganglion cell. (A and B from Cajal, 1893/1973.)

displaces these structures and causes the basilar membrane in turn to vibrate, with different tones producing vibration in different places. It was, however, von Békésy's actual observations of the movements of the basilar membrane that provided the first experimental evidence for the mechanism of auditory sensation in the mammalian ear (see Chapter 6 and von Békésy, 1960).

The visual pigments of the eye were also first discovered in the nineteenth century (an excellent summary of this early work can be found in Brindley, 1960), and Kühne showed that the molecule rhodopsin, or *sehpurpur* as he called it, changes color (bleaches) when exposed to light. These observations eventually led to the discovery by George Wald and colleagues that it is not the protein component of rhodopsin that absorbs light but rather a relative of vitamin A called 11-*cis* retinal, which is covalently bound to the protein (see Chapter 9 and Wald, 1968).

Some of the first electrical recordings of the responses of sensory receptors were made by E. D. Adrian, who dissected away the axons of single touch receptors from the skin and placed them over a wire electrode to record action potentials (Adrian, 1928, 1931, 1947). A typical result from Adrian's experiments is illustrated in Figure 1.3A. Pressure applied to the skin causes the frequency of action potential firing to increase (upper trace). Pricking the skin (lower trace) is also an effective stimulus but evokes action potentials in more than one kind of mechanoreceptor (note the different amplitudes of the spikes recorded by the electrode). Adrian concluded that action potentials from these receptor cells are communicated to the brain and form the basis of our sensation of touch. Using a similar technique, Hartline recorded action-potential discharges from the compound eye of the horseshoe crab *Limulus* (Hartline and Graham, 1932) and showed (Figure 1.3B) that the frequency of action potential firing depended both on the intensity and duration of the light stimulation (Hartline, 1934). These were the first single-cell responses recorded from an eye, though we now know that they were not produced by the photoreceptors themselves but rather by a cell called the eccentric cell, which receives direct synaptic input from the photoreceptors.

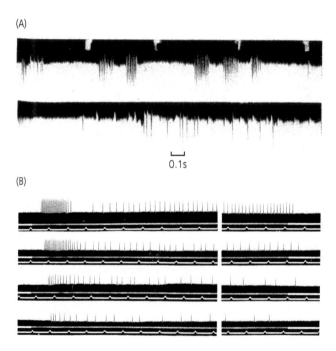

(A)

0.1s

(B)

Figure 1.3 Early electrical recordings of sensory responses. (A) Action potentials recorded from single axons dissected from the cutaneous nerve of a frog. (B) Action potentials from the lateral eye of the horseshoe crab *Limulus*. Each trace gives the response to a different light intensity, which was systematically increased by an additional factor of ten from dimmest (bottom) to brightest (top). (A from Adrian, 1947; B from Hartline and Graham, 1932.)

The method of dissection of single nerve fibers is difficult and tedious and was soon replaced by recording with fine metal microelectrodes. These electrodes are made from tungsten or platinum wire exposed and sometimes gold-plated at the tip but insulated along the rest of the length with glass or plastic resin. These metal electrodes can be inserted directly into the tissue to record the small extracellular currents produced by action potentials of single cells. Metal-electrode recording from the nerve coming from the ear established many of the basic properties of auditory responses, such as their time course and dependence on the frequency of the sound (see Kiang, 1965; Evans, 1975). The recordings in Figure 1.4 were collected from a single axon from the ear and show action potentials as a function of sound intensity on the ordinate, with the

frequency (pitch) of the sound on the abscissa. As the sound was made progressively weaker, the range of frequencies to which the axon responded became progressively more restricted. The nerve fiber showed greatest sensitivity to a tone near 10 kilohertz (kH), since at this frequency (called the *characteristic frequency*) a response could still be recorded even when the sound was made very weak indeed. Recordings of this kind showed that different axons in the auditory nerve have different characteristic frequencies, spanning the entire range of perceptible sound. The axons are therefore *labeled lines*, each carrying information about a different range of sound frequencies. These experiments showed that the ear must have some way of responding to sounds of different frequencies, so that the different auditory receptors can be tuned each to its own characteristic frequency.

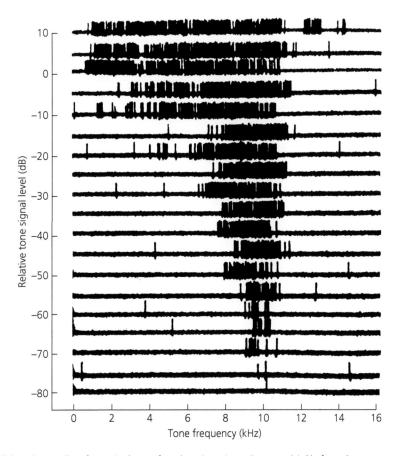

Figure 1.4 Extracellular spike recordings from a single axon from the guinea pig ear. Frequency (pitch) of sound was systematically swept from low to high for a range of different sound intensities. Frequency is plotted on abscissa and intensity is plotted on ordinate in a log scale of decibels (dB). An increase of 20 dB is equivalent to a 100-fold increase in intensity. (From Evans, 1972.)

The first extensive study with metal microelectrodes from olfactory receptors produced a completely different result (Gesteland et al., 1965). There seemed to be no consistent pattern to the responses, with many receptors responding to the same chemicals, sometimes with excitation, sometimes with inhibition. Later recordings confirmed some but not all of these conclusions. They showed that vertebrate olfactory receptors all appear to respond with excitation, producing an increase in spike frequency to stimulation with an odor. Single cells do nevertheless appear to be able to respond to a wide variety of odors. Thus olfactory receptors seem *not* to be labeled lines, at least not in the way originally supposed. I return to this matter in Chapter 7, after I have described the mechanism of olfactory transduction in detail.

These early recordings indicated that receptor cells signal the arrival of sensory stimuli by producing a change in electrical activity. What is the nature of this electrical signal? Is it produced by some change in the cell membrane potential? If so, what is the mechanism that converts the sensory stimulus into an electrical response?

A powerful tool for the investigation of these questions became available with the invention of the intracellular microelectrode in the late 1940s

(Ling and Gerard, 1949). An intracellular microelectrode is made from a piece of glass tubing typically 1 mm in diameter. The tubing is melted and pulled to a fine point, in early studies by pulling the glass by hand over a Bunsen burner, but later by placing the glass in a mechanical device that heats the middle of the tubing and pulls at either end to form two electrodes, each with a fine glass tip. The bore of the electrodes is then filled with a concentrated salt solution such as 3M KCl.

The first intracellular recordings from sensory receptors were made by Hartline and collaborators, again from the compound eye of the horseshoe crab *Limulus* (Hartline et al., 1952). Figure 1.5A is from the later study of Millecchia and Mauro (1969b), also from *Limulus*. Light produces a positive-going change in membrane potential, called a depolarization. Similar depolarizing responses were recorded from many other types of sensory receptors, including mechanoreceptors (Eyzaguirre and Kuffler, 1955; Loewenstein and Altamirano-Orrego, 1958) and chemoreceptors of the nose (Getchell, 1977). It came therefore as a great surprise when Tomita and collaborators first showed that the response of a vertebrate photoreceptor to light is a negative-going hyperpolarization (Figure 1.5B and Tomita, 1965).

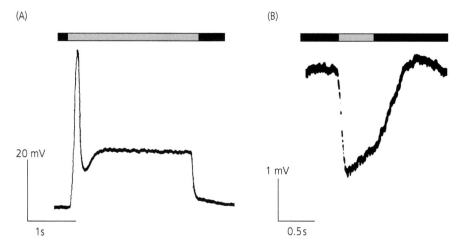

Figure 1.5 Intracellular recordings from sensory receptors. Bars above recordings show timing and duration of light flashes. (A) Depolarizing voltage response from photoreceptor of *Limulus* ventral eye. (B) Hyperpolarizing voltage response from photoreceptor (cone) of a fish. This is the first published recording of the response of a vertebrate photoreceptor. (A from Millecchia and Mauro, 1969b; B after Tomita, 1965.)

Cracking the problem: molecular physiology

Although the important observations of neuroanatomists of the nineteenth and twentieth centuries and the first extracellular and intracellular recordings from receptor cells provided many clues about the early steps in sensory processing, they told us very little about transduction; that is, about the way the electrical signal is generated by light or odor or sound. The physical stimulus received by the sense organ is somehow translated into a change in membrane potential, which is then transmitted into the central nervous system (CNS). The nature of this process remained for a very long time a complete mystery. This puzzle has now been substantially solved for most of the senses in a variety of organisms, providing a fairly clear picture of how sensory signals are produced. These striking advances were greatly facilitated by many years of patient biochemical and electrophysiological investigation, but they were then rather suddenly accelerated by the discovery of the technique of patch-clamp recording and of methods for cloning proteins and expressing their activity.

The invention of the patch electrode by Neher and Sakmann (1976) first made possible direct recordings from the molecules responsible for the electrical activity of nerve cells (see Sakmann and Neher, 1995). A patch electrode is made from fine glass tubing like an intracellular electrode, but the tip of a patch pipette is made very smooth, either by a specialized pipette puller (Brown and Flaming, 1977) or by polishing the end of the pipette with heat under a microscope. The pipette is then pressed against the soma of a cell and slight suction is applied, usually by mouth (Figure 1.6A). The glass of the pipette may then adhere to the cell membrane to form a very tight seal, sometimes called a gigaseal, with a resistance often of the order of 10 gigaohms (10^{10} Ω) or greater. The very high resistance of this seal reduces the electrical noise of the recording and makes it possible to distinguish the opening and closing of single channels in the membrane within the orifice of the pipette (Figure 1.6B). Single-channel responses first of acetylcholine receptors (Neher and Sakmann, 1976) and then of the Na⁺ channels of axons (Sigworth and Neher, 1980) were studied

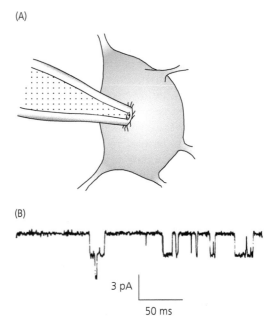

(A)

(B)

3 pA

50 ms

Figure 1.6 Patch-clamp recording from single channels. (A) The tip of a patch pipette is pushed against the cell body of a cell and slight suction is applied to form a seal. (B) Single-channel currents recorded from muscle acetylcholine receptors. The pipette contained 0.3 μM acetylcholine. Downward deflections indicate channel opening. At least two channels were present in this membrane patch. (B from Trautmann, 1982.)

with patch-clamp recording. In a very short time, recordings were obtained from many of the principal kinds of channel molecules of the cells of the nervous system, including those of sensory receptor cells.

Recordings made with patch electrodes sealed to the surface of the plasma membrane as in Figure 1.6 are called on-cell or cell-attached recordings. The extracellular surface of the membrane is exposed to the solution inside the pipette, and the intracellular surface is exposed to the cytosol. If the pipette is sealed in this way and then gently lifted off the cell, the plasma membrane often remains attached to the pipette, forming an excised or inside-out recording (Figure 1.7), so-called because the inside surface of the membrane now faces the outside bathing solution. Inside-out recording makes possible the study of channels that are opened or closed by the binding of some intracellular substance to the cytoplasmic side of a channel protein, such as Ca^{2+}, cyclic nucleotides, and other putative second messengers. As we shall see, inside-out recording provided crucial evidence establishing the identity of the intracellular second messengers mediating vertebrate visual (Fesenko et al., 1985) and olfactory (Nakamura and Gold, 1987a) transduction.

If, on the other hand, a pipette is sealed onto a cell and additional pressure or a brief voltage pulse is applied, the membrane within the pipette can often be made to break, establishing a direct connection between the inside of the pipette and the inside of the cell. This method of recording is called whole-cell (Figure 1.7, left middle) and is useful for introducing small-molecular-weight molecules from the pipette into the cell. The whole-cell mode of patch clamp is also extensively used to voltage clamp small cells. I describe the method of voltage clamping in more detail in Chapter 3. Whole-cell recording has revolutionized cellular physiology, greatly facilitating the study of electrical responses of a variety of neurons such as pyramidal cells in the cortex and granule cells in the cerebellum, as well as many types of sensory receptor cells, including photoreceptors, auditory hair cells, and the chemosensory receptor cells of the nose and tongue.

A pipette in the whole-cell mode can also be lifted off the cell. As Figure 1.7 shows (lower right), the membrane will often then flip around and reseal, leaving a small patch of excised membrane whose outside surface faces the outside solution. This is called an outside-out recording. The outside-out

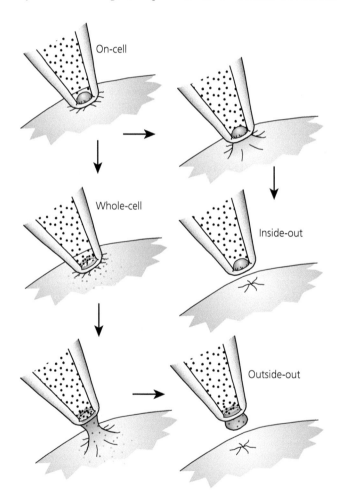

Figure 1.7 Different configurations of recording with patch-pipette technique. On-cell, whole-cell, inside-out, and outside-out recording techniques as described in the text.

mode of the patch-clamp technique has been especially useful for studying ligand-gated channels like those at synapses. These channels have an extracellular binding site for a small-molecular-weight transmitter molecule. A putative transmitter can be added to the bathing solution, and its effect on channel opening can be examined directly.

The revolution of molecular biology

The search for the mechanism of transduction was also greatly facilitated by the development of the techniques of molecular biology. Many of the most important molecules of sensory cells are integral membrane proteins, including the sensory receptor proteins of the nose and tongue, as well as enzymes of second-messenger cascades and the channels that ultimately produce electrical responses. These proteins are firmly embedded in the plasma membrane and difficult to extract and study. In the 1960s and early 1970s, the first attempts were made to isolate these important molecules from neurons and sen-

sory cells and to sequence and study them. In a few favorable cases, it was possible to extract enough of a protein in this way to obtain its complete amino acid sequence (for example rhodopsin, see Artamonov et al., 1983; Hargrave et al., 1983; Hargrave, 2001). In most cases, however, only a very small amount of protein could be extracted—too little to be studied in detail, but enough in many cases to allow the gene of the protein to be cloned.

Many membrane protein genes were first cloned in a similar fashion (Figure 1.8). A small amount of the protein was first extracted and purified, generally with chromatography or electrophoresis. It was then digested with a protease, and a few small-molecular-weight peptides were isolated and sequenced. From these peptides, synthetic nucleotide sequences were synthesized and used to screen a library of clones, made from tissue of the animal from which the protein was originally extracted. Alternatively, an antibody was made to an isolated peptide and used to screen an expression library. From the DNA sequence of the clone, it was possible to infer the

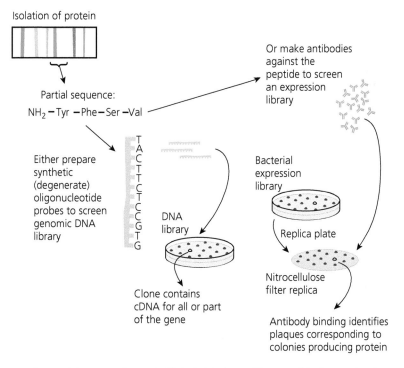

Figure 1.8 Cloning a gene from partial sequence of a protein. Method of cloning used for many of the first proteins whose genes were cloned from the nervous system. The method begins with isolation of partial sequence of a protein, which is then used to prepare oligonucleotides for screening tissue DNA libraries.

amino acid sequence of the protein. It has also been possible to identify families of related proteins within an organism and from organism to organism, by examining complete genomic sequences. We now have complete sequences of the genomes of many model organisms such as the fruit fly *Drosophila*, the mouse, and the zebrafish, as well as of hundreds of other species, from sponges to *Homo sapiens*.

Ultimately, the identification of a DNA sequence as that of a functional protein rests upon the demonstration that the DNA in question can direct the synthesis of a molecule with biological activity. This task can be done by *expressing* the protein. The DNA of the identified clone can be used to make complementary RNA (cRNA), which is then inserted, for example, into an oocyte of the frog *Xenopus* (Figure 1.9A). The oocyte can then be used for voltage-clamp studies of the expressed protein. Alternatively, and now more commonly, the DNA from the clone can be incorporated directly into the DNA of a cultured cell by a process called transfection (Figure 1.9B). DNA packaged into a plasmid or viral vector can be introduced into the cell by a variety of methods, for example by exposing a cell to lipid vesicles containing the DNA, or by giving high-voltage pulses of electricity to pierce holes in the cell membrane. The DNA can then become incorporated into the genome of the cell, and the cells are cultured to select those expressing the DNA of interest. If properly linked to promoters or other regulatory elements, the DNA is transcribed into RNA, which is in turn translated into protein. A stable population of cells may be produced in this way expressing the protein of interest. Transfection is often more convenient than RNA expression in oocytes, because cultured cell lines provide an excellent starting point for producing large quantities of expressed protein for structural or other studies, as well as a convenient preparation for patch-clamp recording.

From the amino acid sequences of the proteins we had our first clues about the structure of the molecules. Many of the most important proteins mediating sensory transduction are integral membrane proteins with extensive sequences lying within the hydrophobic interior of the lipid bilayer. From the sequence alone, reasonable guesses can be made about which amino acids lie within the membrane

and which are more likely to face the cytoplasmic or extracellular solution (Figure 1.10). Some amino acids (such as valine and isoleucine) are hydrophobic and much more likely to be surrounded by lipid or other protein, whereas others (such as aspartate and lysine) are hydrophilic or even charged and much more likely to be surrounded by water. By a process known as hydropathy analysis, the sequence of amino acids can be used to make inferences about how the protein folds, indicating the parts of the sequence that are integrated into the membrane and those that are exposed to the intracellular or extracellular solution. Antibodies to specific sequences can then be used to localize parts of the protein on one side of the membrane or the other. Sequences can be identified as substrates for glycosylation or phosphorylation, or can be shown actually to be glycosylated or phosphorylated. These identifications are often helpful in indicating regions that are intracellular or extracellular, because glycosylases and protein kinases only add sugar groups or phosphates at sites accessible to one side of the membrane or the other.

Finally, membrane proteins either isolated or expressed as in Figure 1.9 can be used to form crystals suitable for X-ray crystallography, from which the complete three-dimensional structure of the protein can be determined. Membrane proteins are in general more difficult to crystallize than soluble proteins, but with continued effort crystals were obtained at about the same time for ion channels and G-protein receptors (Doyle et al., 1998; Palczewski et al., 2000). These methods have given us extensive information about mechanisms of ion movement through potassium (Jiang et al., 2003b; Long et al., 2005a, 2005b, 2007) and sodium (Payandeh et al., 2011; Catterall, 2012; McCusker et al., 2012) channels, as well as protein conformation changes producing activation in rhodopsin (Deupi et al., 2012) and other G-protein receptors (see Erlandson et al., 2018).

Structures of near atomic resolution can also be obtained by a newer method called cryogenic electron microscopy (cryo-EM). The protein of interest is expressed in either bacteria or a cell line, and it is then isolated and distributed onto an EM grid. The grid is plunged into liquid ethane and then into liquid nitrogen, to freeze the sample rapidly so as to prevent the formation of ice crystals. The regularities

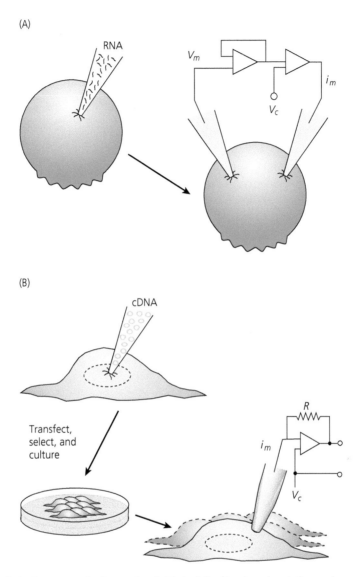

Figure 1.9 Common methods of gene expression for recording electrical activity of ion channels and other membrane proteins. (A) Injection into a *Xenopus* oocyte, which can then be studied by voltage clamping. (B) Transfection. DNA incorporated into a plasmid or viral vector is introduced into the cell by electroporation, Ca^{2+} shock, or direct injection in the nucleus (as shown). The cell line may then be used for patch-clamp recording. V_m, Membrane potential; V_c, command potential; R, feedback resistance of patch amplifier; i_m, membrane current.

of the structure of the frozen particles can then be used to determine the structure of the protein. Large proteins, proteins in solution, and proteins with significant structural heterogeneity can now be visualized with this method.

The techniques of molecular biology can also provide essential information about the function of sensory proteins. Experiments of this kind have been especially informative for receptors that use second-messenger cascades, such as those in the eye and nose. The cloning, for example, of the genes of the family of receptor proteins mediating olfactory transduction in the nose (Buck and Axel, 1991) has led to remarkable insight into the organization and mechanism of transduction in this tissue, which I describe in considerable detail in Chapter 7. Similar

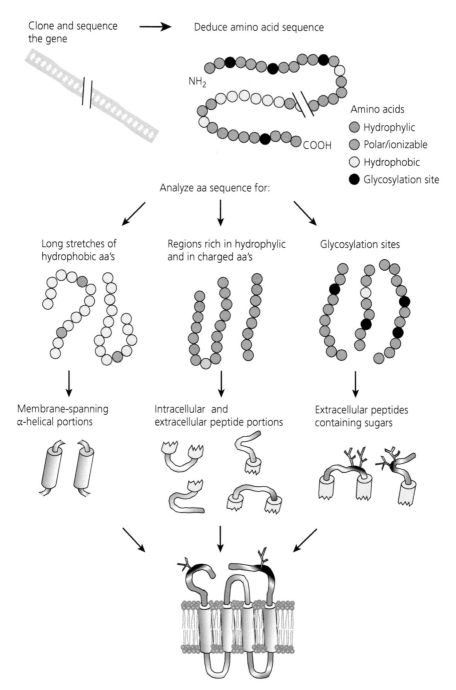

Figure 1.10 Analysis of hydropathy and the folding of membrane proteins. The amino acid sequence of a membrane protein can be used to make inferences about protein structure, as described in the text.

discoveries for the receptors of taste cells (Liman et al., 2014; Roper and Chaudhari, 2017) promise to bring important advances in our understanding of gustatory transduction.

Molecular techniques have provided powerful methods for studying protein structure and function. Site-directed mutations of single amino acids can be made in selected positions in a protein to test the role of specific sequences in substrate binding or catalysis. The DNA can be altered at predetermined locations to produce specific deletions of part of the protein sequence. In this way, whole regions of the protein can be excised. Regions of sequence can even be exchanged between related proteins to produce chimeric molecules, containing part of one protein and part of another. These experimental approaches have been greatly facilitated by new methods of altering the genome, including those based on CRISPR/Cas9 (Doudna and Charpentier, 2014; Doudna and Sternberg, 2017). The CRISPR/Cas9 system is part of a bacterial defense mechanism against viruses, which can be exploited in higher animals including mammals to alter genomic DNA so as to modify or knock out selected genes much more easily and rapidly than was previously possible. These techniques can be combined with structural information to provide remarkable insight into the physiology of the body.

Piezo proteins: channels mediating touch

The power of these methods can be illustrated by the recent isolation and characterization of the piezo proteins (Murthy et al., 2017), which are now known to be essential components of mammalian touch sensation and proprioception (see Chapter 5). These proteins were discovered in the following way. Bertrand Coste in the laboratory of Ardem Patapoutian began to look for channels that might be responsible for tactile sensation. He and his colleagues started screening several commercially available cell lines (Coste et al., 2010), recording from single cells from each of the lines with the whole-cell method of patch clamp while depressing the membrane of the cell with a second pipette (Figure 1.11A). In this way, they discovered that membrane indentation of a Neuro2A cell produces a rapidly decaying inward current, caused by an increase in membrane permeability to cations, principally Na^+. They then isolated messenger RNA from Neuro2A cells and screened for highly expressed membrane proteins, finding nearly 100 candidate sequences. For each of these sequences they constructed a small interfering RNA (siRNA). An siRNA is a nucleotide sequence of twenty to twenty-five base pairs that is complementary to the sequence of the mRNA of the candidate protein. The siRNA binds to the candidate mRNA sequence to inhibit its translation.

The siRNAs were introduced one by one into the Neuro2A cells and the responses of the cells were recorded as in Figure 1.11A, until one of the siRNA sequences was shown to produce a dramatic decrease in the mechanosensitive current. In this way, Coste and colleagues identified a protein called Piezo1 (Figure 1.11B), named from the Greek verb πιέζω, meaning "to press." A second protein called Piezo2 (Figure 1.11C) was identified by its sequence homology and was quickly recognized to be expressed in somatosensory neurons. In a subsequent paper (Ranade et al., 2014), this laboratory used molecular techniques to knock out the *Piezo2* gene in adult mice and showed that it mediates an important component of touch sensitivity (see Chapter 5). Later observations revealed that humans with mutations in *Piezo2* show a selective loss of proprioception and touch perception (Chesler et al., 2016).

Piezo1 and Piezo2 are extremely large multimeric proteins, whose three subunits each have over 2000 amino acids. Hydropathy analysis (Figure 1.12) indicated that each subunit should have nearly forty α-helical hydrophobic regions spanning the membrane (Coste et al., 2015), many more than most ion channels (see Chapter 3). The identification of these regions was confirmed by marking extracellular sites accessible to binding by antibodies against Myc tags, which are small peptide antigens containing ten amino acids. The Myc tags were introduced into the protein structure at selected locations by inserting the Myc sequence into the protein coding region, and the altered protein sequence was then expressed in a cell line. Intracellular sites were identified as sites of phosphorylation with mass spectrometry. Site-directed mutagenesis of single amino acids, followed by expression of the channels as in Figure 1.9, showed that the ion channel of the protein was located near the carboxyl terminus—to the right in Figure 1.12.

Because of its unusual size and sequence, a piezo protein should have quite an interesting structure;

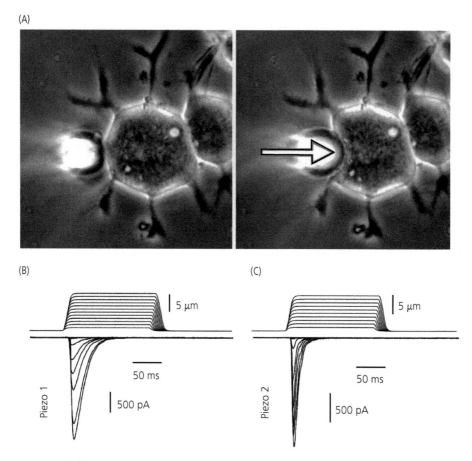

Figure 1.11 Responses of cells transfected with piezo proteins. (A) Method of stimulation. A pipette was mechanically moved against plasma membrane of an N2A mouse neuroblastoma cell during whole-cell recording, visibly deforming the cell membrane. The recording pipette is not shown for clarity. Responses of N2A cells transfected with genes for Piezo1 (B) and Piezo2 (C) to pipette movements of increasing distance in 1-μm increments. Magnitude and waveform of stimuli are shown in upper traces. Lower traces give currents superimposed for each of the stimulus magnitudes, recorded under voltage clamp at a holding potential of −80 mV. (From Coste, 2011.)

but because these proteins are so large, they would be nearly impossible to crystallize for X-ray studies with present techniques. Several groups therefore turned to cryo-EM (Ge et al., 2015; Guo and MacKinnon, 2017; Saotome et al., 2018; Zhao et al., 2018). The protein was expressed in a bacterium or cell line, isolated and solubilized with detergent, and placed on a grid for electron microscopy. Figure 1.13A (from Guo and MacKinnon, 2017) shows an image from such an experiment. Individual proteins are clearly observable, randomly distributed over the grid. Protein images were grouped into classes depending upon their orientation, and the

images were then averaged. Representative averages at two different orientations are shown in Figure 1.13B. The averages were refined under software control and assembled into a model to provide a 3D structure of the protein.

These structures have several interesting features. When viewed from above (Figure 1.13C), the channel resembles a propeller with three blades contributed by the three subunits (in red, green, and blue), meeting in the center to form the ion channel. When viewed from the side (Figure 1.13D), the subunits can be seen to have numerous α-helical regions running perpendicular to the plasma membrane and

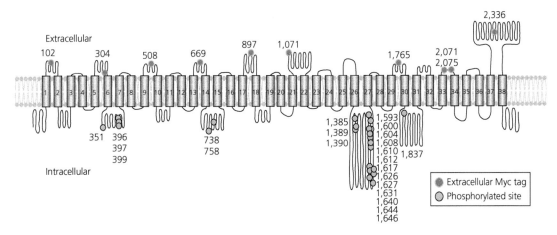

Figure 1.12 Membrane topology of Piezo1 protein. The amino acid sequence of Piezo1 was subjected to hydropathy analysis as in Figure 1.10 in order to predict regions of α helix and provide a preliminary membrane topology of the protein. This preliminary sequence was then tested by identifying regions of the protein that are extracellular or intracellular. Extracellular regions were identified by altering the protein sequence to include Myc tags, which are sequences of ten amino acids inserted in many different places along the protein sequence. Sequences with Myc tags were then expressed one by one in HEK293T cells (a cell line) and exposed to fluorescent antibody against the Myc tag. Cells were only stained if the Myc tag was extracellular and accessible to the antibody. Positions where staining was successful are shown together with the number of the amino acid in the sequence where the Myc tag had been introduced. Intracellular regions of the protein were localized by identifying phosphorylated peptides with mass spectroscopy. Phosphorylated amino acids are shown together with the number of the amino acid in the protein sequence. (From Coste et al., 2015.)

spanning it from one side to the other. Because these regions lie at an angle to one another, they would force the membrane to bend, producing tension. A small cap region on the extracellular side of the membrane lies just above the ion channel, which in the cryo-EM preparations was closed.

How then does membrane tension gate channel opening? The propeller blades of the three subunits are essential, because deletion of any of the groups of helices making up this extended part of the protein produced either a greatly attenuated response or no response at all. The propeller blades could act as a lever, transmitting pressure on adjacent plasma membrane into the middle of the protein to gate the channel (Saotome et al., 2018; Zhao et al., 2018). Alternatively, pressure on the membrane could flatten the channel and its surrounding lipid, which are under tension from the curved shape of the protein (Figure 1.13D). Channel flattening could then produce the change in conformation required to open the channel (Guo and MacKinnon, 2017). Further experimental work may help to distinguish these two alternative explanations.

The code deciphered: sensory transduction

It is remarkable to be able to say that we now understand, at least in outline, how sensation occurs in all of the major senses of the body. This development is, I think, one of the major achievements of modern neuroscience. Though many details of importance still have not been elucidated, it is nevertheless now clear that mechanoreceptors have ion channels like the piezo proteins specialized for the perception of membrane pressure and directly responsible for producing the electrical response of the cell. Vertebrate photoreceptors and olfactory receptors, on the other hand, use complicated signal transduction cascades. Light or odor produces a change in the concentration of one of the cyclic nucleotides which acts as a second messenger and binds to channels to produce the electrical response. An enormous amount of detail has been learned about the proteins responsible for sensation. We have paradoxically much more information about vertebrates for the rather complicated structures of their eyes, ears,

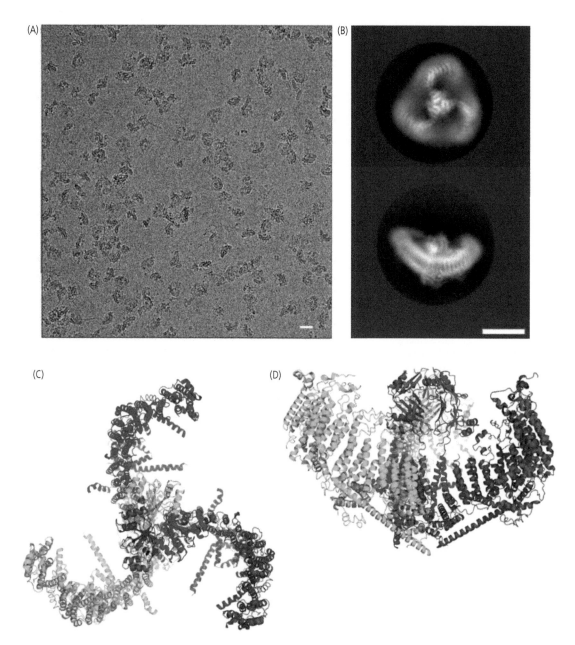

Figure 1.13 Cryo-EM structure of Piezo1. The gene for the Piezo1 protein was expressed in a cell line and purified. The protein was suspended on an electron-microscope grid and rapidly frozen by plunging the grid first into liquid ethane and then into liquid nitrogen. (A) Representative raw micrograph of the protein on the grid; scale bar is 200 Å (20 nM). (B) Protein images like those in (A) were separated into groups according to their orientation, first manually to produce templates and then automatically under computer control. Images in each class were averaged, and representative averaged classes are shown viewed from the top (upper image) and side (lower image); scale bar is 100 Å (10 nM). (C, D) Atomic model of the trimeric channel at an overall resolution of 3.7 Å shown as a ribbon diagram, viewed from the top (C) and side (D). The three subunits have been given different colors. (From Guo and MacKinnon, 2017.)

and noses than for invertebrates and their comparatively much simpler ocelli and sensory hairs. Our understanding of the detection of temperature has made remarkable progress, whether for the hot and cold receptors on the surface of the skin or for the infrared detectors that rattlesnakes use to hunt in darkness. Some species have now been shown to respond to the magnetic field of the earth, though the fundamental nature of sensory transduction in magnetoreception remains a mystery.

The purpose of this book is to describe these major discoveries as well as present areas of uncertainty. Together with an extensive treatment of sensation in mammals, I have included experiments on bacteria and protozoans as well as worms and arthropods. I also describe some mechanisms of sensation like electroreception, for which we as a species have no experience. The focus of this book is the cellular mechanism of transduction: the ion channels, G proteins, enzymes, and second messengers that produce the responses of sensory cells. I also attempt to summarize what we know about the modulation of transduction during adaptation. Where possible, I try to compare the mechanisms used by different sensory receptors. It is remarkable, for example, that adaptation in photoreceptors, olfactory receptors, and hair cells in every case seems to require the calcium ion, though what Ca^{2+} does in each cell is still not clear.

In the first part of the book, I attempt to outline some features common to many kinds of sensory cells, such as mechanisms of elaboration and renewal of sensory membrane, the organization of sensory organs, and the general features of electrical responses.

Chapters 3 and 4 provide background information about electrical activity in the nervous system and metabotropic cascades, with sensory cells used to emphasize general principles of cellular physiology. The remaining chapters provide a review of the literature of sensory transduction for the major senses. It would have been possible to write an entire book just on photoreceptors or hair cells, and perhaps even on olfactory receptor cells. It seemed to me much more useful to summarize in one volume the major features of sensory transduction for each of the receptors. In this way, we may more easily appreciate the many similarities in structure and mode of operation of the cells in the different sense organs.

It comes as an enormous source of satisfaction to me personally to be able to write this book in the way now possible. My own laboratory first began working on photoreceptors when I became a member of the faculty at UCLA in the fall of 1975. For nearly as long, I have taught sensory physiology to undergraduates. When I first lectured about the nose, there was almost nothing to say, apart from the anatomy of the receptor cells and the remarkable ability of the sensory epithelium to regenerate. Now it is easily possible to lecture on this subject for many hours without exhausting the fascinating body of knowledge that has since been acquired. Nearly the same could be said for touch and taste. I hope I can successfully convey the excitement of past discovery, and the anticipation I myself feel for the many surprises the future will doubtless bring about these fascinating organs, our senses.

CHAPTER 2

Mechanisms of sensation

Sensory receptor cells have many common characteristics. All have the same task: to convert a stimulus into an electrical signal in a form that can be used to alter behavior. Protozoans such as *Paramecium* accomplish this entire task within the membrane of a single cell. A *Paramecium* can sense a mechanical stimulus, generate a change in membrane potential, and use this electrical signal to alter the direction of movement of the cilia, ultimately changing the orientation of the organism (Chapter 5). The receptor cells of higher animals, on the other hand, are specialized for just one aspect of this process: the conversion of a sensory stimulus into a change in membrane potential that can be communicated to other cells in the nervous system. This process of conversion is called sensory transduction. An external signal such as a sound or an odor in some way produces a change in the conformation of an integral membrane protein called a sensory receptor molecule. Different sense organs contain different kinds of receptor molecules, each specialized for the detection of a particular stimulus.

Transduction in a sensory receptor can occur in two fundamentally different ways. The change in conformation of a sensory receptor molecule may produce a change in membrane potential directly (Figure 2.1A). Some receptor molecules are themselves ion channels whose opening and closing are directly gated by stimulation. The sensory receptor molecules of mechanoreceptors like the piezo proteins of Figures 1.11–1.13 are specifically designed to be sensitive to changes in membrane pressure. As the membrane of the receptor is deformed by touch

or vibration, the channel of the protein opens or closes. Temperature can produce channel opening or closing in a similar way (see Chapters 3 and 10). Direct modulation of channel gating of this kind is called ionotropic sensory transduction.

Sensory transduction in photoreceptors or vertebrate olfactory receptor cells is much more complex (Figure 2.1B). The change in the conformation of the sensory receptor molecule causes the binding of another protein, usually a heterotrimeric G protein, which in turn interacts with an effector molecule, ultimately producing a change in the concentration of a second messenger such as Ca^{2+} or a cyclic nucleotide. It is then the second messenger that controls the opening and closing of an ion channel. In this kind of transduction, the channel—also a molecule embedded in the plasma membrane—is entirely distinct from the receptor molecule that initially receives the stimulus. Sensory transduction of this kind is said to be metabotropic, and the molecules used to detect the stimulus are closely related to many hormone and metabotropic synaptic receptors, sharing both a common structure and a common mechanism of signal transduction.

Not all sensory receptors have receptor proteins. Some species of fish have electroreceptors, which can detect voltage gradients in water with very high sensitivity. They detect external gradients by sensing the voltage change across their plasma membrane. Though these cells may be highly specialized to sense voltage (see Chapter 10), no receptor protein is required. For other sensory receptors, detection needs a protein, but this protein may not be

Sensory Transduction. Second Edition. Gordon L. Fain, Oxford University Press (2020). © Gordon L. Fain 2020.
DOI: 10.1093/oso/9780198835028.001.0001

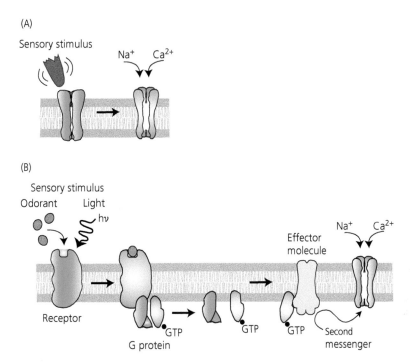

(A)

Sensory stimulus

Na^+ Ca^{2+}

(B)

Sensory stimulus

Odorant   Light

$h\nu$

Effector
molecule

Na^+ Ca^{2+}

Receptor

GTP   GTP   GTP   Second
messenger

G protein

Figure 2.1 Mechanisms of sensory transduction. (A) Ionotropic transduction. The stimulus directly gates an ion channel that is part of the receptor molecule. (B) Metabotropic transduction. The receptor is not itself a channel but activates a heterotrimeric G protein that initiates a transduction cascade. Channels in both cases are usually (though not always) non-selectively permeable to cations.

specialized for sensory transduction. The perception of salt by the tongue in some species is thought to use (at least in part) Na^+ channels similar to those found in many epithelial cells (Chapter 8). The permeability of these channels to Na^+ is thought to produce a change in membrane potential directly, as the concentration of Na^+ in the surrounding medium is increased by salt ingestion.

Sensory membrane

Because receptor proteins are integral membrane proteins, the process of sensory transduction requires specialized regions of membrane. This membrane is often continuous but distinct from the rest of plasma membrane, though in some cells it is enclosed in organelles within the cell cytoplasm. Sensory membrane is generally quite abundant because, as a rule, the more the membrane, the greater the number of receptor proteins and the greater the sensitivity of transduction. The higher the concentration of olfactory

receptor proteins, for example, the smaller the concentration of odorant that can be reliably detected.

To have a large number of receptor proteins, the cell must fabricate a large quantity of sensory membrane to accommodate them. The receptor proteins are usually specifically targeted to the sensory membrane of the cell, and it is common for the transduction machinery to be organized in some way into units of detection, containing accessory proteins and molecules that participate in the transduction cascade.

Sensory membrane is elaborated in most sensory receptors in one of two ways. In hair cells of the ear and vestibular system, olfactory receptors in the vomeronasal organ of the mouse nose, and arthropod photoreceptors the sensory membrane is formed from microvilli (Figure 2.2A). A microvillus is a highly structured membrane evagination, generally containing 40 or so parallel fibers of actin and other structural proteins such as fimbrin and villin, which are thought to play an important role in the

(A)

(B)

(C)

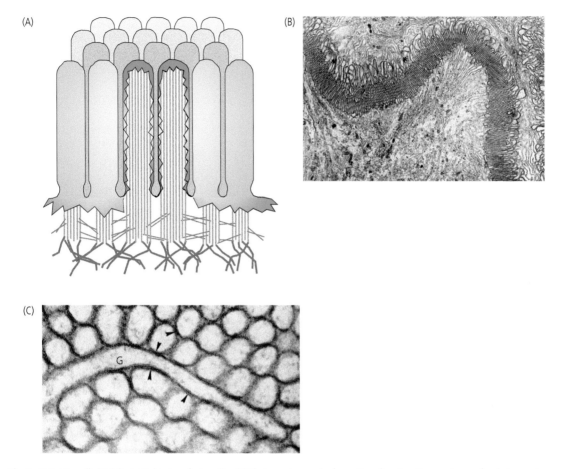

Figure 2.2 Microvilli. (A) Schematic drawing of microvillus. (B) Electron micrograph of a portion of a microvillar membrane of a photoreceptor of the horseshoe crab *Limulus*. Magnification 3,230×. (C) Higher power electron micrograph showing a cross-section of microvilli from a *Limulus* photoreceptor. G, Process of adjacent glial cell. Magnification 80,000×. (B and C from Clark et al., 1969.)

rigidity and maintenance of microvillar shape. These proteins may serve as attachment sites for proteins and probably assist in the organization of the transduction machinery. The base of the microvillus has a specialized structure also containing actin and often myosin, perhaps required to maintain the microvillus in a fixed position. Microvilli in a sensory receptor are generally concentrated in a specific region of the cell (Figure 2.2B). The intracellular space within the microvillus is usually rather restricted (Figure 2.2C).

In vestibular receptors of mollusks (Coggeshall, 1969), mechanoreceptors of insects, olfactory receptors of the principal nasal epithelium of the vertebrate nose, and photoreceptors of verte-

brates and some invertebrates (see Eakin, 1965) the sensory membrane is elaborated in association with the growth of a cilium. Cilia are also highly structured membrane evaginations (Figure 2.3A) but have quite a different internal organization (Figure 2.3B). The inside of the cilium contains microtubules formed from the protein tubulin, generally in a characteristic pattern of nine microtubule doublets ringing the circumference. In many cilia, there are also two single microtubules in the center of the cilium, in what is generally referred to as a nine-plus-two arrangement. These central microtubules are not present in some cells; they are, for example, absent in vertebrate rods and cones. In addition

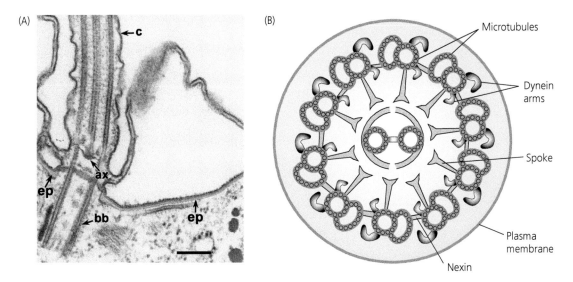

Figure 2.3 Cilium. (A) Structure of a *Paramecium* cilium. ax, Axosome at base of cilium; bb, basal body; c, cilium; ep, epiplasm (dense cytoskeletal formation underneath plasma membrane). Scale bar, 0.2 μm. (B) Schematic drawing of a cross-section of cilium. (A courtesy of R. D. Allen.)

to the microtubules, other proteins such as nexin and dynein help to hold the cilium together or function in cilium movement.

Cilia are produced from structures called basal bodies (Figure 2.3A), formed from centrioles which, in a dividing cell, are essential for the organization and separation of the chromosomes. In vertebrate photoreceptors, one of the basal bodies acts as a growth point for a cilium, which then organizes the formation of the sensory membrane (Steinberg et al., 1980; see also Burgoyne et al., 2015; Ding et al., 2015; Volland et al., 2015). The cilium appears to initiate the out-pocketing of membrane (Figure 2.4), which then grows away from the cilium, gradually reaching the diameter of a mature disk. In rods, the cilium also appears to direct the fusion of the disk rim, so that the disks pinch off to form vesicles in the cytoplasm separate from and enclosed within the plasma membrane. Each rod outer segment may have as many as 500–1000 of these membrane enclosures. In cones, membrane fusion occurs more slowly and is never complete, so that the membrane lamellae never entirely seal off. This leaves cone sensory membrane continuous with the plasma membrane and facing the extracellular medium.

Organization of membrane and sensory protein

In most sensory receptor cells, the sensory membrane forms a distinct part of the cell, specialized for transduction. In a vertebrate photoreceptor, the disks elaborated by the cilium form a structure called the outer segment, which contains all of the receptor proteins, transduction enzymes, and channels required for producing the electrical response to light. In a similar way, the cilium of an olfactory receptor cell contains the receptor molecules as well as most of the other proteins required for the detection of odor. Stretch receptors and many tactile receptors in the skin have neither microvilli nor cilia but nevertheless appear to target mechanoreceptive channels to particular regions of the plasma membrane. Thermoreceptive proteins are probably targeted in a similar way to the dendritic terminals of receptor cells responding to increases or decreases in temperature.

Within the specialized part of the cell containing sensory membrane, the transduction proteins are often organized in a way optimally suited to producing the electrical response. All cells have scaffolding proteins, which are molecules associated

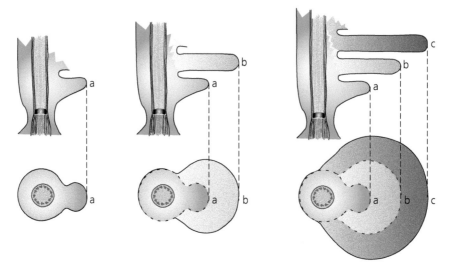

Figure 2.4 Formation of disks of a rod photoreceptor. Disks are initiated at the base of the rod outer segment adjacent to a cilium. (After Steinberg et al., 1980.)

with the cytoplasm that tether membrane proteins and even some cytosolic proteins to particular regions of the cell. A scaffolding protein called inactivation-no-after-potential D protein (INAD) in the photoreceptors of *Drosophila* contains five protein-binding domains (called PDZ domains), each of which seems able to bind to a different protein (see Hardie and Postma, 2008). In this way, INAD brings together several of the different components of the cascade, including effector proteins and ion channels (Figure 2.5). Such a closely organized cluster of transduction proteins has been called a transducisome (Tsunoda and Zuker, 1999) or signalplex (Montell, 1998). The juxtaposition of proteins of the sensory cascade minimizes the time for collision within the membrane and reduces the distance of diffusion of second messengers and other metabolites between the receptor proteins and the other molecules of transduction, including the ion channels. This process can increase the speed of the electrical response. The INAD also serves to maintain essential proteins within the fly photoreceptor at a high concentration with a fixed stoichiometric ratio.

In the disks of vertebrate photoreceptors, the receptor protein rhodopsin is not tethered by scaffolding proteins but is free to diffuse within the plane of the membrane, as if floating on a sea of olive oil. Other proteins of the transduction cascade, such as the G protein transducin and the effector molecule cyclic-GMP phosphodiesterase, are probably also mobile. As Poo and Cone (1974) first observed, free diffusion on a two-dimensional surface such as the plasma membrane greatly facilitates the collision of the molecules of the cascade. Activated rhodopsin can diffuse to find G protein, and G protein can diffuse to find effector. Some time is required for this process, but the minimum latency of the photoreceptor response is only about 7 ms (Cobbs and Pugh, 1987; Hestrin and Korenbrot, 1990). There is some evidence that rhodopsin may exist in the disk at least partially as a dimer (Liang et al., 2003; Zhang et al., 2016), which may also in some way facilitate the rate of transduction, but the role of dimerization remains unclear (Chabre et al., 2003; Edrington et al., 2008).

Membrane renewal

The sensory stimulus that activates transduction is often deleterious to the cell. Light, especially in the ultraviolet, can damage DNA; continuous activation of transduction in the photoreceptor can cause cell death (Fain, 2006). Olfactory receptor cells and taste buds are directly exposed to the external

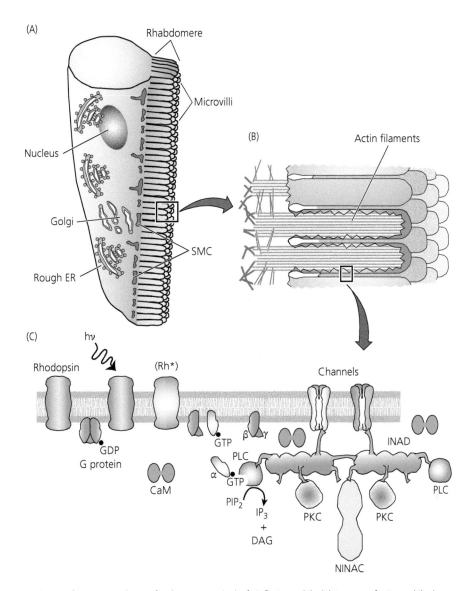

Figure 2.5 Organization of sensory membrane of a photoreceptor in the fruit fly *Drosophila*. (A) Anatomy of a *Drosophila* photoreceptor. The sensory membrane forms a structure called a rhabdomere, composed of about 30,000 microvilli. (B) The membrane of the microvillus is highly organized by actin filaments. (C) A scaffolding protein called INAD binds to proteins in the cytosol and plasma membrane. PLC and PKC are proteins shown as if cytosolic but likely to be at least peripherally associated with the plasma membrane. CaM, Calmodulin; DAG, diacylglycerol; ER, endoplasmic reticulum; GDP, guanosine diphosphate; GTP, guanosine triphosphate; IP$_3$, inositol 1,4,5-trisphosphate; NINAC, neither inactivation nor afterpotential C protein, a form of myosin; PIP$_2$, phosphatidylinositol 4,5-bisphosphate; PKC, protein kinase C; PLC, phospholipase C; Rh*, activated form of the photopigment rhodopsin; SMC, submicrovillar cisternae.

world, to perfume and soufflé but also to sulfur dioxide and three-alarm chili. For this reason, many sensory cells have mechanisms for the replacement and renewal of sensory membrane.

For photoreceptors in insects and other arthropods, light exposure can produce a rapid breakdown of the microvilli of the sensory membrane (see Blest, 1988). The membrane at the end of the

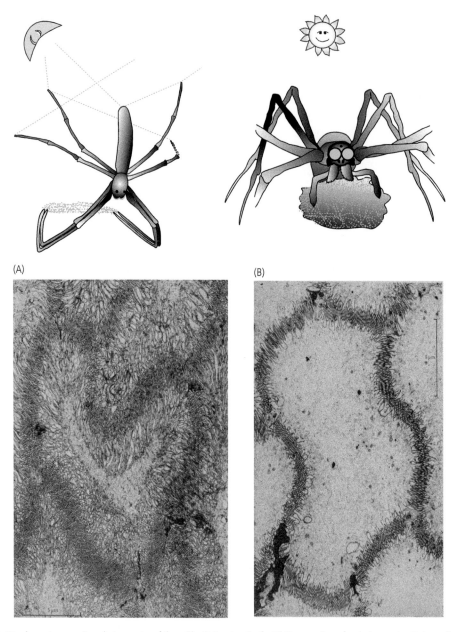

Figure 2.6   Membrane turnover in a photoreceptor of the spider *Deinopis subrufa*. (A) Cross-section of a photoreceptor that was fixed 2 hours after sunset. Magnification 2,850×. (B) Cross-section of a photoreceptor that was fixed 3 hours after sunrise. Magnification 3,850×. (From Blest, 1978.)

microvillus pinches off, forming multivesicular bodies that coalesce and are metabolized by the cell. In some species, new membrane is immediately resynthesized even in the light (Chamberlain and Barlow, 1979; Herman, 1991). In others, the reforma-

tion of new membrane seems to be accelerated once the animal is returned to darkness (White and Sundeen, 1967; White and Lord, 1975).

This process of light-dependent membrane turn-over is especially dramatic in the spider *Deinopis subrufa* (Blest, 1978). This animal emerges at dusk and spins a small web, which it holds as a net with its four anterior legs to drop on prey passing underneath. Because the spider hunts in dim light, it needs to have highly sensitive photoreceptors with a high concentration of sensory membrane and many microvilli. During the night the microvilli may occupy as much as 90 percent of the volume of the receptor (Figure 2.6A). At dawn, there is a massive destruction of this membrane, so that during the day the microvilli come to occupy only about 15 percent of the cell volume (Figure 2.6B).

Vertebrate rods have evolved an even more complicated procedure (Young, 1976). Each day, 5–10 percent of the disks in the outer segment are fabricated by the mechanism outlined in Figure 2.4. As these new disks are produced, they push the older disks upward. At dusk, the reduction in the intensity of the light acts as a signal that triggers the pinching off of 5–10 percent of the disks at the top of the outer segment, which are then phagocytozed by an adjacent cell layer, called the retinal pigment epithelium. If the photoreceptors are given a brief exposure to radioactive amino acid, protein (mostly rhodopsin) is first labeled in the inner segment, which contains the endoplasmic reticulum and Golgi apparatus (Figure 2.7). The label then migrates to the base of the outer segment and is incorporated

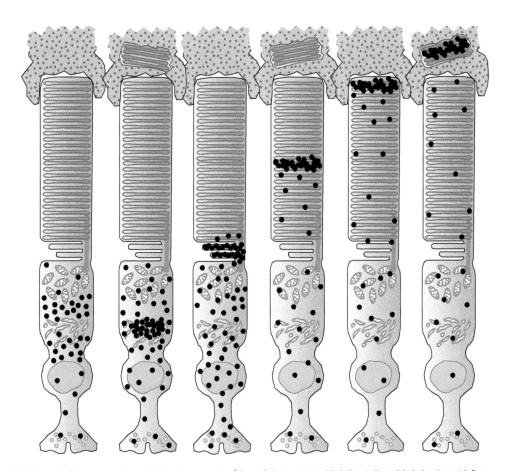

Figure 2.7 Renewal of sensory membrane in the outer segment of the rod photoreceptor. Black dots indicate labeled amino acid, first incorporated into protein in the inner segment, then transported to the outer segment as components of the disk (largely as rhodopsin). Synthesis of new disks pushes label upward until, after 10–14 days, the labelled disks are shed by the outer segment and phagocytosed by the cells of an adjacent cell layer, called the retinal pigment epithelium. (After Young, 1976.)

into the membrane of newly elaborated disks. Formation of subsequent disk membrane pushes labeled disks up the outer segment, so that a band of radioactivity marches up the outer segment over a period of 10 days to 2 weeks until disks at the tip are pinched off and lost from the cell.

In the nose, not only the sensory membrane but also the whole of the receptor cell is replaced. Olfactory receptor cells in mouse have a half-life of about 90 days. The olfactory epithelium contains a population of stem cells called basal cells that are constantly dividing and producing more receptor cells (Chapter 7). What is so remarkable about this process is that each receptor cell is highly differentiated and seems to express only a single type of the thousand or so olfactory receptor molecules in the mouse genome. A sensory cell containing a particular receptor protein has an axon that enters the CNS and appears to terminate at a very few specific locations in the olfactory bulb, coded according to the receptor protein the cell contains. As each olfactory receptor cell is replaced, the progenitor cell must not only express the appropriate receptor molecule, but also generate an axon that finds its way to the correct sites of termination in the brain. How the progenitor cells accomplish this task for every different kind of receptor cell in the epithelium is still not understood. There seem to be guidance clues within the olfactory bulb that persist into adulthood, so that newly formed receptor cells can terminate in their proper locations (Gogos et al., 2000). The discovery of the mechanism of this remarkable phenomenon may provide important insight into the renewal of cells not only in the nose but also elsewhere in the nervous system.

External specializations

The sensory membrane of the receptor cell is often surrounded by a covering or enclosed in some external structure that can make an important contribution to the detection of the stimulus. A fly, for example (see Dethier, 1976), has chemosensory bristles or hairs on the bottom of its feet (Figure 2.8A). These bristles are part of the exoskeleton of the insect composed mostly of chitin, a carbohydrate polymer similar in structure to cellulose. The receptors inside the bristles can sense the composition of chemicals in food because bristles have one or more tiny holes at their tip (Figure 2.8B). Inside the bristle are several chemosensory neurons, which elaborate microvilli called dendrites whose membrane contains taste receptor molecules. The space between the cuticle and the microvillus is filled with fluid, and the pores in the sensory bristles must be large enough and numerous enough to let chemicals in, but small enough to prevent the fluid from evaporating too rapidly.

The skin of mammals, including humans, contains a variety of mechanoreceptors. Among the most interesting are the Pacinian corpuscles, generally found buried within the skin and responding to deep pressure (see Loewenstein, 1971; Abraira and Ginty, 2013). The nerve ending of a Pacinian corpuscle is surrounded by a gelatinous structure formed from plasma membrane, wrapped round and round the nerve, much like the leaves of an onion (Figure 2.9A). Pressure applied to the corpuscle pushes on the layers of this structure but must ultimately be detected by the membrane of the nerve, buried deep within.

The many layers of membrane wrapped around the nerve have a dramatic effect on the response produced by applied pressure. Think of the corpuscle as a water balloon. Pressure placed at one point on the surface will push the balloon inward. This pressure will at first be transmitted into the center of the balloon, but the applied force will be rather quickly absorbed and the balloon will bulge outward. When the pressure on the balloon is removed, the balloon will snap back into shape as the energy stored in the elasticity of the structure is released.

For the Pacinian corpuscle, much the same process will occur. Deep pressure is felt only briefly and transiently, as it is first applied, and then again when it is removed. When recordings are made from the corpuscle (Figure 2.9C), pressure produces only a brief depolarization at the beginning and end of the stimulus, even when pressure is applied continuously. The corpuscle acts therefore to sense only changes in pressure. If, on the other hand, much of the thickness of the corpuscle is dissected away from the nerve ending (Figure 2.9B) and pressure is again applied (Figure 2.9D), the response is much more sustained (Loewenstein and Mendelson, 1965).

(A)

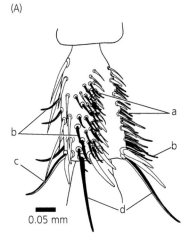

0.05 mm

(B)

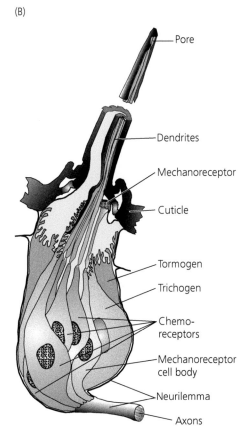

— Pore

— Dendrites

— Mechanoreceptor

— Cuticle

— Tormogen

— Trichogen

— Chemo-
receptors

— Mechanoreceptor
cell body

— Neurilemma

— Axons

Figure 2.8 Taste receptors of the blowfly. (A) Chemosensory taste bristles (hairs) on the tarsus of a blowfly. The tarsus is the most distal part of the leg in contact with the substrate. Letters indicate different anatomical classes of hairs (type a, type b, etc.; see discussion in Chapter 8). (B) Structure of a chemosensory taste bristle. In addition to two to four chemoreceptors, the bristle contains a single mechanoreceptor. Trichogen and tormogen cells are accessory cells, which secrete the hair and bristle socket. (A from Grabowski and Dethier, 1954.)

In addition to encasements covering the sensory membrane, sensory cells may be surrounded by associated structures that may play a large role in stimulus detection. In many mollusks and crustaceans, there are specialized organs of equilibrium called statocysts, which sense the position of the animal in the field of gravity; that is, the statocysts can tell the animal whether it is right-side up or upside-down. In many species they also provide information about movement velocity and acceleration.

In decapod crustaceans such as the crayfish or lobster, the statocysts are located on antennae on either side of the mouth (Figure 2.10A). The inside of the statocyst resembles a hollow sphere that contains 400 or so mechanoreceptor bristles projecting into the interior (Figure 2.10B and Cohen, 1960; Takumida and Yajin, 1996), with each bristle containing at least one mechanoreceptive neuron. The interior of the statocyst is filled with fluid and contains a structure called a statolith, composed of sand grains cemented together into a compact mass. When the lobster changes position, the statolith rolls over the bristles of the mechanoreceptors, stimulating different cells to varying degrees depending upon the position of the statolith and the location of the receptor cells inside the organ. If the statolith is removed, the response to position disappears (Cohen, 1955).

The gustatory hairs on the feet of house flies and the Pacinian corpuscles in the skin of mammals are relatively simple, consisting of single cells surrounded by an external specialization. Statocysts contain several cell types. There are mechanoreceptors, sometimes (as in crustaceans) encased in bristles, sometimes without bristles but projecting numerous cilia into the statocyst lumen. There are specialized cells that secrete the statolith, and accessory cells that encase the mechanoreceptors and enclose the whole of the organ.

Even rather simple eyes contain a large number of different kinds of cells. The eye of the scallop (Dakin, 1928; McReynolds and Gorman, 1970a), for example (Figure 2.11A), has a lens formed from specialized lens cells and two completely separate retinas in distinct layers, called distal and proximal. Photoreceptors in the distal retina contain sensory membrane elaborated from a cilium like vertebrate photoreceptors, whereas those in the proximal retina have microvilli (Figure 2.11B). Since the classic

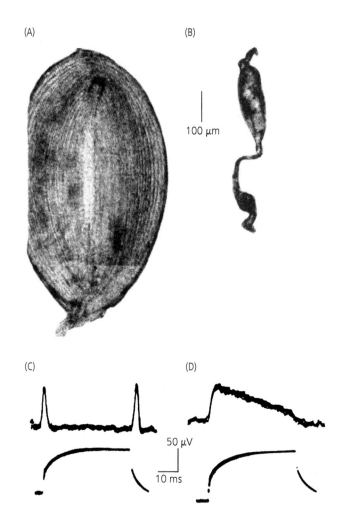

(A)

(B)

100 μm

(C)

(D)

50 μV

10 ms

Figure 2.9 Anatomy and physiology of the Pacinian corpuscle. (A) Phase-contrast micrograph of a normal Pacinian corpuscle isolated from the cat mesentery. (B) Pacinian corpuscle decapsulated by stripping away layers of membrane. (C) Electrical response recorded extracellularly from a normal corpuscle after blocking action potentials with procaine or tetrodotoxin. (D) Electrical response of a decapsulated corpuscle recorded as in part (C). Lower traces in (C) and (D) give the time course of the mechanical stimulus. (A and B from Loewenstein and Rathkamp, 1958; C and D from Loewenstein and Mendelson, 1965.)

work of Hartline (1938), we have known that these two retinal layers produce responses of different polarity, and we now know that the distal and proximal photoreceptors use completely different mechanisms of phototransduction (see Chapter 9). In addition, there are pigment cells, a reflecting layer called a tapetum, and a variety of accessory cells that probably assist in maintaining the shape of the eye and in transporting nutrients into and out of the eye interior. A scallop has fifty to a hundred eyes of

this sort, each 1 mm or so in diameter, which sit on small stalks and line the interior of the mantle of the shell.

The most complicated organs of sensation are certainly the vertebrate eye and ear, which both contain many different cells and an abundance of accessory structures. In addition to the photoreceptors, the retina of the eye has at least a hundred distinct types of neurons, which process the visual signal. There are several sorts of lens cell: a retinal

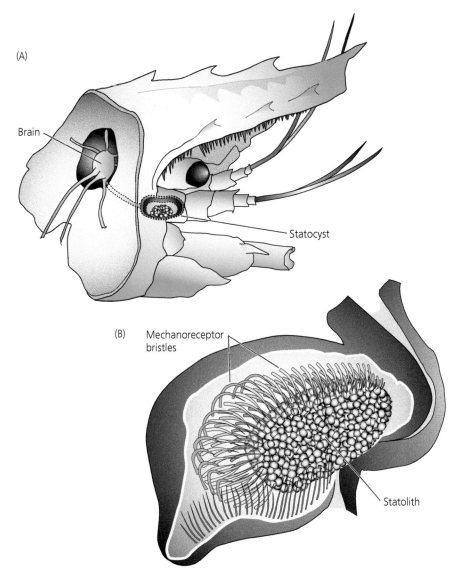

(A)

Brain

Statocyst

(B) Mechanoreceptor
bristles

Statolith

Figure 2.10 Structure of a statocyst in the lobster *Homarus americanus*. (A) Location of the statocyst in the lobster showing the nerve connecting to the brain. Dashed lines indicate that statocyst and its nerve are encased within the cuticle. (B) Internal organization of mechanoreceptive bristles and the statolith of lobster statocyst. (After Cohen, 1955, 1960.)

pigment epithelium, which as we have seen (Figure 2.7) phagocytoses disks from the tips of rod and cone outer segments and has other additional functions in physiology and nutrition; the ciliary body, which secretes a fluid (called the aqueous humor), bathing the cells of the eye interior; and the pupil and cornea, which again contain cells quite specialized in their anatomy and physiology. The ear is

similarly complex, as described in some detail in Chapter 6.

Detection of the stimulus

The activation of sensory receptor protein directly or indirectly causes a change in the conformation of membrane channels (Figure 2.1). This conformational

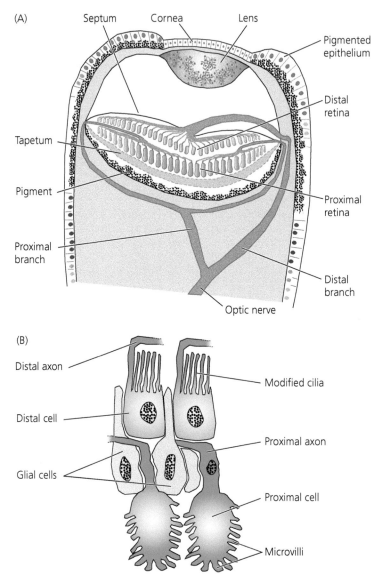

Figure 2.11 Organization of the eye of the scallop *Pecten*. (A) Anatomy of the eye. (B) Schematic drawing of the two photoreceptor layers. (A after Dakin, 1928; B after McReynolds and Gorman, 1970b.)

change increases or decreases the flow of ions through the channels and is responsible for the voltage response of the cell (Chapter 3). All sensory cells have channels specifically gated in some way by a sensory stimulus, but these channels are of considerable variety and can respond to stimulation in different ways, sometimes opening and sometimes closing. In some cells, both can occur: for hair cells of the ear and vestibular system, movement of

microvilli in one direction causes channels to open, and movement in the other causes them to close. The change in membrane potential produced by stimulation is similarly variable; in some receptor cells it is positive or depolarizing (Figure 1.5A), and in others, negative or hyperpolarizing (Figure 1.5B).

In spite of these differences, the electrical responses of sensory cells have many common features. For most sensory receptors, the change in membrane potential produced by stimulation, called the receptor potential, has a characteristic waveform (see for example Figure 1.5A). When the receptor cell is presented with a constant, maintained *step* of stimulation, such as a steady light or a steadily presented concentration of a taste or odorant, there is an initial change in potential which reaches a transient peak, and the response then declines to a steady level, often called a *plateau*. The decrease in the amplitude of the response during constant stimulation reflects a decrease in the sensitivity of the receptor, called adaptation.

Adaptation is a nearly universal feature of sensory receptors. The change in sensitivity during adaptation decreases the effect of background stimulation, such as constant noise or an ever-present odor. At the same time, adaptation adjusts the dynamic range of the response so that it is appropriate to the mean level of stimulation. This modulation of the response lets the receptor detect *changes* in odor concentration or sound amplitude over a very wide range of stimulus intensities. In the Pacinian corpuscle, the adaptation of the response (Figure 2.9) is mostly the result of the physical properties of the accessory structure surrounding the nerve ending. In photoreceptors or olfactory receptors, on the other hand, adaptation occurs mostly within the receptor cell itself and may require a multiplicity of biochemical processes to modulate the proteins of the transduction cascade.

Primary and secondary receptor cells

Some sensory cells, like the statocyst mechanoreceptors in Figure 2.10 or the scallop photoreceptors in Figure 2.11, have axons and generate action potentials. The change in membrane potential produced by the sensory stimulus depolarizes or hyperpolarizes the axon hillock of the cell, changing the frequency of action potentials conducted into the CNS of the animal.

Other sensory cells, such as vertebrate photoreceptors and vestibular hair cells, do not generate action potentials but communicate with another cell, which may be an afferent nerve fiber or another neuron, sometimes called a second-order neuron. The afferent terminal or second-order neuron receives synaptic input from the sensory receptor and responds with a change in membrane potential, which may be either positive or negative. It may then produce action

potentials, which are conducted into the CNS. In some sensory organs such as vertebrate retina, second-order neurons also lack action potentials but synapse onto yet other third-order neurons that do spike, and it is these cells (in vertebrate retina called ganglion cells) that convey the sensory signal to the CNS.

The voltage response of a secondary sensory receptor is a *graded* signal, varying continuously in value. Signals of this sort cannot be transmitted very far, because they decay in amplitude and become progressively slower in waveform as they move down the processes of a neuron. For this reason, these signals must eventually be converted into action potentials, which are all or none and do not decay as they propagate. The sensory code is then no longer represented by the amplitude of the signal, which for an action potential is reasonably constant, but rather by frequency, that is by the number of spikes per unit time.

Because the accuracy of the conversion of the receptor response into a frequency code is so important for the transmission of the sensory signal, many sensory receptors have evolved specialized structures at their synapses. In the fly retina, the second-order neuron receiving input from the photoreceptor is called a laminar cell, and the synapse between the photoreceptor and the laminar cell has a specialized anatomy (Figure 2.12A). The cytoplasm of the photoreceptor contains a curious "T"-shaped structure (Trujillo-Cenoz, 1965; Burkhardt and Braitenberg, 1976), consisting of a bar surrounded by a regular arrangement of synaptic vesicles. The bar is capped by a presynaptic plate. The function of this "T body" is probably to direct synaptic vesicles in an organized fashion to vesicle release sites, where transmitter is exocytosed into the synaptic cleft.

Vertebrate photoreceptors, hair cells in the ear and vestibular system, and electroreceptors all have presynaptic densities analogous to the T bodies of fly photoreceptors. In electroreceptors, for example (Figure 2.12B), there is a prominent dense structure called a synaptic ribbon surrounded by synaptic vesicles. These vesicles appear to be attached to the ribbon by fine filaments. As for the T body of insects, the ribbon is probably directing vesicles to release sites on the presynaptic membrane.

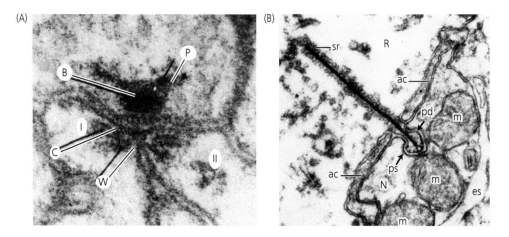

Figure 2.12 Presynaptic specializations at synapses of sensory receptors. (A) Electron micrograph of a synapse between a photoreceptor and second-order laminar cells in the housefly. P, Presynaptic plate; B, synaptic bar; I and II, processes of laminar cells; C and W, postsynaptic specializations. Magnification 115,000×. (B) Electron micrograph of a synapse between a skate electroreceptor (R) and an afferent nerve fiber (N). R, Receptor cell; N, nerve process; ac, accessory cell process; es, extracellular space; m, mitochondrion; pd, presynaptic density of receptor cell; ps, postsynaptic density of nerve process; sr, synaptic ribbon with associated vesicles. Magnification 33,000×. (A from Burkhardt and Braitenberg, 1976; B from Waltman, 1966.)

These presynaptic specializations are likely to make an important contribution to the sensitivity of synaptic transmission. In a vertebrate photoreceptor, which has synaptic ribbons like those in electroreceptors (see Chapter 9 and Figure 9.15), the minimum voltage change required to signal the detection of light is at most 10 µV, a deviation of less than one-tenth of 1 percent of the photoreceptor resting membrane potential (Fain et al., 1977). In the electroreceptors of elasmobranchs, the minimum voltage required to produce a significant change in synaptic transmission may be even smaller (see Chapter 10). In photoreceptors, hair cells of the ear, and many electroreceptors the release of transmitter at the synapse occurs continuously, even at rest, so that a stimulus need only modulate the frequency of release about a steady level. This basal release rate may also contribute to the very high sensitivity of transmission at these synapses.

Sensitivity of transduction

In addition to specialized mechanisms of synaptic transmission, transduction itself in many sensory receptor cells makes a large contribution to the sensitivity of detection. Many cells fabricate large amounts of membrane to accommodate a high concentration of sensory receptor protein. The gain of transduction may also be very large. In photoreceptors, for example, the activation of one rhodopsin molecule by a single photon of light produces a response that is easily detectable both by the organism and by a physiologist with a recording electrode. In Figure 2.13A, flashes at the same dim intensity activating on average 0.5 pigment molecules were given repeatedly, and the response of a monkey rod was recorded, as described in more detail in Chapter 9. The response of the cell was *quantized* (Baylor et al., 1979a; Field and Rieke, 2002), with some flashes giving no response, some a response that in this cell averaged about 2 pA, and some a response of about twice this value, or 4 pA. These values represent responses to 0, 1, and 2 photons. Since a response of 2 pA is equivalent to a change in the influx of many millions of sodium ions across the rod plasma membrane, the gain of transduction can be seen to be quite high. In a vertebrate rod, some of this gain occurs at the very first step: a single activated rhodopsin molecule can stimulate many G proteins, perhaps ten to fifteen in the small rods of a mouse but a hundred to two-hundred in a larger amphibian rod (Reingruber

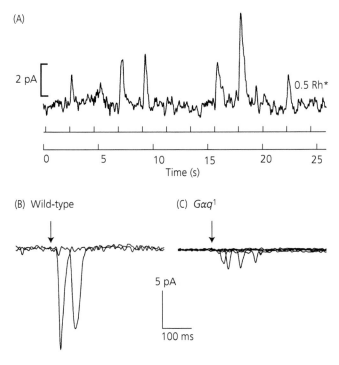

Figure 2.13 Responses to single photons of light. (A) Suction-electrode recording from a monkey rod (see Chapter 9 for method). Flash timing is indicated by the upward ticks of the trace below the recording. (B, C) Voltage-clamp recordings of current responses to single photons (quantum bumps) from *Drosophila* photoreceptors of wild-type (B) and *Gαq¹* mutants (C) containing only about 1 percent of the normal amount of G protein. (A from Field and Rieke, 2002; B and C courtesy of Roger Hardie.)

et al., 2015). Each of these G proteins then interacts with other proteins and initiates the rest of the transduction cascade.

In an invertebrate photoreceptor, the gain of transduction is also very large, but less of this gain comes from the first step of the cascade. The recordings in Figure 2.13B and C (from Hardie et al., 2002) were made from photoreceptors from the eye of the fruit fly *Drosophila*, and the large downward deflections are currents produced by the opening of many membrane channels following the absorption of a single photon of light. These responses are called quantum bumps and were first detected with intracellular microelectrodes from the horseshoe crab *Limulus* (Yeandle, 1958), where they were shown to be produced by excitation of single rhodopsin molecules (Fuortes and Yeandle, 1964).

The quantum bumps in Figure 2.13B were recorded from cells of a wild-type *Drosophila*, whereas those in Figure 2.13C (labeled *Gαq¹*) are from an animal whose photoreceptors contained only 1 percent of the normal amount of G protein. The amplitude of the bumps is about three to four times smaller. The most likely interpretation of this finding is that in the *Gαq¹* flies, the bumps are produced by activation of only a single G-protein, but in wild-type flies, something like three to four G proteins are stimulated. Most of the gain that produces the quantum bump must therefore be occurring in later stages in transduction (see also Kirkwood et al., 1989).

It is of some interest that much more light is required to produce the bumps in Figure 2.13C than in Figure 2.13B. The reason for this difference is presumably as follows. An activated rhodopsin molecule must find a G protein molecule in the membrane of the photoreceptor during the interval of time before the rhodopsin molecule is inactivated. If the amount of G protein is greatly reduced, the probability of rhodopsin molecules finding a target during this time interval is decreased, and many rhodopsins must be stimulated each by a single photon before one G protein is successfully encountered. Once the G protein is turned on, transduction proceeds, but with a smaller amplitude in mutants than in wild-type animals, because only one G protein has been activated within the lifetime of the activated rhodopsin molecules.

Noise

If a sensory receptor is to detect a weak stimulus, the gain must be high, but the noise of transduction must also be low. A sensory receptor molecule that changes its conformation in response to a single photon of light will also occasionally change its conformation spontaneously, as the result of random motion of the atoms of the receptor protein. These random motions can produce a response that the receptor cell and CNS would be unable to distinguish from the response produced by illumination. The larger the number of receptor molecules in the membrane of the cell, the greater the probability that a spontaneous response will occur. Noise can also be produced from the random opening and closing of the ion channels of the cell, or indeed from the activity of any of the enzymes along the transduction cascade. Transduction and channel noise, if large, would limit the ability of the organism to detect a sensory stimulus (Reingruber et al., 2015).

Sensory cells use several strategies to limit noise. In many cases, the sensory receptor molecules have been designed so that spontaneous changes in conformation occur very rarely, greatly reducing the probability of a spontaneous event. We shall encounter a dramatic example of this phenomenon for vertebrate photoreceptors in Chapter 9. In addition, it is very common for the CNS to average the signals of many receptors. The simultaneous occurrence of the same response in more than one receptor cell can greatly increase the reliability of detection of the sensory event.

Sex pheromone detection in the male moth

One of the most impressive examples of a sensory system operating at the physical limit of sensitivity is the detection of sex pheromones by male moths.

Pheromones are chemicals released into the environment that evoke specific behavioral responses. The first sex pheromone to be identified was an aliphatic compound called bombykol extracted from the secretion glands of female silk moths (*Bombyx mori*). No more than 1000 molecules of bombykol per cubic centimeter are required to elicit a behavioral response from a male moth. Males have large antennae (Figure 2.14) with a surface area as great as 1 cm². In the silk moth it is estimated that about a quarter of all the molecules of bombykol entering the space occupied by the antennae are absorbed onto the olfactory receptors. Detection elicits a complicated pattern of flight that brings the male moth to its amatory goal.

The antennae of the male moth contain hundreds of thousands of tiny bristles, called sensilla, and something like 40 percent of these bristles are specifically involved in the detection of sex pheromone. Like the taste receptors of flies (Figure 2.8B), these sensilla contain one or more olfactory receptor neurons surrounded by a hard shell of cuticle. In male moths, pheromone detection seems to occur primarily in large male-specific trichoid sensilla (Lee and Strausfeld, 1990), which form arches on the surface of the antennae (Figure 2.15A). Within the wall of the cuticle are numerous tiny pores leading into narrow channels, which allow odors from the air to enter (Figure 2.15B). As explained in more detail in Chapter 7, the odorant then binds to receptor molecules in the membrane of the receptor cells. Since these cells are primary sensory receptors, the

Figure 2.14 Full-frontal view of male silk moth. (Courtesy of Muriel Williams.)

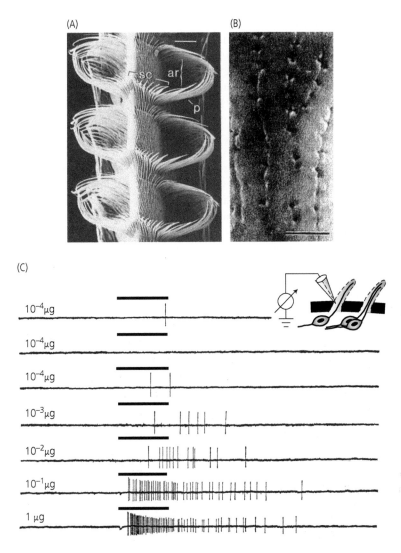

Figure 2.15 Chemosensory receptors of the male silk moth respond to single molecules of sex pheromone. (A) Long male-specific trichoid sensilla on the upper (right) and lower (left) surfaces of the antenna of the moth *Manduca sexta*. Sensilla form arches (ar), whose proximal elements (p) are connected by a group of shorter elements to form a scoop (sc) extending around the leading edge of the formation. Scale bar, 100 μM. (B) Linear array of pores on the bristle surface. Magnification 25,000×. (C) Response of a male silk moth olfactory receptor to bombykol. Bars indicate period of stimulation at concentration shown to left. See text. (A from Lee and Strausfeld, 1990; B from Riesgo-Escovar et al., 1997; C from Kaissling and Priesner, 1970.)

detection of the odor produces action potentials, which are conveyed to the brain of the moth.

In the antennae of the male silk moth, the sensitivity of detection is so great that a single sensillum can apparently detect one molecule of pheromone (reviewed in Boeckh et al., 1965; Kaissling, 1971). In Figure 2.15C, an electrode was inserted into a sen-

sillum to record the action potentials of a receptor axon. Bombykol was placed on a piece of filter paper in an air stream which could be puffed at the antenna. A clear response was produced by 10^{-3} μg bombykol, which was estimated to have released seven molecules of bombykol onto the sensillum during the 1-second period when the air flow was activated (indicated by the bars above each of the responses in the figure). The sensillum could respond even to a concentration of 10^{-4} µg bombykol. Not every puff of air at this concentration produced a response from this sensillum (Figure 2.15C), but most did, suggesting that only a single molecule is sufficient to elicit the response.

A single molecule of bombykol binding to a single sensillum would not, however, alter the behavior of the moth. When the male begins to fly toward the female, in some species from a distance of several kilometers, many of the thousands of sensilla of the antennae would each be detecting one or more molecules of sex pheromone. The behavioral response of a male moth in many species may even require the detection of more than one *type* of pheromone (Hildebrand, 1995). The sensing of these chemicals by many cells allows the moth to distinguish sex hormones from the myriad of other compounds in its environment, as well as from the spontaneous noise of transduction and random channel opening.

Summary

Sensory cells convert a stimulus into an electrical response by a process known as sensory transduction. Most sensory cells contain a specific population of receptor molecules, which are integral membrane proteins localized to a specialized region of the cell. The sensory stimulus produces a change in conformation of the sensory receptor that triggers transduction, ultimately leading to the opening and closing of channels in the receptor plasma membrane.

Sensory receptor molecules are usually targeted to specialized membrane in the cell. In many cells, this membrane is elaborated either as microvilli or as a modification of a cilium to produce elongated structures providing an extensive membrane surface for the localization of the receptor proteins. Sensory membrane may contain other molecules important for transduction, such as enzymes or ion channels, and in some receptor cells many of these molecules are held together in close proximity by scaffolding proteins. Sensory membrane must be constantly renewed. Renewal may occur by the regular and organized replacement of the membrane itself, or by the regeneration of the whole of the cell, including its axon and terminations within the CNS.

Sensory cells are often surrounded by external specializations that protect the cell or modify its response to external stimuli. There are often accessory cells that secrete external encasements or internal structures important for the mechanism of sensation. In the most complicated of the sensory organs, such as the vertebrate eye or ear, there are a great variety of different cell types which each make some contribution to the homeostasis of the organ or the detection of the stimulus.

The change in conformation of the sensory receptor molecule ultimately produces channel gating and a change in membrane potential. Despite the great variety of channel mechanisms and voltage responses, all sensory signals have common characteristics. The voltage change produced by the cell, called the receptor potential, usually shows a pronounced decline from a peak to a steady plateau during constant stimulation, reflecting a decrease in the sensitivity of the cell. This phenomenon is called adaptation. For primary receptor cells, the change in receptor potential produces action potentials communicated directly to the CNS. For secondary receptor cells, action potentials are not generated by the receptor cell itself. Instead, a synapse that is often highly specialized in structure links the receptor either to an afferent nerve terminal or to another cell, called a second-order neuron. For many sensory receptors, mechanisms have evolved to reduce the noise of transduction and increase its gain, to maximize the probability of detection. This enhanced sensitivity is nowhere more pronounced than in photoreceptors, which are able to detect single photons of light; and in the olfactory receptor cells on the antennae of male silk moths, which appear to be able to respond to a single molecule of female sex attractant.

CHAPTER 3

Channels and electrical signals

The electrical responses of sensory cells are generated by ions flowing through membrane channels. Ion channels are a large and diverse group of membrane proteins present in every cell in the body. These proteins have much in common. They are all integral membrane proteins; that is, they all contain extensive sequences of amino acids that extend from one side of the membrane to the other, embedded in the hydrophobic environment of the lipid bilayer. All contain a specialized sequence, which for many channels is called the P loop or P region. This part of the protein forms a pore or channel through which ions such as Na^+, K^+, and Cl^- can pass from one side of the membrane to the other. For most ion channels the opening of the pore is gated. That is, the protein is predominantly in a conformation with the pore closed and blocked to the passage of ions. Spontaneous openings at room temperature in most gated channels occur only rarely and briefly. In order for the pore to open, some signal must cause a change in conformation sufficiently large that some amino acids occluding the pore are moved to one side to allow ions to pass. This conformation must be sufficiently stable over a period of at least a few milliseconds so that a population of open channels can produce a change in membrane permeability large enough to produce a significant change in the membrane potential of the cell.

Structure and function of ion channels

In the first chapter, I briefly recounted the revolution in our understanding of membrane proteins brought about by the cloning of their genes, the recording of single-channel activity with patch-clamp electrodes, and the discovery of the structure of these proteins with X-ray crystallography and cryo-EM. To describe some of these discoveries in more detail, I begin with perhaps the most famous ion channel of all, the *Shaker* potassium channel. The gene for this protein was first cloned from the fruit fly *Drosophila* by taking advantage of the large chromosomes of this species and mutant animals called *Shaker*, which were originally isolated because mutant flies shake their legs uncontrollably when anesthetized with ether (Papazian et al., 1987; Tempel et al., 1987).

The *Shaker* gene codes for a protein that forms the principal subunit of an ion channel. The conformation of this protein is illustrated in Figure 3.1. Part A shows the result of a hydrophobicity plot and other earlier structural studies. There are six membrane-spanning regions, with both the amino and carboxyl termini facing the cytoplasm. Figure 3.1B gives the amino acid sequence of the P loop, which dips into the membrane between the fifth and sixth membrane-spanning sequences and forms the channel through which ions move. A similar pore sequence has now been described for a large number of different kinds of ion channels, including those used by many sensory receptor cells.

A functioning Shaker channel is formed from four protein subunits (called α subunits) like the one in Figure 3.1, which are non-covalently bound and face one another so that the P loops come together to produce the physical pore through the

Sensory Transduction. Second Edition. Gordon L. Fain, Oxford University Press (2020). © Gordon L. Fain 2020.
DOI: 10.1093/oso/9780198835028.001.0001

(A)

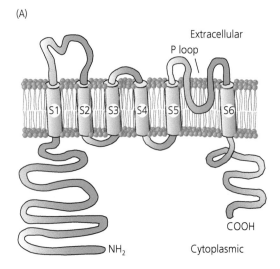

(B)

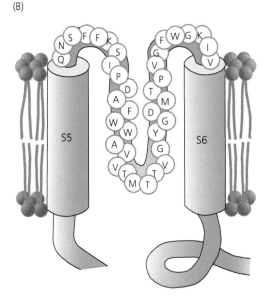

Figure 3.1 The Shaker potassium channel. (A) Membrane topology of the Shaker protein. Both amino and carboxyl termini are intracellular. S1–S6, Transmembrane domains; P loop (pore-forming domain). (B) Amino acid sequence of the P loop of the Shaker potassium channel. Individual amino acids in the sequence are indicated with single letter code (Tempel et al., 1987). (A after Isom et al., 1994; B after Brown, 1993.)

membrane. *Drosophila* contains three other genes similar to the *Shaker* gene, which each make proteins of similar structure and also form channels with a similar selectivity for potassium ions. Mam-

mals, including humans, have 17 genes for K⁺-channel α subunits similar to *Shaker* (Gutman et al., 2005). Each α protein variant has somewhat different properties, providing an enormous variety in the functioning of K⁺ channels in the cells of the body. Most mammalian K⁺ channels are thought to contain at least one additional species of structural subunit called β, which can also be coded by several different genes producing subtle changes in the properties of the assembled channel.

Although the structure of the complete *Drosophila* Shaker protein has not been solved, Figure 3.2 (from Long et al., 2005a) shows the X-ray crystallographic structure of a closely related mammalian protein called $K_V2.1$. Part A shows the channel in the plane of the membrane, and part B is the structure of the channel looking down onto its extracellular surface. Each of the four α and β subunits has been given a different color. The ion pore formed from the fifth and sixth membrane-spanning sequences (S5 and S6) is in the center of the tetramer. The other four membrane-spanning sequences (S1–S4) are located in the periphery and are largely responsible for the voltage-dependent gating of the channel, which I describe later in the chapter.

Structure of the pore

The first structural studies on ion channels were not done on *Shaker* or $K_V2.1$ but rather on the simpler $K_{CS}A$ channel from the bacterium *Streptomyces lividans* (Doyle et al., 1998). This channel has only two membrane-spanning sequences (Figure 3.3A) and was consequently easier to crystallize in these groundbreaking studies. Though this bacterial protein is simpler than most other K⁺ channels, it nevertheless has a P loop like the one in *Shaker* and $K_V2.1$, and the channel is also selectively permeable to K⁺. The structure of the $K_{CS}A$ channel provided the solution to a puzzle of long standing: how can a K⁺ channel be more permeable to K⁺ than to Na⁺, even though the diameter of the K⁺ ion is *larger* than the diameter of the Na⁺ ion? If K⁺ can permeate the channel, shouldn't Na⁺ also pass through?

A stick model of the atomic structure of the $K_{CS}A$ P loop is show in Figure 3.3B (from Alam and Jiang, 2011). The membrane-spanning surface of the channel is lined with oxygen atoms, shown as red

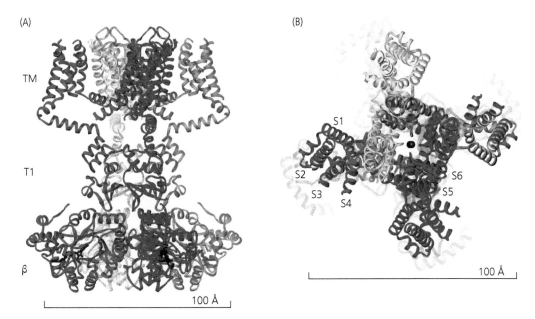

Figure 3.2 Structure of rat brain $K_V1.2$ voltage-gated potassium channel with β2 ancillary subunit. (A) Side view of the channel, with the extracellular surface above and intracellular surface below. Each of the four subunits of the channel has been given a different color. TM, Transmembrane domains; T1, tetramerization domains of each of the α channel subunits providing docking sites for the β ancillary subunits, which lie just below in the cytoplasm. (B) Top view of the channel looking down onto the extracellular surface. Each of the subunits has been given a different color, and the positions of each of helices S1–S6 are indicated for the "red" subunit. (From Long et al., 2005a.)

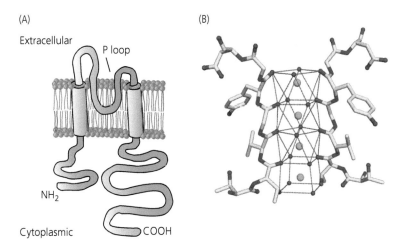

Figure 3.3 The pore of an ion channel. (A) Membrane topology of the simpler potassium channel of the bacterium *Streptomyces lividens*. (B) Magnified view of the selectivity filter; just two of the four subunits are shown. The potassium ions in the filter (green) are surrounded by a cage formed by ion-binding sites provided mainly by carbonyl oxygens (red bars and red spheres for out-of-plane oxygens). (A after Doyle et al., 1998; B from Alam and Jiang, 2011.)

spheres and bars in the figure, which are mostly coming from carbonyl oxygens of P-loop amino acids. Within the P loop there are four binding sites for potassium ions, with each ion shown in green. Each K^+ is surrounded by eight oxygen atoms from the structure of the protein, which collectively produce a complete hydration cage around each K^+ ion like the one formed by water in solution.

The hydration cage formed by the carbonyl oxygens of the four protein subunits is the secret to permeation. A potassium ion in solution is fully hydrated by water and loses its shell of water molecules with difficulty, because the energy of hydration is rather large. A potassium ion can nevertheless enter into the channel pore, because the water molecules in the solution are replaced by the carbonyls of the amino acids lining the channel. The carbonyls are positioned the appropriate distance from the K^+ ion to mimic the hydrating water molecules, and the K^+ ion can therefore lose its hydrating shell and gain hydrating amino acid carbonyls with a minimal change in free energy. Furthermore, once an ion enters the channel, it can move freely and rapidly from one binding site to the next, and it can leave the pore and enter the external medium by regaining its shell of water molecules. A Na^+ ion can pass through the selectivity filter much less readily because it is too large to enter with its shell of water molecules, and it is too small to replace its water of hydration with the oxygens of the backbone carbonyls. The center of the Na^+ ion is too far away from the channel carbonyls to be easily hydrated.

Similar P loops are present in other channels, including those selectively permeable to Na^+ and Ca^{2+}, and it is likely that the structure of the P loop in all of these channels is similar. The pore of Na^+ channels may be narrower, making possible the coordination of the sodium ion presumably again with carbonyls of amino acids within channel.

Gating

The *Shaker* and $K_v2.1$ K^+ channels are two of a large number of proteins known to be gated by a change in voltage. Voltage-dependent gating is largely the result of charged amino acids within the S1–S4 sequences (Figure 3.2B), which are positioned in the membrane and move when the voltage across the

membrane changes (Figure 3.4A). The S4 sequences of both *Shaker* and $K_v2.1$ contain positively charged amino acids every third residue, forming a gentle spiral of positivity down the S4 α helix across the membrane. When the potential of the inside of the cell becomes more positive (depolarizes), these amino acids can be repelled by the increase in positivity inside the cell and move outward. Other amino acids (such as glutamate and aspartate) which are negatively charged can be attracted and can move inward. The motion of amino acids produces a change in the conformation of the protein which, if sufficiently large, opens the channel. A similar mechanism is responsible for the gating of voltage-gated Na^+ and Ca^{2+} channels.

Voltage-gated Na^+ and K^+ channels produce action potentials and are found in the axons and somata of cells of the CNS. They are also present in primary sensory cells (Chapter 2) and provide the rapid conduction of signals from sense organs into the brain. Many cell types, including, for example, secondary sensory receptor cells, contain voltage-gated channels selective for Ca^{2+}, which regulate the entry of Ca^{2+} into the presynaptic terminal and control the rate of release of synaptic transmitter.

In addition to voltage gating, there are other common mechanisms of channel gating caused by the binding of small-molecular-weight ions or metabolites to the channel, either directly or in association with a soluble binding protein. In one large class of channels, the binding site is located externally (Figure 3.4B), and the small-molecular-weight molecule diffuses to its binding site through the extracellular medium. These channels include a group of proteins called ligand-gated channels, which are gated by synaptic transmitters such as acetylcholine, glutamate, glycine, or GABA released into the extracellular space. The binding of these molecules to their binding sites triggers a change in the conformation of the receptor to open the channel pore.

For a third group of channels, the binding site is intracellular (Figure 3.4C). These channels are called second-messenger-gated channels, because the opening of the channel is gated by a second messenger diffusing either within the cytoplasm or within the membrane itself. Second-messenger-gated channels that bind the cyclic nucleotides cAMP and cGMP are essential components of the transduction

(A)

(B)

ACh

(C)

cGMP

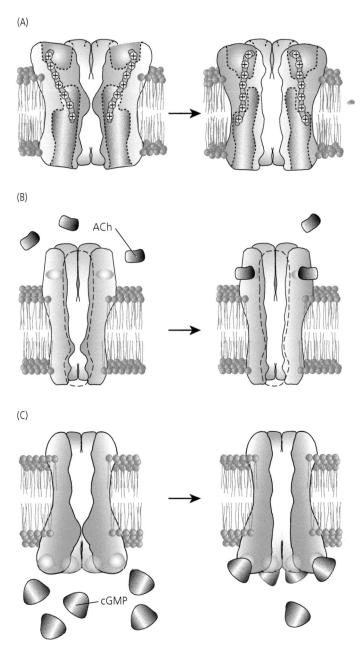

Figure 3.4 Gating of ion channels. (A) Voltage gating. (B) Gating by binding of a small-molecular-weight molecule such as the synaptic transmitter acetylcholine (ACh) to an extracellular binding site. (C) Gating by binding of a small-molecular-weight second messenger, such as cGMP, to an intracellular binding site.

mechanism of vertebrate photoreceptors and olfactory receptors. Many cells including sensory receptors also contain channels gated by other second messengers such as Ca^{2+}, which can bind to the channel either by itself or in association with the small-molecular-weight protein calmodulin. A channel permeable to Cl^- that is gated by intracellular Ca^{2+} has an important role in producing the electrical response of vertebrate olfactory receptors.

Ionotropic receptor molecules

Channels can also be directly gated by a sensory stimulus. An important group of sensory receptor cells called mechanoreceptors underlie our senses of touch, proprioception, hearing, and equilibrium (Chapters 5 and 6). These cells are all thought to contain proteins like the piezo channels of Chapter 1, specialized for the detection of pressure or deformation of the plasma membrane.

Pressure-sensitive channels were discovered not long after the development of patch-clamp recording, in experiments placing patch-clamp electrodes directly on the surface of a skeletal muscle fiber to record single-channel activity (Brehm et al., 1984). When pressure was applied to the membrane through the bore of the pipette, either by gentle sucking or by blowing, the probability of opening of a channel was dramatically altered (Figure 3.5A). Because similar recordings could be obtained in a membrane patch completely excised from the cell (Figure 3.5B, from Guharay and Sachs, 1984), the opening of the channel must be produced directly by pressure to the membrane, that is ionotropically as in Figure 2.1A, without the necessity of transduction enzymes or second messengers.

Although we are uncertain why muscle cells contain channels of this kind, we have excellent evidence that mechanoreceptors can also have them. Some of these channels (like the piezo channels) are able to respond to pressure even in artificial lipid membranes (Syeda et al., 2016), which is evidence that the proteins are inherently mechanosensitive. Other mechanoreceptor channels seem to require intracellular or extracellular accessory proteins attaching the channel to the structure of the inside and outside of the cell. I describe some of these adaptations in more detail in Chapter 5.

Another important group of ionotropic receptor molecules can mediate sensitivity to temperature. The first protein of this kind to be identified was shown to respond primarily to noxious heat at temperatures above 40°C. This channel can also be gated by capsaicin, which is the active ingredient in hot pepper. This feature turned out to be the key to

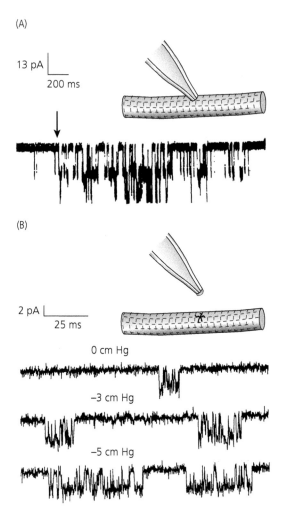

Figure 3.5 Stretch-activated channels. (A) On-cell recording from skeletal muscle fiber from the tail of a *Xenopus* tadpole. Negative pressure was applied to the patch pipette at the arrow and produced an increase in the frequency of channel opening. (B) Recording from an excised, inside-out patch from embryonic chick skeletal muscle. The numbers above the traces give the value of suction applied to the patch pipette in units of centimeters of mercury (Hg). (A from Brehm et al., 1984; B from Guharay and Sachs, 1984.)

identifying the protein. Clones from a cDNA library (as in Figure 1.8) were transfected into cultured cells (as in Figure 1.9) and tested for responses to capsaicin. The protein identified in this way belonged to the class of transient-receptor potential (TRP) proteins and is now usually referred to as TRPV1.

The sequence of TRPV1 indicates that it is quite similar in structure to the *Shaker* protein, with six transmembrane domains, a putative P loop between S5 and S6, and cytoplasmic amino and carboxyl termini (Figure 3.6A). Near the intracellular amino terminus, there are six ankyrin repeats, which are stereotypic sequences of thirty-three amino acids containing binding sites for ATP, calmodulin, and other modulators of protein activity (Li et al., 2006;

Lishko et al., 2007). Like the piezo proteins, the TRPV1 channel can respond to heat even in an artificial lipid membrane (Figure 3.6B) and seems therefore to be intrinsically sensitive to elevated temperature. Results from site-directed mutagenesis of the TRPV1 sequence suggest that regions around the P loop are essential for temperature

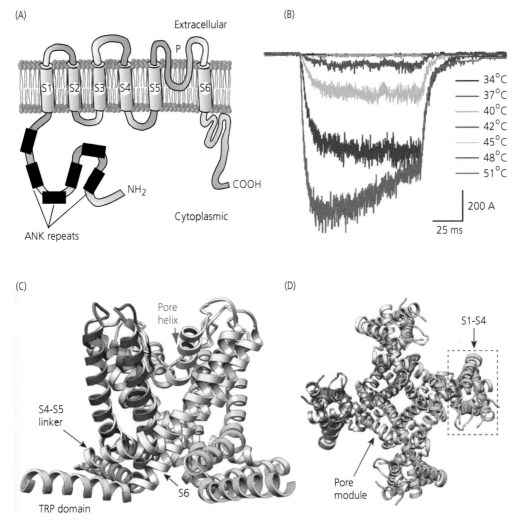

Figure 3.6 Temperature-sensitive channel. (A) Membrane topology of TRPV1 protein inferred from hydropathy analysis and structural and labeling techniques. Both amino and carboxyl termini are intracellular. S1–S6, Transmembrane domains; P, P loop forming channel pore; ANK, ankyrin repeats. (B) Patch-clamp recordings of inside-out patches taken from artificial lipid membranes (liposomes) containing reconstituted TRPV1 protein. The patches were rapidly heated for 100 ms with an infrared laser to the temperatures shown for each trace. Elevations of temperature produced channel opening and inward (negative-going) currents from an increase in permeability to cations. (C, D) Cryo-EM structure of the TRPV1 protein. The assembled channel is a tetramer whose four subunits are indicated with different colors. The channel is shown from the side with the extracellular surface above (C), and from the bottom looking onto the intracellular surface (D). (A after Julius, 2013; B from Cao et al., 2013b; C and D from Liao et al., 2013.)

sensitivity (Grandl et al., 2010; Yang et al., 2010), but other regions of the protein near the amino terminus seem also to contribute (Yao et al., 2011; see Zheng and Ma, 2014). I return to the mechanism of channel gating when I describe thermoreception in more detail in Chapter 10.

The three-dimensional structure of the TRPV1 protein (Figure 3.6C and D) is quite different from that of the piezo protein and bears a striking resemblance to the structure of $K_V2.1$ and other voltage-gated channels. The assembled TRPV1 is a tetramer, whose S5 and S6 sequences together with the pore loop form the center of the channel protein and provide the pathway for ion conduction. The S1–S4 sequences are located on the periphery. In $K_V2.1$ these sequences contain charged amino acids responsible for voltage-dependent gating, but the S1–S4 sequences of the TRPV1 receptor lack these amino acids. Their primary role may be structural (Cao et al., 2013a).

Membrane potentials

When stretch-sensitive or temperature-sensitive channels open, they produce a change in the permeability of the membrane to ions, which in turn alters the voltage of the cell. These changes in membrane voltage are the signals produced by the sensory stimulus, which are communicated to the CNS where they are processed so as to be recognized, understood, and remembered. Because changes in membrane potential are so important to the physiology of sensory cells, it is essential to understand how the voltage across a cell membrane is produced, and how it is altered when channels open and close.

Cells at rest in the absence of a sensory (or neural) stimulus have a resting membrane potential determined both by the concentration of ions on the two sides of the membrane and by the resting permeabilities of the cell to these ions. Most cells in the body, including neurons and most sensory receptor cells, have a much higher concentration of K^+ in the cytoplasm than in the extracellular fluid. The Na^+ concentration is also asymmetric, with the concentration inside the cell considerably smaller than that outside (Figure 3.7). These asymmetries produce driving forces which, with everything else being

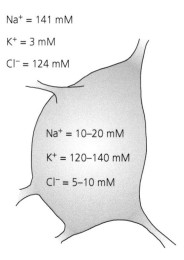

Na⁺ = 141 mM

K⁺ = 3 mM

Cl⁻ = 124 mM

Na⁺ = 10–20 mM

K⁺ = 120–140 mM

Cl⁻ = 5–10 mM

Figure 3.7 Intracellular and extracellular concentrations of Na⁺, K⁺, and Cl⁻ in a typical neuron. Note asymmetric concentrations of Na⁺ and K⁺ inside and outside the cell.

equal, would tend to move K^+ out of the cell and Na^+ inside.

The differences in the concentrations of Na^+ and K^+ on the two sides of the membrane are produced by a very important protein called the Na^+/K^+ ATPase. This protein is a multimer, whose largest subunit (called the α subunit) is responsible for the actual transport of ions (see Morth et al., 2011; Toyoshima et al., 2011). There are two additional, smaller subunits, which can regulate pump activity and are essential for assembly and targeting of the protein.

The mechanism of ion transport by the α subunit is as follows (see Figure 3.8 and Kuhlbrandt, 2004). The Na^+/K^+ ATPase can assume two basic conformations, called E_1 and E_2. When three Na^+ ions and ATP bind to E_1 on the intracellular side of the protein forming $E_1 3Na \cdot ATP$ (upper left), the protein can catalyze the phosphorylation of an internal aspartate residue to form $E_2 3Na \cdot P$, with the ion binding sites now facing the extracellular space. This change in conformation happens because phosphorylation of the protein places the negative charges of a phosphate group directly onto the structure of the protein, pushing other charged amino acids toward or away from the phosphate group. This conformational change alters the affinity of the ion binding sites, so that they have a lower affinity for Na^+ and a higher affinity for K^+. The Na^+

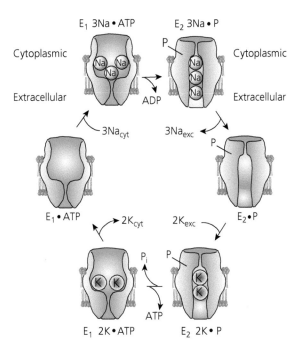

E_1 3Na • ATP E_2 3Na • P

Cytoplasmic   Cytoplasmic

Extracellular   Extracellular

ADP

3Na$_{cyt}$ 3Na$_{exc}$

E_1 • ATP E_2 • P

2K$_{cyt}$ 2K$_{exc}$

P$_i$

ATP

E_1 2K • ATP E_2 2K • P

Figure 3.8 Pumping cycle of the Na$^+$/K$^+$ ATPase.

ions leave the protein and diffuse into the extracellular space, forming E_2•P and exposing the binding sites to K$^+$. The binding of two K$^+$ ions and release of the phosphate changes the conformation of the protein back to E_1, followed by rebinding of ATP to form $E_1$2K•ATP. The K$^+$ ions are released into the cytoplasm, and the cycle starts all over again.

The Na$^+$/K$^+$ ATPase, then, catalyzes the movement of three Na$^+$ out of the cell for two K$^+$ inward with the expenditure of the energy of a single high-energy phosphate bond of ATP. The pumping is electrogenic: three Na$^+$ are moved outward for every two K$^+$ inward, producing a net movement of one positive charge out of the cell with each cycle of the pump. There is also a net transport of one osmotically active particle out of the cell, which is essential for the maintenance of osmotic equilibrium—not only for neurons and sensory cells, but for every cell in the body.

The Nernst equation

Imagine now a cell in which the Na$^+$/K$^+$ ATPase has transported enough Na$^+$ and K$^+$ so that there is con-

siderably more K$^+$ inside than outside (Figure 3.9A). Suppose also that the membrane of the cell at rest is permeable only to K$^+$, even though other ions such as Na$^+$ and Cl$^-$ are also present in the cytoplasm and extracellular space. At a molecular level, we can think of this resting permeability as a population of ion channels in the cell membrane, which are not voltage-dependent like K$_V$2.1 but are open spontaneously at the resting membrane potential. Neurons have K$^+$-selective channels with some of these properties, which are called two-pore K$^+$ (K2P) channels because each channel monomer has two P loops. The channel pore is still formed by four P loops like K$_V$2.1 and TRPV1, because K2P proteins assemble as dimers rather than as tetramers (Miller and Long, 2012). These channels may be largely responsible for the resting membrane potential of neurons in the CNS.

Our hypothetical cell has more K$^+$ inside than outside, so K$^+$ will tend to leave the cell through K$^+$ channels by diffusion. At a molecular level, the potassium ions inside and outside the cell occasionally collide with the pore of one of the spontaneously open channels, and some of the ions colliding

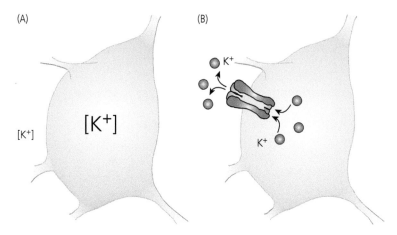

Figure 3.9 Membrane potential of a cell that is permeable only to potassium ions: the Nernst equation. (A) Gradient of K+ concentration. Size of letter indicates relative concentration, which is greater inside the cell as in Figure 3.7. (B) Concentration gradient will tend to move K+ out of the cell, making the inside of the cell negative.

with the pore then pass across the membrane. Since the concentration of K+ is greater inside the cell than outside, the probability that K+ will collide with the pore and pass across the membrane will be greater for K+ inside the cell than outside. This process of diffusion will accordingly produce a net exit of a small number of K+ ions from the cell. The departure of K+ will in turn generate a net decrease in the amount of positive charge inside the cell, with the result that the voltage inside will become negative. The more K+ that leaves, the greater the negativity of the inside of the cell.

The electrical negativity inside the cell will produce a difference in voltage across the plasma membrane, tending to prevent K+ from leaving. If this voltage difference becomes sufficiently large, it prevents any further net movement of K+ out of the cell. It is important to understand that the deficit of potassium ions required to produce this voltage difference is extremely small in comparison to the total amount of K+ in the cytoplasm and has almost no effect on the concentration of K+ inside the cell. Small changes in the concentration of charged particles inside the cell can produce large changes in voltage, because the capacitance of the lipid bilayer of the plasma membrane is so large. This small exit of K+ can nevertheless produce a negative voltage sufficiently large to restrain K+ efflux. And if the cell membrane is permeable only to K+, the effects of

voltage and diffusion can exactly cancel, leaving the cell in equilibrium. Potassium ions can still enter and exit the cell through the K+ channels, but the number of ions entering exactly equals the number leaving, and there is no net flux across the membrane.

For a cell like the one in Figure 3.9 permeable only to K+, the membrane potential at equilibrium can be shown to be given by the Nernst equation:

$$V_m = \frac{RT}{F} ln \frac{[K]_o}{[K]_i} \qquad (3.1)$$

where R is the gas constant, F is the Faraday (also a constant), T is the absolute temperature, $[K]_o$ is the concentration of K+ outside the cell, $[K]_i$ is the concentration inside the cell, and V_m is the membrane potential, inside minus outside. At room temperature, RT/F is approximately 0.025 volts (V) or 25 millivolts (mV). For a typical nerve cell with a resting $[K]_o$ of 120 mM and a $[K]_i$ of 3 mM (as in Figure 3.7), the Nernst equation predicts a membrane potential of approximately –0.09 volts, or –90 mV at room temperature, with the inside of the cell negative to the outside.

In the more general case, for an ion c whose concentrations are given by $[c]_o$ outside and $[c]_i$ inside, the Nernst equation is

$$V_m = \frac{RT}{zF} ln \frac{[c]_o}{[c]_i} \qquad (3.2)$$

where z is the valence (or charge) of c.

Ion homeostasis

The Nernst equation gives the membrane potential of an ideal cell permeable only to a single ion, in our example K^+. For real cells such as neurons in the CNS or sensory receptor cells in the nose or ear, life is not so simple. These cells at rest are often permeable mostly to K^+ but always have some permeability to other ions. They are, for example, usually permeable to Na^+, either because Na^+ moves with low probability through resting K^+ channels, or

because there are other channels open at rest and selectively permeable to Na^+.

Imagine now a cell mostly permeable to K^+ but with a small additional permeability to Na^+. Suppose again that there is considerably more K^+ inside than outside, but also (as in Figure 3.7) more Na^+ outside than inside. For such a cell (Figure 3.10A), the dominant K^+ permeability might be thought to produce a resting membrane potential much as before, by K^+ moving out of the cell. This situation can, however, never occur in practice. Although K^+

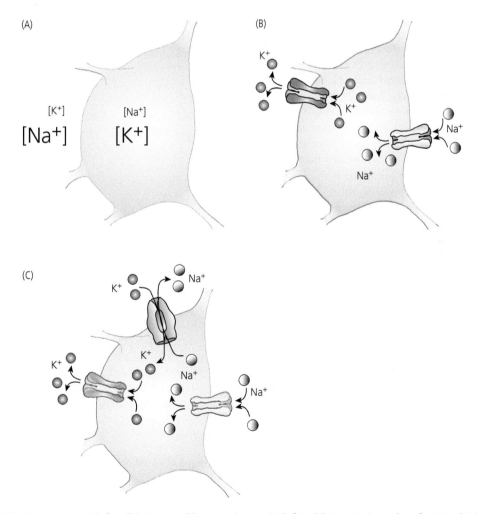

Figure 3.10 Membrane potential of a cell that is permeable to more than one kind of ion. (A) Concentration gradients for Na^+ and K^+ in a typical cell at rest. (B) Predicted movements of Na^+ and K^+ resulting from their concentration gradients at rest. (C) Steady state, produced by passive movement of Na^+ and K^+ through membrane channels and energy-dependent pumping by the Na^+/K^+ ATPase.

will move down its concentration gradient, the exit of positive charge produced by the outward movement of K^+ can be compensated by the inward movement of Na^+ (Figure 3.10B). Na^+ will move inward because there is more Na^+ outside the cell than inside, and also because the charge imbalance produced by the exit of K^+ will tend to make the inside of the cell negative, and this potential difference will attract Na^+. In the absence of active ion transport, K^+ and Na^+ would continue to move across the membrane, eventually nearly equalizing the concentrations of both ions inside and outside. For a cell permeable *only to a single ion*, the effects of diffusion and voltage can exactly balance one another, producing a true chemical equilibrium. In a real cell, equilibrium of this sort can never occur.

To the rescue comes the Na^+/K^+ ATPase (Figure 3.10C). As Na^+ and K^+ enter and leave the cell passively through membrane channels, this protein pumps them back again. The cell never reaches a true equilibrium but does achieve a steady state, at which there is no net movement of either Na^+ or K^+ into or out of the cell. Ion movement passively through membrane channels is exactly balanced by active transport by the pump, so that there is no change with time in the charge, voltage, or ion concentration on the two sides of the membrane. It is this steady state that is responsible for the membrane potential of a real cell at rest.

The Goldman voltage equation

Because a resting cell is permeable to more than one ion, we cannot use the Nernst equation to calculate the value of the resting membrane potential but must instead use the Goldman voltage equation, derived from a consideration of the forces acting on ions across the membrane rather than from the laws of chemical equilibrium (for a more detailed explanation see, for example, Fain, 2014). This equation takes into consideration the concentrations and permeabilities of all of the ions capable of moving through the membrane and has the form

$$V_m = \frac{RT}{F} ln \frac{P_{Na}[Na]_o + P_K[K]_o + P_{Cl}[Cl]_i}{P_{Na}[Na]_i + P_K[K]_i + P_{Cl}[Cl]_o} \quad (3.3)$$

where P_{Na}, P_K, and P_{Cl} are the permeabilities of the cell to Na^+, K^+, and Cl^-, and the terms $[Na]$, $[K]$, and

$[Cl]$ are concentrations inside or outside the cell. A term for Cl^- has been included in Eq. (3.3) for those cells showing a significant Cl^- permeability. For many neurons and sensory cells, the Cl^- permeability at rest is so small that it can be ignored, giving

$$V_m = \frac{RT}{F} ln \frac{P_{Na}[Na]_o + P_K[K]_o}{P_{Na}[Na]_i + P_K[K]_i} \quad (3.4)$$

If we let α be the ratio of the Na and K permeabilities (P_{Na}/P_K) and divide the numerator and denominator of Eq. (3.4) by P_K, we have

$$V_m = \frac{RT}{F} ln \frac{\alpha[Na]_o + [K]_o}{\alpha[Na]_i + [K]_i} \quad (3.5)$$

Suppose now that $[K]_o$ is 3 mM and $[K]_i$ 120 mM, as before; $[Na]_o$ is 140 mM and $[Na]_i$ 15 mM; and α is 0.02, a typical value for a resting neuron. The resting potential becomes about –75 mV at room temperature instead of –90 mV, closer to the resting potential actually measured from an olfactory receptor cell or a pyramidal cell in the cerebral cortex.

For most neurons and sensory cells at rest, the membrane potential is the result of a steady state produced by the passive movement of Na^+ and K^+ through membrane channels and active transport by the Na^+/K^+ ATPase. The value of the resting potential is negative inside but can vary from –35 mV to –90 mV, depending upon the magnitude of the resting Na^+ permeability. Nearly every cell in the body also actively removes Ca^{2+}, whose concentration is a few millimolar in the blood and cerebrospinal fluid but of the order of 100 nanomolar or 10^{-7} M inside the cell—four orders of magnitude lower than the concentration outside. Cells extrude Ca^{2+} in part with a plasma membrane Ca^{2+} ATPase molecularly similar and closely related to the Na^+/K^+ ATPase. Most cells also have a transport protein that exchanges Na^+ for Ca^{2+} (or Na^+ for Ca^{2+} and K^+), using the inwardly directed concentration gradient for Na^+ to transport Ca^{2+} outward.

Driving force and voltage change

The asymmetric movement of ions across the membrane produced by the Na^+/K^+ ATPase and by other ion transport proteins generates a concentration gradient, which, together with the membrane potential, can produce a driving force tending to move ions into or out of the cell. The driving force has two components: a diffusive component produced by the difference in ion concentration, and an electrophoretic component produced by the voltage difference across the membrane. When the ion channels in the membrane open or close, the membrane permeability can change, producing a change in membrane potential. In a sensory cell, this change in potential is the signal that ultimately produces our conscious perception.

We can calculate the value of the change in membrane potential from the Nernst and Goldman equations I have just introduced. Let us take a simple example. Imagine a neuron (say an olfactory receptor cell in the vertebrate nose) that has concentration gradients resulting from the pumping of the Na^+/K^+ ATPase, so that there is more K^+ inside and more Na^+ outside, much as I have described. Let the resting membrane be considerably more permeable to K^+ than to Na^+, so that the value of the resting potential of the cell is -75 mV, a typical value. Imagine now that a very large number of channels in the membrane of the cell suddenly open and that these channels are permeable to Na^+, as actually occurs during the response of this cell to an odor. What will happen to the membrane potential?

The answer can be derived from Eq. (3.5),

$$V_m = \frac{RT}{F} ln \frac{\alpha [Na]_o + [K]_o}{\alpha [Na]_i + [K]_i}$$

where α is again the ratio of the Na^+ to K^+ permeabilities, P_{Na}/P_K. At a resting potential of -75 mV, α is of the order of 0.02, as we have seen. If channels open and the permeability of the membrane to Na^+ increases, P_{Na}/P_K will increase. *Because $[Na]_o$ is greater than $[Na]_i$, any increase in P_{Na}/P_K (that is, in α) will cause the membrane to become more positive*, provided of course that there is no significant change in the concentrations of the ions, as is very likely to be the case. As α increases, the numerator of Eq. 3.5 becomes progressively larger than the denominator. Such a positive change in membrane potential is called a depolarization. If the increase in α is large enough, the membrane will act as if it is permeable only to Na^+, and Eq. 3.5 becomes very nearly the Nernst equation for Na^+,

$$V_m = \frac{RT}{F} ln \frac{[Na]_o}{[Na]_i}$$

At the concentrations in Figure 3.8, V_m would be between $+50$ and $+60$ mV, which is about the value of the membrane potential at the peak of a Na^+-dependent action potential.

In primary sense cells, a depolarization produced by the opening of channels permeable to Na^+ can stimulate action potential firing directly. In secondary sense cells, a depolarization can increase the rate of release of synaptic transmitter onto second-order cells, whose action potential firing will be accentuated. If, on the other hand, α were to decrease, because either channels permeable to Na^+ closed or channels permeable to K^+ opened, the membrane potential would become more negative, called a hyperpolarization. In secondary sense cells, a hyperpolarization would decrease the rate of release of synaptic transmitter onto second-order cells, whose action potential firing would then be attenuated. In this way, changes in channel opening and closing produce the signals sensory cells use in order to communicate to the rest of the nervous system.

The voltage response of hair cells

To show for a real cell how channel opening produces a voltage change, I take as my example the hair cell. Hair cells are sensory receptor cells used by the vestibular system to monitor balance and equilibrium, by the ear to detect sound, and by the lateral line organs of fish and aquatic amphibians to sense movement in water. As described in Chapter 6, considerable evidence indicates that hair cells are ionotropic sensory receptor cells—cells that produce a sensory signal with a protein that is itself an ion channel. They are also secondary sense cells, which do not generate action potentials. Instead they synapse onto the terminals of the eighth cranial nerve, which can convey information about hair cell responses into the CNS. Because hair cells do not produce spikes, they provide a particularly simple system to explore how the membrane potential is changed by sensory stimulation.

The response of a hair cell is generated by the movement of microvilli like those in Figure 2.2 but arranged in a characteristic hair bundle (Figure 3.11A). In addition, there is often (though not always) a single large cilium, called a kinocilium. The tips of the kinocilium and microvilli are usually embedded in some external structure, which can often be removed by the experimenter to allow direct manipulation of the microvilli and recording from the cells.

This method was exploited by Hudspeth and Corey (1977) for hair cells of the bullfrog sacculus, one of the organs of the bullfrog inner ear. The hair cells in the sacculus are normally covered by a dense otolithic membrane, but this membrane was peeled away so that a fine intracellular microelectrode could be inserted directly into the hair cell to record the membrane potential. Another pipette

with a larger-diameter tip (labeled *Stimulus probe* in Figure 3.11A) was slipped over the tips of the microvilli and moved in a precise fashion, to mimic the normal mode of stimulation of the cell. In these experiments, the outside of the cell was perfused with a typical amphibian saline solution containing 113 mM Na^+ and 2 mM K^+, not very different from the concentrations of these ions in mammalian blood or cerebrospinal fluid (Figure 3.7).

When the stimulus probe was moved in a direction toward the kinocilium and the largest of the microvilli, the hair cell membrane potential depolarized from its resting potential. When the probe was moved in the opposite direction, the membrane potential hyperpolarized, reaching a value a few

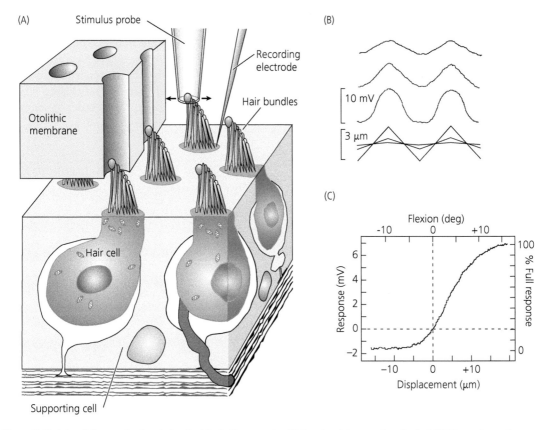

Figure 3.11 Intracellular recording from hair cells of the bullfrog sacculus. (A) Experimental preparation. See text. (B) Change in membrane potential. The three superimposed traces at the bottom of the figure show triangle waves (10 Hz) for three amplitudes, used as the signal to move the stimulus probe. Positivity indicates movement towards the kinocilium. The upper traces show the responses of the hair cell to these three stimuli. Each trace is the average of thirty-two trials. (C) Curve relating the change in membrane potential to the amplitude of hair bundle movement. Zero displacement is the resting position of the hair bundle. Positive displacements are toward the kinocilium; negative displacements are away from the kinocilium. Note the asymmetry of the curve (see text). (After Hudspeth and Corey, 1977.)

millivolts below the resting potential (Figure 3.11B). The amplitude of the voltage change increased with increasing movement of the hair bundle but was asymmetric: depolarizations were larger than hyperpolarizations (Figure 3.11C). When the stimulus probe was moved sideways, there was no response.

The technique of voltage clamping

The results of Figure 3.11 are by themselves consistent with two very different explanations of the hair cell response. On the one hand, the deflection of the hair bundle in the direction toward the largest microvilli and kinocilium might open channels (Figure 3.12A), increasing the permeability of the membrane to Na^+ to produce a depolarization as I have previously explained. If the channels are partially open at rest, deflection in the opposite direction might close Na^+ channels, decreasing P_{Na}/P_K and sending the membrane potential to a more hyperpolarized value. The curve in Figure 3.11C would then suggest that about 20 percent of the Na^+ channels are open in the absence of any stimulation.

Alternatively, we can imagine that motion in the opposite direction against the kinocilium could open channels permeable to K^+ (Figure 3.12B). Such a change in permeability would decrease P_{Na}/P_K and produce a negative-going hyperpolarization. Motion toward the kinocilium might then close such channels, increasing α and depolarizing the cell in a way entirely consistent with the results in Figure 3.11B and C. According to this hypothesis, 80 percent of the channels would be open at rest.

The key difference between these two hypotheses is that, in one case, movement toward the kinocilium increases membrane conductance (Figure 3.12A);

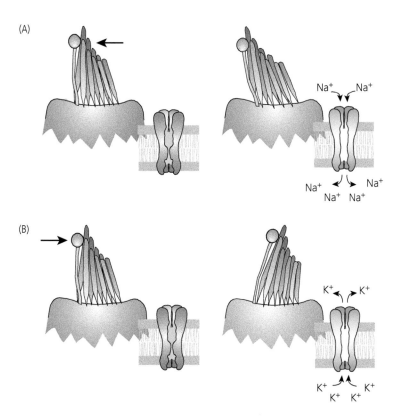

Figure 3.12 Two possible mechanisms for production of hair cell response. (A) Movement of hair bundle toward the largest microvilli and kinocilium opens channels permeable to Na^+. (B) Movement of hair bundle in the opposite direction opens channels permeable to K^+.

and in the other case, movement in this direction causes the membrane conductance to decrease. These alternative explanations can be distinguished with the technique of voltage clamping. A voltage clamp is an electronic device that uses a feedback circuit to set the voltage of the cell at a predetermined value, called the command or holding potential. The circuit then keeps the membrane potential at this value, even when ion channels open or close and the permeability of the membrane changes.

To understand the significance of voltage clamping, it is helpful first to realize that an ion channel is a kind of electrical conductor. As for any conductor, charge moves through the channel in the presence of a voltage gradient. For a simple conductor such as a wire, the amount of current that flows is equal to the product of the voltage difference across the conductor and the value of the conductance, g:

$$i = gV \qquad (3.6)$$

Conductance is measured in Siemens (abbreviated S). Since conductance is the inverse of resistance ($g = 1/R$), Eq. (3.6) is just Ohm's law, $V = iR$.

For ion channels in cell membranes, the current cannot be calculated from an expression as simple as Eq. (3.6). For a wire, the flow of current is determined only by the voltage difference across the conductance. For a channel, the flow of current is a result of the electrophoretic force produced by the membrane potential but also the diffusive force generated by the concentration difference of ions inside and outside the cell. Instead of Eq. (3.6), we must use

$$i_m = g\left(V_m - E_{rev}\right) \qquad (3.7)$$

where i_m is the membrane current, g is the conductance of the membrane, V_m is the membrane potential, and $(V_m - E_{rev})$ is the total magnitude of the driving force.

E_{rev} is called the reversal potential. It is the membrane potential at which no current flows through the channel even when the channel is open. No current flows because when E_{rev} is equal to V_m, the driving force is zero and i_m is zero regardless of the value of g. The value of E_{rev} is determined by the concentration gradients and ion permeabilities of the membrane and can be calculated from the

same equations we used for the resting membrane potential. Now, however, it is not the channels at rest for which we make this calculation, but the channels gated by the sensory stimulus.

Equation (3.7) says that net current can flow through the channels of the hair cell only when the membrane potential differs from the reversal potential, and the larger the difference, the greater the current through the channel. The sign of the driving force determines the sign of the current. When V_m is positive of E_{rev}, the current is positive and outward, from the cytoplasm into the extracellular medium. When V_m is negative of E_{rev}, the current is negative and inward. The current therefore reverses sign for voltages on either side of the reversal potential. We will see this effect more clearly when we examine actual recordings from voltage-clamped cells.

Equation (3.7) makes apparent the principal advantage of voltage clamping. A voltage clamp holds the membrane potential V_m at a constant value. Provided there is no change in the concentrations of ions during the voltage-clamp recording, E_{rev} and therefore ($V_m - E_{rev}$) will also remain constant. Then, from Eq. (3.7), the current measured by the voltage-clamp circuit can be seen to be directly proportional to membrane conductance and can therefore be used to measure conductance. The conductance, in turn, depends only on the opening and closing of channels in the membrane. When a sensory receptor cell such as a hair cell is voltage clamped and then stimulated, the net current recorded by the voltage-clamp circuit is produced only by flow of ions through the mechanoreceptive channels. It is equal to the product of the number of channels opened by the sensory stimulus, the conductance of each of the channels, and the value of the driving force set by the experimenter and held constant by the voltage-clamp circuit.

Voltage clamp of the hair cell

The first voltage clamp recordings from hair cells were made by inserting two intracellular microelectrodes into a single cell (Corey and Hudspeth, 1979a), one to measure voltage and the other to apply current. In Figure 3.13A, the voltage electrode on the left provided one of the inputs to a voltage-recording amplifier (labeled V_m). The other input to this amplifier was connected to an electrode that was inserted into the solution bathing the cell. The potential of the solution is usually referred to as "ground" (in the USA) or "earth" (in the UK) and is represented in the diagram by the open triangle connected to the lower input of the amplifier labeled V_m. This amplifier therefore measures the difference in potential between the inside and outside of the cell, which is equal to the membrane potential of the cell, inside minus outside. The output of this amplifier was led into one of the inputs of a feedback amplifier, labeled *FBA* in Figure 3.13A. The other input into this amplifier is the command voltage (V_c), which was set by the experimenter.

The key to understanding the operation of the voltage clamp is the feedback amplifier. This amplifier senses the difference in the voltages V_m and V_c at its two inputs and injects a current i_m into the current electrode that passes across the membrane of the cell. The sign and amplitude of i_m are automatically adjusted by the feedback amplifier to bring V_m as close to V_c as the gain of the feedback circuit will permit. V_m and V_c in an actual experiment typically differ by no more than 0.1 percent. A well-designed feedback amplifier achieves this result so rapidly and accurately that for all practical purposes the membrane potential of the cell can be assumed to remain very nearly equal to the value of the command potential during the response to the stimulus. The circuit can therefore give the value of the membrane current that flows through the mechanoreceptive channels at the membrane voltage set by the command potential, and the value of this current is a direct measure of the membrane conductance of the cell.

Let us now suppose that the hair cell has been voltage clamped at a command potential equal to the resting membrane potential, which for the hair cell in the preparation of Figure 3.13A was approximately –60 mV. At the resting membrane potential, the cell is at steady state, and there is no net current across the membrane of the cell. Thus for the cell at rest and held at the resting membrane potential, no current would need be injected into the cell by the voltage-clamp circuit, and i_m is kept at zero. Any departure of i_m from zero would move the membrane potential away from the resting potential. But because the circuit is designed to keep the membrane potential at the command potential, and because (in this case) the command potential *is* the resting membrane potential, i_m remains unchanged.

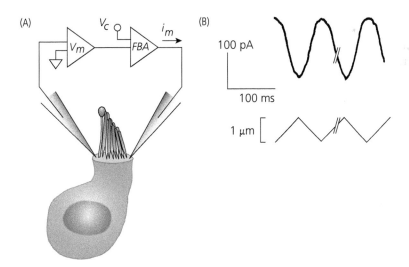

Figure 3.13 Voltage clamping the hair cell. (A) Method of voltage clamping. The hair bundle was moved with a stimulus probe as in Figure 3.11A. *FBA*, Feedback amplifier of voltage clamp circuit; V_m, membrane potential of cell; V_c, command or holding potential; i_m, membrane current. (B) Voltage-clamp current recorded as in (A) from bullfrog saccular hair cell. The lower trace shows movement of the stimulus probe as in Figure 3.11B, positivity indicating movement toward the kinocilium. The upper trace gives the membrane current at a holding potential of –60 mV, near the normal resting potential of the cell. Hashes indicate that two parts of the original record have been spliced together. (After Corey and Hudspeth, 1979b.)

Suppose we now stimulate the cell by moving the hair bundle as in Figure 3.11. What current will flow across the membrane of the cell? If the voltage clamp circuit keeps the voltage constant, the membrane current will be given by a modified form of Eq. (3.7),

$$\Delta i_m = \Delta g \left(V_c - E_{rev} \right) \qquad (3.8)$$

where g has been replaced with Δg, the change in conductance; i_m with Δi_m, the change in membrane current; and V_m by V_c, the command potential of the voltage-clamp circuit. The current injected by the circuit is the current that must flow through the ion channels at the command voltage set by the experimenter. Provided ion concentrations do not change, (V_m-E_{rev}) is constant, and the current Δi_m is directly proportional to the change in conductance of the channels.

Voltage clamping, though technically more difficult than simple voltage recording, produces an enormous simplification in our understanding of what the cell is doing. If the sensory stimulus modulates the opening of only a single type of channel (as seems to be true for hair cells), Δg in Eq. (3.8) is exactly equal in sign and magnitude to the aggregate change in the conductance of these channels. The voltage-clamp current therefore gives a direct measure of the time-dependent change in the opening or closing of the channels activated by the sensory stimulus. I have already shown voltage-clamp recordings of currents for piezo channels in Figure 1.11 and TRPV1 channels in Figure 3.6, and this technique will be used extensively in the rest of the book.

Voltage-clamped response of the hair cell

Let us now return to the problem we originally posed: does bending of the hair bundle open channels or close them, and what are the properties of these channels? These questions were answered by voltage-clamp experiments (Corey and Hudspeth, 1979a; Ohmori, 1985; Holton and Hudspeth, 1986; Howard and Hudspeth, 1987; Crawford et al., 1989). Typical recordings are shown in Figure 3.13B from the paper of Corey and Hudspeth (1979a). When

the cell is clamped to the resting membrane potential, the sinusoidal bending of the hair-cell bundle first toward the kinocilium and then back again produces a sinusoidal current, first negative (inward) and then positive (outward). This current is responsible for generating the voltage response of the cell.

The experiment in Figure 3.13B shows that movement of the hair cell bundle toward the kinocilium produces a negative or inward current Δi_m through the mechanoreceptive channels, but this information is not by itself sufficient to determine whether Δg is positive or negative, that is whether channels are being opened or closed. This is because, from Eq. (3.8),

$$\Delta g = \frac{\Delta i_m}{\left(V_m - E_{rev} \right)}. \qquad (3.9)$$

The sign of the conductance change can be seen to depend upon the sign of the driving force as well as the sign of the current.

To determine the sign of the driving force, we must measure E_{rev}, the voltage at which the current Δi_m reverses sign. We can measure E_{rev} by systematically varying the command potential of the voltage clamp circuit to change the membrane potential of the cell. The sign and amplitude of the current then change. A typical result is given in Figure 3.14 (from Ohmori, 1985), in experiments in which the hair cell was voltage clamped with whole-cell patch recording instead of with two electrodes. For hair bundle movement toward the kinocilium, the inward current measured at the resting membrane potential becomes smaller as V_m is depolarized toward zero. At a membrane potential a little positive of zero, Δi_m is zero, and for further depolarization to membrane potentials even more positive, the current reverses direction and becomes positive or outward. The reversal potential of the current is therefore at about 0 mV or perhaps somewhat positive of this value.

Let us now return to the recordings in Figure 3.13B. At the resting membrane potential of the cell, in this case –58 mV, movement of the hair cell bundle toward the kinocilium produces a Δi_m that is negative. The driving force is equal to $V_m - E_{rev}$, which, after inserting the appropriate values (and taking E_{rev} as zero), can be calculated to be (–58–0) = –58 mV. So the driving force is also negative. The conductance change is therefore positive, and the

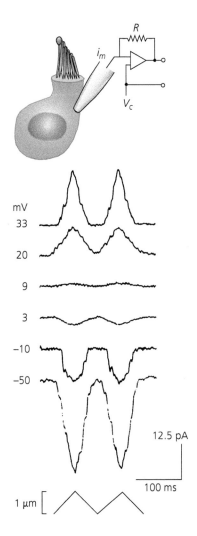

Figure 3.14 Reversal potential of hair-cell current. A chick vestibular hair cell was voltage clamped with the whole-cell patch technique (see Figure 1.7). The cell was stimulated by pushing against the hair bundle with a glass rod. The stimulus waveform is given by the lowermost trace, with upward deflections indicating movement toward the kinocilium. Numbers to the left of each trace give the membrane potential, held by the voltage-clamp circuit at the command potential set by the experimenter. Note the reversal of response polarity somewhat positive of 0 mV. (From Ohmori, 1985.)

bending of the hair cell bundle toward the kinocilium produces an increase in conductance and opening of the mechanoreceptive channels. Since movement in the opposite direction produces a current of opposite sign but with the same reversal potential (see Figure 3.14), it must close these channels.

Ion selectivity

Voltage-clamp recordings are useful in determining the sign of the conductance change, but they can also tell us something about the nature of the channels opened by the sensory stimulus. It would be interesting to know what sort of ions permeate the channels, because this information would tell us something about the size and structure of the ion pore.

For mechanoreceptive channels of hair cells in the experiments I have described, the reversal potential of the current is near zero. The value of the reversal potential is given by the Nernst equation for a channel permeable to only a single ion species, or by the Goldman voltage equation for a channel permeable to more than one kind of ion. Since for hair cells the reversal potential of the current is near zero, the channels are clearly not selectively permeable either to Na^+ or to K^+, because the Nernst potentials for Na^+ and K^+ are quite different from zero, one rather positive and the other quite negative. It is also unlikely that the channels are selective for anions, because for most neurons the Nernst potential for Cl^- is generally rather negative and near the resting membrane potential. The channels might, however, be non-selectively permeable to both Na^+ and K^+. If, for example, we assume that P_{Na} and P_K are equal, E_{rev} can be calculated from Eq. (3.4) or (3.5) to be 1–2 mV more positive than zero at room temperature, probably within experimental error of the value obtained from the data of Figure 3.14.

A more systematic way to study the ion selectivity of the channels is to change the composition of the extracellular medium bathing the receptor cell. Suppose, for example, the cell is bathed in a solution whose only cation is Na^+. The reversal potential of the current can be calculated from the Goldman equation to be

$$E_{rev}^1 = \frac{RT}{F} ln \frac{P_{Na}[Na]_o}{P_{Na}[Na]_i + P_K[K]_i} \tag{3.10}$$

Chloride has been omitted, because a large body of evidence shows that hair cell mechanoreceptive channels have a negligible permeability to anions. There is only one term in the numerator, because there is no potassium in the external solution and K_o is therefore zero.

Suppose we now replace the medium bathing the cell with one for which all of the Na^+ has been replaced with some other ion, say the lithium ion Li^+. The reversal potential is now

$$E_{rev}^2 = \frac{RT}{F} ln \frac{P_{Li}[Li]_o}{P_{Na}[Na]_i + P_K[K]_i} \quad (3.11)$$

If we then assume that when we change solutions bathing the cell, we do not change the composition of the ions inside the cell, we can subtract Eq. (3.10) from Eq. (3.11) to give

$$E_{rev}^2 - E_{rev}^1 = \frac{RT}{F} ln \frac{P_{Li}[Li]_o}{P_{Na}[Na]_o} \quad (3.12)$$

The terms for the ion concentrations inside the cell cancel out (remember, $ln\ x - ln\ y = ln\ x/y$). If we use the same concentrations of Li^+ and Na^+ in our experiment (or better, the same *activities*), this equation becomes even simpler:

$$E_{rev}^2 - E_{rev}^1 = \frac{RT}{F} ln \frac{P_{Li}}{P_{Na}} \quad (3.13)$$

Equation (3.13) shows that it is possible to estimate the relative permeability of the mechanoreceptive channel for Li^+ with respect to Na^+ simply by replacing Na^+ with Li^+ in the bathing medium and measuring the change in the reversal potential of the current. With this method, the ion selectivity of the hair cell mechanoreceptive channel has been extensively studied (Corey and Hudspeth, 1979a; Ohmori, 1985). The channel can be permeated by many different monovalent cations including Na^+, Li^+, and K^+, as well as by much larger ions such as the choline ion and tetrametylammonium (TMA^+). Even divalent cations such as Ca^{2+}, Sr^{2+}, and Ba^{2+} can flow through the channels. As we shall see (Chapter 6), the ion selectivity of the channel and in particular its permeability to Ca^{2+} can have important consequences for the response of the cell to bending of the hair bundle.

Summary

Ions move across cell membranes through channels, which are integral membrane proteins forming pores whose opening is gated. Gating can occur from a change in voltage across the membrane or the binding of a small ion or molecule to the intracellular or extracellular surface of the protein. Channels can also be gated by a sensory stimulus such as temperature or mechanical pressure, and channels of this sort produce the responses of ionotropic sensory receptor cells, including hot and cold receptors in the skin and hair cells in the ear.

Neurons including sensory cells have concentration gradients for K^+ and Na^+ produced by energy-dependent transport by the Na^+/K^+ ATPase. These concentration gradients, together with the membrane potential of the cell, provide a driving force for ion entry and exit. At the resting membrane potential, the passive movements of K^+ and Na^+ through channels spontaneously open at rest are exactly balanced by transport by the Na^+/K^+ pump. When stimulation of the cell causes additional channels to open, the membrane permeability of the cell can change, producing a change in the potential across the membrane.

The membrane channels of hair cells of the ear and vestibular system are directly gated by movement of a microvillar sensory hair bundle. Movement of the bundle toward the largest of the microvilli produces a depolarization, and movement in the opposite direction produces a hyperpolarization. Voltage-clamp experiments have shown that depolarization is produced by the opening of channels non-selectively permeable to cations including Na^+ and K^+ and even Ca^{2+}. The methods of intracellular recording and voltage clamping introduced in this chapter are important techniques used extensively in the investigation of the electrical responses of sensory receptor cells and indeed for all the cells in the nervous system.

CHAPTER 4

Metabotropic signal transduction

For many sensory receptor cells, the receptor protein activates a heterotrimeric G protein that triggers a series of enzymatic reactions, resulting in most cases in a change in the concentration of a second messenger (Figure 2.1B). The second messenger then gates the opening of ion channels. Transduction of this sort can have a much higher gain than mechanisms where the receptor protein is itself a channel, because one receptor protein can activate many G proteins and cause a significant change in second-messenger concentration, opening or closing a large number of channels.

The mechanisms of signal transduction of this more complicated variety can be called metabotropic, because they are similar to those used by many hormone receptors and metabotropic synaptic receptors in the CNS. Metabotropic sensory transduction has now been described for vertebrate and invertebrate photoreceptors and a wide variety of chemosensory cells, including those of the mammalian tongue and nose and other, simpler sensory structures in lower invertebrates such as the worm *Caenorhabditis elegans*. For cells using this more complicated transduction scheme, the sensory receptor proteins so far identified have been shown in nearly every case to be members of the same large family of proteins, sometimes called heptahelical receptors or metabotropic receptors or seven transmembrane-spanning proteins, but which we will call *G-protein-coupled receptors*.

All G-protein-coupled receptors and metabotropic signal cascades work in a similar fashion (Figure 4.1). An integral membrane receptor protein such as a vertebrate olfactory receptor molecule is activated by the binding of an odorant. The receptor protein then changes conformation, facilitating the binding of a heterotrimeric G protein peripherally associated with the intracellular surface of the plasma membrane. This protein has three subunits—$G\alpha$, $G\beta$, and $G\gamma$. $G\alpha$ has a binding site for guanosine nucleotides which, in the absence of stimulation, is occupied by GDP. The binding of the G protein to activated receptor causes the exchange of GTP for GDP at the binding site of $G\alpha$. G-protein subunits can themselves bind to and modulate ion channels, but more commonly they accelerate the activity of a wide variety of proteins called effector molecules. The effectors can then synthesize or degrade second messengers, directly or indirectly modulating ion channels or other membrane and cytosolic proteins.

Some effector molecules and their actions are summarized in Figure 4.2. They include adenylyl cyclase, the synthetic enzyme for cyclic adenosine monophosphate (cAMP); phospholipase C, which produces the two second messengers diacylglycerol (DAG) and inositol 1,4,5-trisphosphate (IP$_3$); phospholipase A$_2$, an enzyme that generates the membrane-bound metabolite arachidonic acid; and phosphodiesterase, a protein that hydrolyzes (degrades) cAMP or cyclic guanosine monophosphate (cGMP). These effector proteins produce changes in the concentration of small-molecular-weight, second messenger molecules such as cAMP, cGMP, or Ca^{2+}, which can themselves modulate the opening or closing of ion channels. Second messengers may also regulate the activity of two very important classes of enzymes known as protein kinases and protein phosphatases. These enzymes

Sensory Transduction. Second Edition. Gordon L. Fain, Oxford University Press (2020). © Gordon L. Fain 2020.
DOI: 10.1093/oso/9780198835028.001.0001

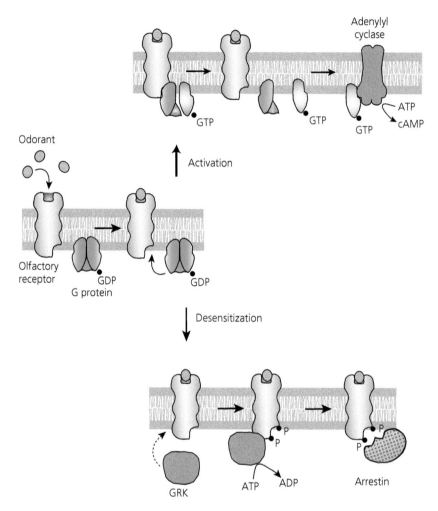

Figure 4.1 Metabotropic sensory transduction. Stimulation with odorant in the vertebrate primary olfactory epithelium produces binding of heterotrimeric G protein to the receptor protein. During activation (upper part of figure), there is an exchange of GTP for GDP on the guanosine-nucleotide binding site of the α subunit of the G protein, producing Gα·GTP. The Gα·GTP then binds to and stimulates adenylyl cyclase. During desensitization (lower part of figure), the receptor protein is phosphorylated by a G-protein receptor kinase (GRK), and arrestin binds to phosphorylated receptor. In *Drosophila* photoreceptors, the arrestin can bind to and desensitize even unphosphorylated receptor.

attach or remove phosphate groups to and from ion channels and other proteins, placing a bolus of negative charge on the structure of the protein. This negative charge attracts positively charged amino acids and repels negatively charged amino acids, altering the conformation of the protein and changing its enzymatic activity. Also shown in Figure 4.2 are the membrane-bound guanylyl cyclases. These enzymes form a special category of proteins that can act as receptors but also as effectors, because the

binding of a signaling molecule to an extracellular site directly stimulates the production of the second messenger cGMP without the need of a heterotrimeric G protein.

The blue arrows in Figure 4.2 follow one particular cascade in detail. When a G-protein-coupled receptor activates a G protein with a Gα subunit from the $G\alpha_{q/11}$ family, the Gα•GTP may stimulate the membrane-bound effector phospholipase Cβ (PLCβ). This enzyme then hydrolyzes a specific

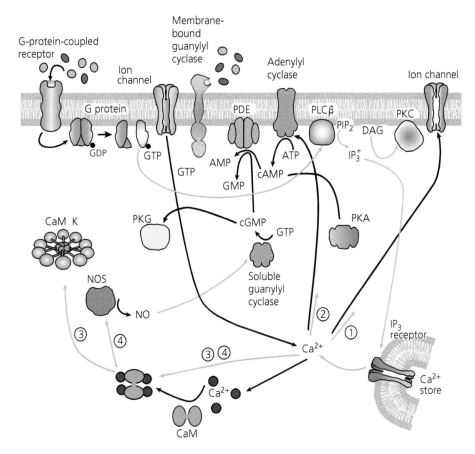

Figure 4.2 Representative pathways of metabotropic signal transduction. G-protein-coupled receptor activates G protein, which in turn can stimulate a variety of effector molecules. Blue arrows and circled numbers indicate one particular transduction cascade, described in more detail in the text. AMP, Adenosine monophosphate; ATP, adenosine triphosphate; CaM, calmodulin; CaMKII, calcium-calmodulin-dependent protein kinase II; cAMP, adenosine 3′,5′-cyclic monophosphate; cGMP, guanosine 3′,5′-cyclic monophosphate; DAG, diacylglycerol; GDP, guanosine diphosphate; GMP, guanosine monophosphate; GTP, guanosine triphosphate; IP_3, inositol 1,4,5-trisphosphate; NO, nitric oxide; NOS, nitric oxide synthase; PDE, cyclic nucleotide phosphodiesterase; PIP_2, phosphatidylinositol 4,5-bisphosphate; PKA, protein kinase A; PKC, protein kinase C; PLCβ, phospholipase Cβ; PKG, protein kinase G.

component of plasma membrane lipid called phosphatidylinositol 4,5-bisphosphate (PIP_2) to produce the two second messengers DAG and IP_3. DAG can diffuse within the membrane and activate protein kinase C (PKC), whereas IP_3 diffuses through the cytoplasm to bind to IP_3 receptors on smooth endoplasmic reticulum (ER), causing the release of Ca^{2+}. The effect of the increase in Ca^{2+} concentration can vary from cell to cell. Among the many possibilities are: (1) the direct gating of ion channels; (2) the activation of certain forms of adenylyl cyclase; (3) direct stimulation of a protein kinase called calcium-calmodulin kinase (CaM kinase); and (4) acti-

vation of the enzyme NO synthase which synthesizes the gas nitric oxide (NO). NO can diffuse widely throughout the tissue and activate soluble guanylyl cyclase, producing an increase in the concentration of cGMP.

Many of the second-messenger pathways in Figure 4.2 are utilized by sensory receptor cells. In some cases, these pathways open or close channels and produce the electrical response of the cell; in other cases, they regulate the response in some way. Cyclic nucleotides directly gate ion channels and are essential second messengers in vertebrate photoreceptors and some olfactory receptors, but these same cells can use Ca^{2+} as a second messenger to modulate the sensitivity of transduction.

As we will see later in the book (Chapter 9), the cascades for cyclic nucleotides and Ca^{2+} can interact to produce negative feedback. In a vertebrate rod or cone, a decrease in cGMP leads to a decrease in Ca^{2+}, and as the Ca^{2+} concentration declines, the activity of a membrane-bound guanylyl cyclase is stimulated. This enzyme then synthesizes more cGMP. As a consequence, the concentration of cGMP first decreases and then partially recovers as the guanylyl cyclase activity is accelerated. Negative feedback pathways of this kind are of crucial importance in the regulation of metabotropic cascades, not only in sensory organs but throughout the body.

G-protein-coupled receptors

There are nearly a thousand G-protein-coupled receptor proteins in humans and an even greater number in mouse, where they amount to 3–5 percent of the total protein-coding genome. In vertebrates these proteins include receptors for many hormones (follicle stimulating hormone, oxytocin, vasopressin), synaptic transmitters (neuropeptides, opiates, dopamine), olfactory receptor proteins, visual pigments, and taste receptors for bitter, sweet, and amino acids. All G-protein-coupled receptors have a similar structure (Figure 4.3A). There are seven mostly α-helical sequences of twenty to twenty-five amino acids, which are primarily hydrophobic and span the lipid bilayer. The carboxyl terminus of the protein is intracellular and contains several sites for phosphorylation which, as described later in the chapter, can regulate the activity of the protein. The carboxyl terminus, together with the intracellular loop domains that interconnect the membrane-spanning domains, can form a binding site for G protein that is essential for activation.

Our understanding of G-protein-coupled receptors was greatly increased by the development of a method for making crystals of one of these proteins, rhodopsin (Okada et al., 2000), and by the solution of the structure of this protein by X-ray crystallography (Palczewski et al., 2000). This report was followed by the solution of the structure of the β_2-adrenergic receptor (Rasmussen et al., 2007) and an ever-increasing number of additional receptors (see Erlandson

et al., 2018). Many G-protein-coupled receptor proteins, including rhodopsin and the β-adrenergic receptor, have a binding site shaped by the folding of the membrane-spanning sequences, which can be occupied by a small-molecular-weight molecule within the lipid bilayer (orange molecules in Figure 4.3A). For rhodopsin this molecule is retinal, which is the part of the visual pigment that actually absorbs the light. For the β-adrenergic receptor, it is epinephrine and norepinephrine (also called adrenaline and noradrenaline), which bind within the volume of the receptor protein. These molecules produce activation by altering the conformation of the receptor (Figure 4.3B), primarily as the result of a shift of the sixth transmembrane domain (TM6) outward toward the cytoplasmic surface of the lipid bilayer, together with a smaller, similar movement of TM5. These movements open up a binding site for G protein on the cytoplasmic side of the receptor and facilitate G-protein activation.

Another group of G-protein-coupled receptors, called the glutamate family, also has seven transmembrane α-helical sequences but in addition a large N-terminal motif called a *Venus flytrap domain*. This extracellular part of the protein forms the binding site for the activating molecules. These receptors seem always to function as dimers. There are relatively few members of this family, but among them are the important metabotropic glutamate and $GABA_B$ receptors of the CNS, as well as the T1R sweet and umami taste receptors of the tongue, whose function is described in detail in Chapter 8.

The change in conformation of a G-protein-coupled receptor not only activates G protein but also exposes residues near the receptor carboxyl terminus and facilitates the binding of a G-protein-coupled receptor kinase (GRK), mediating termination and desensitization of the receptor (Figure 4.1). The GRKs are a family of proteins that catalyze the attachment of phosphate groups from ATP to serine and threonine residues. Phosphorylated receptor can then be bound by a member of another family of proteins, called the arrestins. The binding of an arrestin to phosphorylated receptor prevents any further binding and activation of G protein, effectively terminating the activity of the receptor (Figure 4.1). Phosphorylation and arrestin

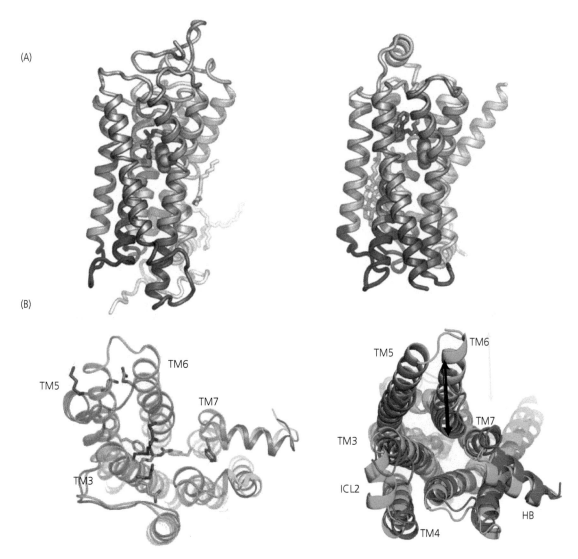

Figure 4.3 Structure and mechanism of activation of G-protein-coupled receptor proteins rhodopsin and β₂-adrenergic receptor. (A) Ribbon models of rhodopsin (left) and β₂-adrenergic receptor (right). Ligands are shown in orange and a conserved tryptophan residue in green. (B) Activation of G-protein-coupled receptors. Left, comparison of partial ribbon models from X-ray crystallographic structures of dark-adapted (inactive) rhodopsin (lavender) and a form of rhodopsin with partial activity (green). TM3–TM7 are α-helical domains of the proteins. Right, comparison of partial ribbon models of active (green) and inactive (blue) β₂-adrenergic receptor. TM5 and TM6 show the largest structural changes. Arrows indicate a 14 Å outward movement of TM6 accompanying activation; TM5 is extended by two helical turns. (A from Hanson and Stevens, 2009; B left from Rosenbaum et al., 2009; B right from Rasmussen et al., 2011.)

binding, first discovered for rhodopsin (Bownds et al., 1972; Kuhn et al., 1973; Kuhn, 1978; Wilden et al., 1986), are together an important mechanism for the inactivation and desensitization of most G-protein-coupled receptor proteins (see Reiter et al., 2012).

Heterotrimeric G proteins

The mammalian genome contains a large assortment of G proteins which, though variable in structure and function, all bind GTP. Some of these molecules, such as *Ras*, *Rho*, and *Rac*, consist of only a single small subunit but are critically important to the growth and differentiation of the cell. The larger, heterotrimeric G proteins have a total molecular weight of about 100 kilodaltons and are the only G proteins that interact with receptors of the G-protein-coupled family.

Heterotrimeric G proteins have three subunits called Gα, Gβ, and Gγ. In mammals there are sixteen genes for Gα, five for Gβ, and twelve for Gγ (Downes and Gautam, 1999; Wettschureck and Offermanns, 2005). The Gα genes can be subdivided into four families, called $G\alpha_s$, $G\alpha_i/\alpha_o$, $G\alpha_{q/11}$, and $G\alpha_{12/13}$. The $G\alpha_s$ family includes $G\alpha_s$ itself, which stimulates the effector adenylyl cyclase (Figure 4.2), leading to the synthesis of cAMP. This family also contains $G\alpha_{olf}$, found in vertebrate olfactory receptor cells. The $G\alpha_i/\alpha_o$ family is the largest and contains the transducins—the Gα proteins of vertebrate rods and cones—as well as gustducin or $G\alpha_{gus}$, found in some taste-bud neurons. This family also includes $G\alpha_o$, the most common G-protein α subunit in the nervous system. The $G\alpha_{q/11}$ family contains the very important group of Gαs coupled to phospholipase C. The function of the two members of the $G\alpha_{12}/\alpha_{13}$ family is still uncertain. At rest, the Gα subunit is bound to a Gβγ dimer to form a membrane-associated trimer. Cells must have rules dictating which Gβ associates with which Gγ to form these Gβγ pairs, but these rules are still largely unknown.

The Gα subunit of the G protein contains a binding site for guanosine nucleotide that in unstimulated G protein is occupied by GDP, forming a Gα•GDP•βγ complex (Figure 4.4). When a G-protein-coupled receptor protein is activated, this G-protein complex binds to the cytoplasmic surface of the receptor, producing a subtle change in the structure of the Gα subunit. The GDP falls off the guanosine-nucleotide binding site, which then becomes free to bind GTP. The binding of GTP forms Gα•GTP•βγ, an unstable intermediate which dissociates into Gα•GTP and Gβγ. Both parts of the G protein are then released from the receptor binding site and can diffuse within the cytosol or along the surface of the plasma membrane. The Gα•GTP can bind to and activate effector molecules, and it used to be thought that *only* Gα•GTP was active and that Gβγ was simply structural, holding the inactive complex together. We now know, however, that Gβγ

dimers can directly activate many different proteins, including ion channels, PLCβ, and several forms of adenylyl cyclase (see Clapham and Neer, 1997; Dupre et al., 2009).

As soon as Gα•GTP and Gβγ have left the receptor protein binding site, another G protein can bind and become activated for as long as the receptor remains in its active conformation (Figure 4.4). In this way, many G proteins can potentially be transformed by a single activated receptor until phosphorylation and/or arrestin binding prevent further Gα•GTP production. The Gα•GTP in turn can continue to stimulate effector molecules until the bound GTP is hydrolyzed to GDP, which can occur by the intrinsic hydrolytic activity of the Gα subunit itself. The intrinsic rate of hydrolysis is, however, very slow for unassisted Gα but is greatly accelerated by GTPase-activating proteins (GAPs). A group of GAPs called RGS proteins have been shown to accelerate GTP hydrolysis by many fold (Ross and Wilkie, 2000; Kimple et al., 2011). There is also evidence that effector molecules themselves can act as GAPs, accelerating the rate of GTP hydrolysis (see Arshavsky and Bownds, 1992; Cook et al., 2000). Once GTP is hydrolyzed to GDP, the Gα•GDP reassociates with Gβγ to form inactive Gα•GDP•βγ. Thus hydrolysis of Gα•GTP to Gα•GDP quenches the activity not only of Gα but also of the Gβγ dimer.

Effector molecules

There are many important enzymes utilized as effectors by sensory receptor cells and other cells in the body. In vertebrate olfaction and photoreception, the two most important are membrane-bound adenylyl cyclase and cyclic nucleotide phosphodiesterase. Adenylyl cyclase catalyzes the synthesis of adenosine 3',5'-cyclic monophosphate from ATP. Adenylyl cyclase is an integral membrane protein composed of two similar, covalently attached halves both required for catalytic activity (Figure 4.5A). Each half has six membrane-spanning (M_1 and M_2) domains, similar to the Shaker protein (Figure 3.1) but with no significant homology in amino acid sequence. The membrane-spanning domains in each half are followed by large catalytic domains (C_1 and C_2), which together are responsible for the synthesis of cAMP. There are nine genes for

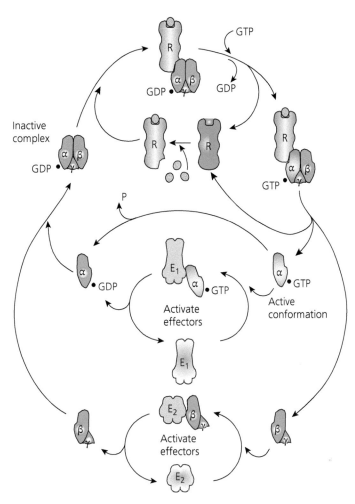

Figure 4.4 Activation and turnoff of heterotrimeric G protein. Activation of G protein by receptor produces Gα•GTP and Gβγ, which in turn activate effector molecules. The hydrolysis of GTP to GDP leads to recombination of Gα•GTP and Gβγ, restoring inactive G protein.

adenylyl cyclases, all of which can be stimulated by Gα$_s$•GTP. Some are inhibited by Gα$_i$•GTP, some stimulated or inhibited by Ca^{2+}, and some regulated by Gβγ. Which cyclase is affected by which G protein may depend upon the distribution of these molecules in the membrane, perhaps in regularly arranged structures directing the G protein to the appropriate effector. Another family of enzymes called guanylyl cyclases can synthesize cGMP from GTP.

Both cAMP and cGMP can be degraded by cyclic nucleotide phosphodiesterase (PDE). Twenty-one different genes for PDE have been described, some specific for cAMP or cGMP. The enzyme of greatest interest for our purposes is the cGMP-specific PDE of vertebrate photoreceptors, which is PDE6 (Figure 4.5B). In vertebrate rods, this protein is a tetramer containing two catalytic subunits (called α and β) and two identical inhibitory subunits (called γ). The direct binding of Gα•GTP of photoreceptor G protein to the PDE6γ subunits displaces these inhibitory subunits from the active sites of the catalytic subunits, releasing the protein from inhibition.

Two more effector molecules of great importance in the physiology of the body are the phospholipases A$_2$ and C, also called PLA$_2$ and PLC (Figure 4.6).

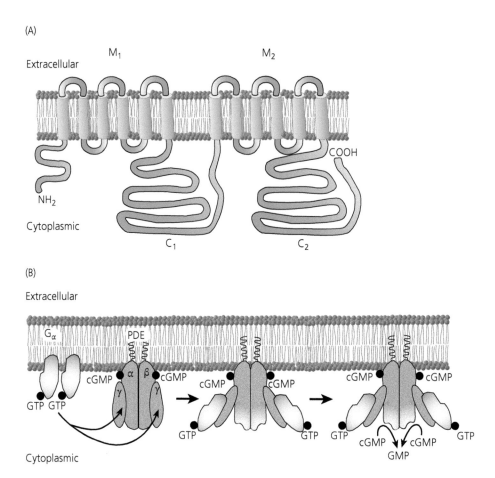

Figure 4.5 Effector molecules: adenylyl cyclase and cyclic nucleotide phosphodiesterase. (A) Membrane topology of adenylyl cyclase. M_1 and M_2, Membrane-spanning domains; C_1 and C_2, catalytic domains that together catalyze cAMP synthesis. (B) Structure and activation of cyclic nucleotide phosphodiesterase (PDE6) from a vertebrate rod. The enzyme is a tetramer, containing two catalytic subunits (α and β) and two inhibitory subunits (γ). The α and β subunits have associated lipid that inserts into the membrane, keeping the enzyme next to the membrane. Also shown are two non-catalytic binding sites for cGMP on the α and β subunits. (A after Birnbaumer and Birnbaumer, 1994; B after an illustration provided by A. S. Polans.)

Phospholipases hydrolyze membrane phospholipids. Phospholipase A_2 releases the membrane lipid arachidonic acid, which diffuses as a second messenger in the plasma membrane and can modulate the activity of ion channels and other proteins. Arachidonic acid can also be converted into a group of biologically active molecules called eicosanoids, important metabolites that can have a wide range of effects in many of the tissues of the body.

The phospholipase C family of enzymes also hydrolyzes lipid, and these enzymes are quite important in many aspects of cell function (Kadamur and Ross, 2012). Especially prominent in sensory cells are the PLCβs, which act as effectors for heterotrimeric G proteins of the $G\alpha_{q/11}$ family. In mammals there are several different kinds of PLCβ linked to specific groups of G-protein-coupled receptors. The PLCβs attack the head group of phosphatidylinositol 4,5-bisphosphate (PIP_2) (Figure 4.6), catalyzing the formation of the two second messengers DAG and IP_3.

Second messengers

Many metabotropic signaling cascades regulate the synthesis or release of small-molecular-weight

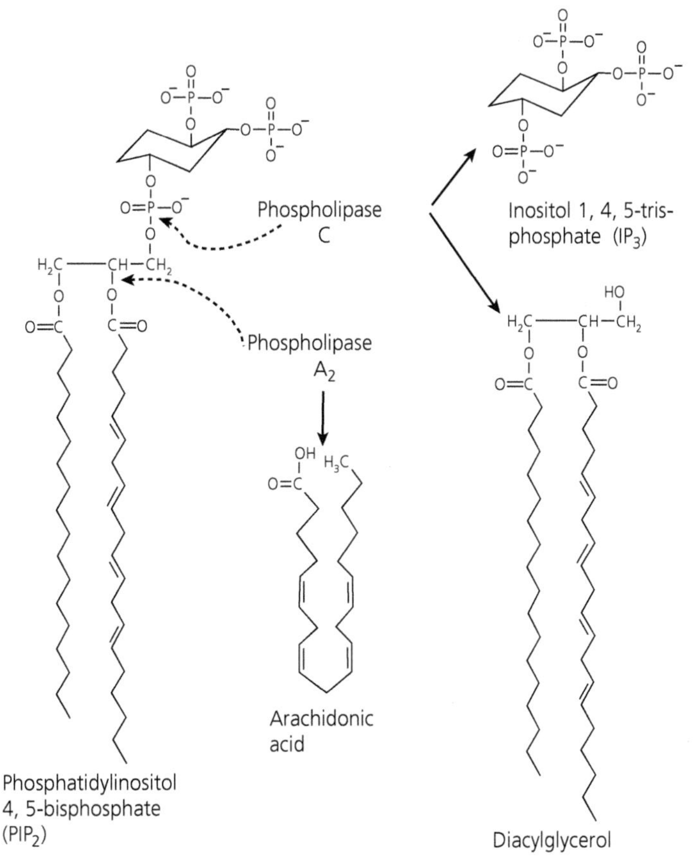

Figure 4.6 Effector molecules: phospholipases. The membrane lipid phosphatidylinositol 4,5-bisphosphate (PIP$_2$) can be cleaved by phospholipase C to form the two second messengers IP$_3$ and diacylglycerol; or by phospholipase A$_2$ to form the second messenger arachidonic acid.

second messengers, compounds that diffuse either within the plasma membrane or in the cytoplasm to activate ion channels or other proteins of functional importance. Cyclic AMP, the first of these compounds to be discovered (Sutherland, 1972), regulates the activity of protein kinase A (PKA), a soluble enzyme that phosphorylates and regulates the activity of many important membrane and cytoplasmic proteins. Similarly, cGMP can bind to and activate cGMP-dependent protein kinase. Both cAMP and cGMP can also bind to and gate an important group of second-messenger-gated channels, as described in more detail later in this chapter.

Diacylglycerol, one of the two messengers generated by PLCβ (Figure 4.6), can stimulate protein kinase C (PKC). The other second messenger IP$_3$ has

an entirely different effect (see Figure 4.2): it diffuses within the cytosol and binds to receptors in the endoplasmic reticulum (ER). Many of the vesicles of the ER are full of Ca^{2+}. An enzyme called the sarco/endoplasmic reticulum Ca^{2+} ATPase (SERCA), similar in structure and mechanism to the Na$^+$/K$^+$ ATPase (Brini and Carafoli, 2009, 2011), pumps Ca^{2+} into the lumen of the ER. Once these vesicles have accumulated Ca^{2+}, they can release it. Many ER vesicles contain membrane channels called IP$_3$ receptors (Figure 4.7), which are large multimeric proteins of total molecular weight exceeding one million daltons. Each of the four receptor monomers has a single binding site for IP$_3$ with several membrane-spanning domains that together form a cationic channel selective for divalent cations. Binding of IP$_3$

releases Ca^{2+} out of the vesicle lumen into the cytoplasm.

Much of the mass of the IP_3 receptor lies within a large amino-terminal region facing the cytoplasm of the cell. This part of the protein, called the regulatory or coupling domain, contains several regions that modulate Ca^{2+} release, including sites that bind Ca^{2+} itself. The probability of opening of the IP_3 receptor channel is increased when the free-Ca^{2+} concentration in the cell cytoplasm increases, provided the Ca^{2+} concentration is rather low. High Ca^{2+} inhibits channel opening. As a result, the release of Ca^{2+} by IP_3 receptors can be initially enhanced as the Ca^{2+} concentration in the cytoplasm begins to grow, but it can then be inhibited as the concentration becomes increasingly elevated.

In addition to IP_3 receptors, another group of proteins called ryanodine receptors are expressed in vesicles of the ER, either together with IP_3 receptors or in a separate fraction of ER membrane (Van Petegem, 2012). These receptors take their name from ryanodine, a plant alkaloid used in their biochemical isolation. They have a structure and function quite similar to IP_3 receptors but are gated by the binding of Ca^{2+} rather than IP_3.

After Ca^{2+} is released, it must be returned back into the ER vesicles. Resequestration is ultimately the task of the SERCA protein, but this process is facilitated by a process called store-operated Ca^{2+} entry (Prakriya and Lewis, 2015). The vesicles of the ER contain a membrane protein called STIM, which has a binding site inside the vesicle for Ca^{2+}. When

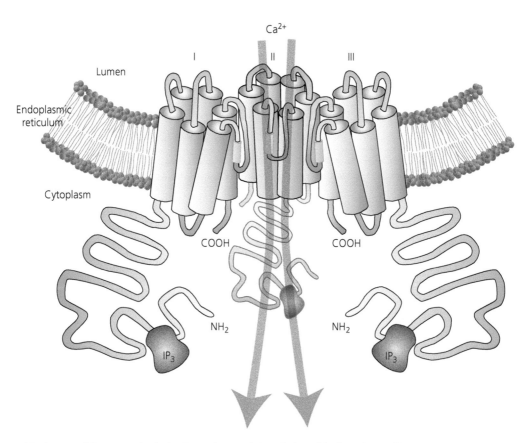

Figure 4.7 Structure of IP_3 receptor of endoplasmic reticulum. For clarity, only three of the four subunits of the tetramer are shown (I–III). The approximate positions of the IP_3 binding sites near the N-terminus of the cytoplasmic domains are indicated. Binding of IP_3 causes release of Ca^{2+} from inside the ER vesicle into the cytoplasm. (After Foskett et al., 2007.)

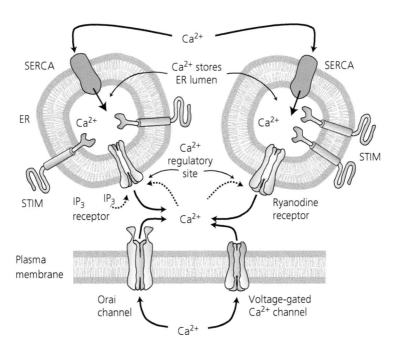

Figure 4.8 Summary of mechanisms of cellular Ca^{2+} release and reuptake. See text.

Ca^{2+} is released from the vesicle and the Ca^{2+} concentration inside the vesicle decreases, Ca^{2+} falls off the STIM binding site. The STIM proteins (probably as dimers) then change conformation and bind to a plasma membrane channel called the Orai channel, which is a thousand times more permeable to Ca^{2+} than to Na^+. These channels gate the entry of Ca^{2+} from outside the cell to the inside. Ca^{2+} can also enter the cell through other membrane channels such as the stretch-sensitive or temperature-sensitive channels I described in Chapter 3, or through voltage-gated Ca^{2+} channels like those at presynaptic membrane.

Mechanisms of Ca^{2+} release and resequestration are summarized in Figure 4.8. Ca^{2+} can be released into the cytoplasm from vesicles of the ER by the opening of channels of IP_3 receptors or ryanodine receptors. To refill these stores, Ca^{2+} must first enter the cell either by store-operated Ca^{2+} entry via the STIM and Orai proteins or by means of some other Ca^{2+}-permeable channel in the membrane. This Ca^{2+} is then pumped back into the vesicle by the SERCA protein.

Measuring cell calcium

Ca^{2+} is an important second messenger, not only in sensory cells but in all the cells of the body. For this reason, many indicator molecules have been developed to measure the Ca^{2+} concentration, of which the most important are undoubtedly the jellyfish protein aequorin and a large family of fluorescent dyes initially synthesized in the laboratory of the late Roger Tsien (see Miyawaki, 2016). Compounds such as indo, fura-2, fluo-3, and their relatives change fluorescence when Ca^{2+} binds (Grynkiewicz et al., 1985). The change in fluorescence is a function of the free-Ca^{2+} concentration and can be calibrated, providing a measure of the relative increase in Ca^{2+}, as well as a less accurate—though often still useful—estimate of the absolute change in the free-Ca^{2+} concentration. These fluorescent dyes have completely revolutionized our ability to understand the role of Ca^{2+} in cell function and have been almost as important as patch-clamp recording for understanding the physiology of the nervous system.

Figure 4.9 is an example of an experiment utilizing one of these dyes to measure Ca^{2+} in a photoreceptor from the primitive cordate amphioxus (Gomez et al., 2009). Amphioxus has four photosensitive organs (Figure 4.9A), though the lamellar body does not persist into adulthood (Lacalli, 2004). A single Joseph cell was loaded with fluo-3 and then stimulated with light. Part B of the figure is a control experiment in which the photoreceptor was bathed in sea water. The images at the top show raw measurements of fluorescence, and the images below give color-coded plots indicating the distribution of the increase in fluorescence (and Ca^{2+}) within the cell. The experiment was then repeated on the same cell this time bathed in sea water lacking Ca^{2+}. The result—shown in part C of the figure—was nearly identical to that in part B. This finding shows that the Ca^{2+} increase of the cell in part B could not have been produced by Ca^{2+} entering the cell from the outside but must have been produced by Ca^{2+} release, probably via $PLC\beta$ and IP_3 (Angueyra et al., 2012; Peinado et al., 2015).

One of the reasons Ca^{2+} release has been so extensively studied is that Ca^{2+} can affect so many processes within the cell (see Figure 4.2). In sensory cells, calcium binding either to the small-molecular-weight protein calmodulin or to a variety of other Ca^{2+}-binding proteins can regulate many important molecules, including transduction enzymes and ion channels essential for the generation of the response. There will be much to say about Ca^{2+} in the remainder of the book.

Channels gated by second messengers

One of the most important roles played by a second messenger in a sensory cell is the direct gating of ion channels. This phenomenon was first discovered by Fesenko and colleagues (1985) for vertebrate photoreceptors. At the time this work was being done, we knew that the light response of vertebrate photoreceptors was regulated by a second messenger. There was evidence for both Ca^{2+} and cyclic nucleotides (see Stryer, 1986). Ca^{2+} had been shown to regulate ion channels (Meech and Strumwasser, 1970; Meech and Standen, 1975), but there was no

previous indication that a cyclic nucleotide could directly gate channel opening.

This uncertainty was resolved by the experiment shown and described in Figure 4.10A. A patch-clamp electrode was sealed onto the outer segment of a frog rod, and the pipette was pulled away from the cell. This maneuver produced an inside-out patch, with the inside of the membrane facing the extracellular solution as in Figure 1.7. Fesenko and colleagues could then perfuse the cytosolic side of the membrane with putative second messengers. They gave a series of 10-mV voltage pulses across the membrane of the patch to measure membrane conductance, which from Ohm's law is equal to the current divided by the voltage (Eq. 3.6). Because the voltage pulses were kept constant, any change in the magnitude of the current indicated a change in membrane conductance. They then perfused cGMP across the cytoplasmic surface of the membrane patch and observed an increase in conductance, which did not require ATP or the addition of a protein kinase. The cGMP seemed to be opening the channels directly. Ca^{2+} had no effect.

Channels gated by cyclic nucleotides were subsequently isolated and cloned (Cook et al., 1987; Kaupp et al., 1989), and they were discovered to have a surprising resemblance to *Shaker* and other voltage-gated potassium channels (Figure 4.10B). This resemblance is even more striking at the level of atomic structure. The cryo-EM in Figure 4.11 is for TAX-4, a CNGA channel from *C. elegans*. This cyclic-nucleotide-gated channel lacks a β ancillary subunit like $K_V2.1$ in Figure 3.2, but the structure of the membrane-spanning sequences is very similar. Moreover, both cyclic-nucleotide-gated channels and K^+ channels are tetramers, and both are commonly heterotetramers in which the four subunits do not all have the same amino acid sequence (see for example Chen et al., 1993; Korschen et al., 1995; Jan and Jan, 2012). There is, however, one important difference between the two kinds of channels: cyclic-nucleotide-gated channels are not selectively permeable to K^+ but are permeable to a wide range of cations, to both Na^+ and K^+ and even to Ca^{2+}. When the pore region of *Shaker* is lined up with that of the rod channel (Figure 4.10C), there are two amino acids present in *Shaker* that are missing in the rod sequence. The deletion of these amino

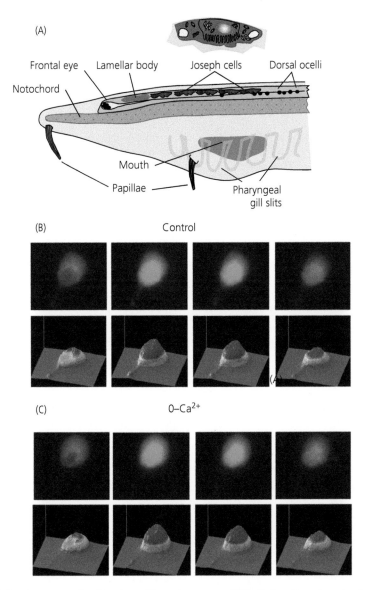

Figure 4.9 Light stimulation elevates cytosolic calcium in amphioxus photoreceptor. (A) Head of amphioxus larva showing four groups of photoreceptors: the frontal eye, lamellar body, Joseph cells, and the dorsal ocelli. A single Joseph cell is shown with microvilli. (B) Upper pictures show the sequence of fluorescence images acquired every 30 ms after illuminating the photoreceptor. The lower pictures give corresponding color-coded surface plots beneath each of the images, with blue indicating lowest level of fluorescence and red the highest. The first frame is basal fluorescence; the increase in fluorescence after the second frame is due to a light-induced increase in intracellular calcium. (C) The same procedure was repeated in the same cell after superfusing the recording chamber with Ca-free artificial sea water. A similar increase in fluorescence was observed, indicating that photostimulation causes the release of calcium from intracellular stores. (A after Lacalli, 2004; B and C from Gomez et al., 2009.)

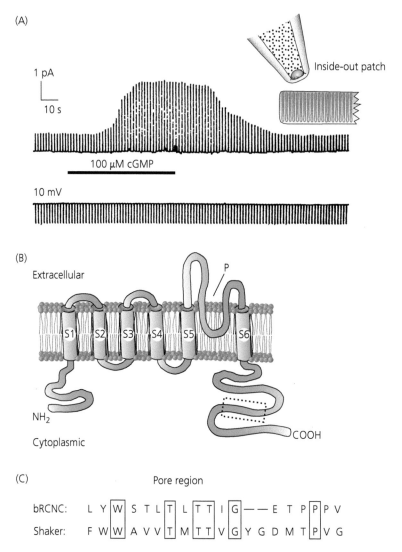

Figure 4.10 Second-messenger-gated channels in rods. (A) Gating of channels. An inside-out membrane patch pulled from the outer segment of a frog rod was voltage clamped and stimulated with 10-mV voltage pulses, whose timing is shown in the lower trace. The upper trace gives the current in response to the constant voltage pulses, which is proportional to membrane conductance. Application of 100 μM cGMP to the cytosolic surface of the patch produced a large increase in conductance, due to the gating of membrane channels. (B) Membrane topology of the photoreceptor cyclic-nucleotide-gated channel. S1–S6, Transmembrane domains; P, pore loop. The dotted box shows the approximate location of the cGMP-binding domain. (C) Comparison of the amino-acid sequence of the pore loop of a rod cyclic-nucleotide-gated channel (bRCNC) and the *Shaker* potassium channel. (A from Fesenko et al., 1985; B after Finn et al., 1996, and Zagotta and Siegelbaum, 1996; C after Finn et al., 1996).

acids has the effect of shifting the positions of all subsequent amino acids in the pore-loop sequence. When these two amino acids are deleted from the *Shaker* gene, the *Shaker* channel becomes non-selectively cationic just like the cyclic-nucleotide-gated channels (Heginbotham et al., 1992).

Mechanism of gating by cyclic nucleotides

Cyclic-nucleotide-gated channels have a binding domain near the carboxyl terminus that has considerable homology to the nucleotide-binding domains of protein kinases PKA and PKG. It is largely

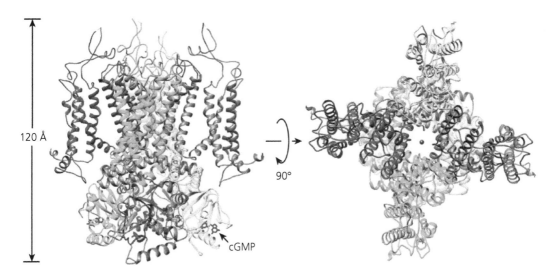

Figure 4.11 Structure of TAX-4, a cyclic-nucleotide-gated channel of *C. elegans*. Cryo-EM electron density map of channel. The structure is viewed on its side parallel to the membrane on the left and from the extracellular surface of the membrane down onto the channel pore on the right. The channel is a tetramer, with each subunit represented by a different color. The cGMP binding site is indicated in the structure to the left, at the bottom right of the protein. (From Li et al., 2017.)

responsible for determining the binding affinity of the channel for cyclic nucleotides (Goulding et al., 1994). This sequence is outlined by the dotted box in Figure 4.10B and is also indicated by the arrow labeled "cGMP" in the lower right of the structure to the left in Figure 4.11. Each of the monomers of the channel has a binding site for cyclic nucleotides, and all four sites are functional.

The mechanism of channel gating was first elucidated in an elegant series of experiments by Ruiz and Karpen (1997, 1999), who used an ingenious technique. They first expressed channels in *Xenopus* oocytes and measured single-channel currents with inside-out recording (as in Figure 4.10A). They then exposed the channels to a photoaffinity analog of cGMP called 8-*p*-azidophenacylthio-cGMP. When the patches in the presence of this compound were illuminated with ultraviolet light, the cGMP analog became tethered to the binding site *covalently*, that is by a chemical bond. The analog then activated the channel just like free cyclic nucleotide but nearly continuously. The longer they exposed the patch to ultraviolet light, the greater the fraction of binding sites covalently bound to analog until, with long light exposure, all four sites became occupied, and the channel was maximally activated.

Ruiz and Karpen (1997) exposed different patches to different light exposures to produce channels with a different number of binding sites covalently occupied. They then tested each patch with *free* cGMP, which could bind only to those sites still unoccupied by the covalently attached analog. By examining the dependence of channel opening on free-cGMP concentration, they could tell for each channel how many sites were still open and available to bind cyclic nucleotide. Having made this determination, they then characterized each of the patches *without* free cGMP present, to record the effect on channel gating when one, two, three, or four of the binding sites were continuously occupied.

Their results are shown in Figure 4.12. Each graph is a histogram, with the ordinate giving the probability of observing a channel opening, and the abscissa the amplitude of the channel current. Let us begin with the graph at the top for recordings from a patch with only a single ligand bound. The histogram shows that the current was zero about 60 percent of the time, and it was quite close to zero for the remaining 40 percent. This result means that the channel with one site occupied was nearly always closed. Very occasional openings were seen but seemed to be spontaneous, since similar openings at about the same frequency were observed when no cyclic nucleotide was present and the binding sites were completely unoccupied.

With two ligands bound (second graph from top), the channel was again mostly closed. The insert shows that there were a few more openings than with only one site occupied, though the average current was only 1 percent of the maximum possible current. With three ligands, the channels opened frequently and produced 33 percent of the maximum current, though the most probable channel current amplitude was still zero, with the channel closed. When all four cyclic

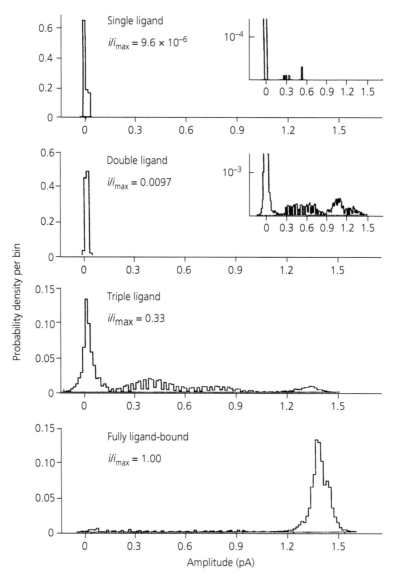

Figure 4.12 Mechanism of opening of cyclic-nucleotide-gated channels. Graphs show the open probability for bovine rod channels with one, two, three, or four cyclic nucleotide binding sites covalently bound to the cGMP analog 8-p-azidophenacylthio-cGMP (ART-cGMP). The ordinate gives probability of occurrence of channel current for the current values given on the abscissa, which have been lumped together into equal-amplitude bins. For channels that bind only a single ligand (top), the most probable value for the current was near zero (closed channel), whereas for a channel with four ligands bound (bottom), the most probable value was between 1.2 and 1.5 pA, the current of a fully open channel. (From Ruiz and Karpen, 1997.)

nucleotides were bound (histogram at bottom), the channels were open nearly all the time.

One curious feature of these recordings is that channels could open to several different states, each with its own conductance, as if the pore of the channel could assume more than one configuration. With two or three ligands bound, the smaller conductance states of the channel were frequently observed, whereas when all four sites were occupied, the largest conductance state was the most probable. The gating of the channel therefore uses all four binding sites and all four channel subunits, but the mechanism of gating is rather complicated and seems to require several intermediary changes in protein conformation before full opening occurs (Ruiz and Karpen, 1999; Mazzolini et al., 2010).

A metabotropic sensory receptor

One interesting example of a metabotropic sensory receptor that uses a complicated signal transduction cascade and a second-messenger-gated channel is the photoreceptor of the lizard parietal eye. Lizard parietal eyes are secondary photosensitive organs located on the top of the head (Figure 4.13A). They seem to function in a simple form of light detection used in thermoregulation and the control of the circadian rhythm. Their photoreceptors resemble other vertebrate photoreceptors with outer segments formed from modified cilia (Ohshima et al., 1999). When the photoreceptor membrane potential was recorded with an intracellular microelectrode, it responded to a flash of light with a depolarization (Figure 4.13B), instead of the hyperpolarization commonly found in photoreceptors of the principal eyes of lizards and all other vertebrates. When, however, the parietal photoreceptor was illuminated with steady light in the green region of the spectrum and then stimulated with a blue flash on top of the continuous green background, the photoreceptor hyperpolarized (Figure 4.13C). The cell is therefore able to produce responses of different polarity with different wavelengths of stimulating light (Solessio and Engbretson, 1993). The depolarization had a peak sensitivity at about 495 nm in the blue-green, whereas the hyperpolarization has a peak sensitivity at about 440 nm in the blue (Figure 4.13D). Both responses could be recorded from single cells isolated from the retina and must therefore be intrinsic to the photoreceptors themselves.

How can a single cell produce two such different responses? It does it, first, with two different photopigments, called parietopsin for the depolarization and pinopsin for the hyperpolarization (Su et al., 2006). These proteins are both related to the rhodopsins of vertebrate rods and cones but have somewhat different properties (Sakai et al., 2012). Moreover, they are coupled to different G-protein α subunits: parietopsin to $G\alpha_o$ and pinopsin to gustducin or $G\alpha_{gust}$, the $G\alpha$ used by some vertebrate taste receptors. Neither uses the transducin of rods and cones.

Though the two responses employ different pigments and G proteins, they seem to gate the same outer-segment channels, which are nearly identical in their properties to the channels of rods and cones (Finn et al., 1998). Inside-out recordings like those in Figure 4.10A show that the outer segments of parietal photoreceptors have a high density of channels selectively opened by cGMP (Figure 4.14A). And like the channels of rods and cones, the parietal-photoreceptor channels are non-selectively permeable to cations. If cGMP is introduced into the cell with a patch pipette, the channels open and the photoreceptor depolarizes (Xiong et al., 1998).

The effector enzyme is also the same as in rods and cones and is a cGMP phosphodiesterase, the enzyme that hydrolyzes cGMP. What seems to be going on is this (Figure 4.14B): In darkness, the phosphodiesterase is spontaneously active at a relatively high rate and keeps the cGMP concentration low. Light in the green stimulates parietopsin and then $G\alpha_o$, which inhibits the phosphodiesterase and allows the cGMP concentration to rise. The increase in cGMP opens the channels, and the photoreceptor depolarizes. Shorter wavelengths have the opposite effect: they stimulate pinopsin and then gustducin, which activates the phosphodiesterase, decreases the cGMP concentration, closes the channels, and produces a hyperpolarization. The parietal photoreceptor is a rare example of a metabotropic sensory cell with two interacting G-protein cascades, which generate responses of opposite polarity.

Summary

For many sensory receptor cells, the receptor protein is not itself an ion channel but triggers a complex series of biochemical reactions, which in turn regulate membrane channels with high gain. The

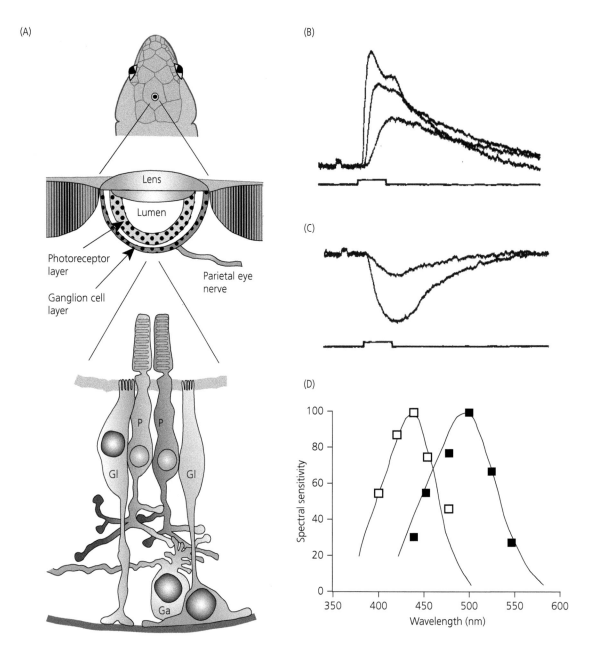

Figure 4.13 Lizard parietal eye. (A) The parietal eye is located under a small aperture between the parietal bones. The retina contains photoreceptors (P), glial cells (Gl), and ganglion cells (Ga), which receive input from photoreceptors and convey this information to the CNS. (B) Depolarizing responses of photoreceptor to 1-s green flashes of increasing intensity. Resting membrane potential was −50 mV. (C) Hyperpolarizing responses recorded to 1-s blue flashes of increasing intensity in the presence of a green adapting background. Resting membrane potential was −36 mV. (D) Relative spectral sensitivities of depolarizing response (closed squares) and hyperpolarizing response (open squares). (From Solessio and Engbretson, 1993.)

(A)

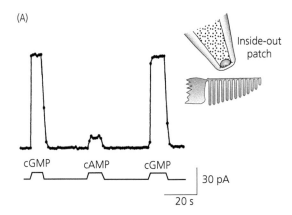

(B)

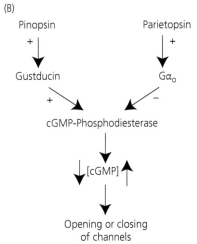

Figure 4.14 Cyclic-nucleotide-gated channels in the lizard parietal eye. (A) Inside-out patch pulled from the outer segment of a parietal eye photoreceptor, voltage clamped as in Figure 4.10A. The concentration of both cGMP and cAMP was 1 mM. (B) Proposed transduction cascades for depolarizing and hyperpolarizing photoreceptor responses. See text. (A from Finn et al., 1997; B after Su et al., 2006.)

receptor protein in these cells is commonly a member of the large G-protein-coupled family that includes many receptors for hormones and synaptic transmitters. All of these proteins use a similar transduction cascade. Activated receptor provides a binding site for one of a family of heterotrimeric G proteins having three subunits: Gα, which has a guanosine-nucleotide binding site, and Gβ and Gγ, which are closely associated with one another and normally act as a Gβγ dimer. After G protein binds to receptor, there is an exchange of GTP for GDP on the nucleotide binding side of Gα. This exchange causes the release of Gα•GTP and Gβγ, which both diffuse either close to or within the plasma membrane and can bind to and activate channels and effector molecules.

Metabotropic cascades can use many different effector molecules. Some effectors, like adenylyl cyclase and cyclic-nucleotide phosphodiesterase, synthesize or degrade second-messenger molecules such as cAMP and cGMP. Other effectors metabolize lipid, producing the second messengers DAG and IP_3. IP_3 diffuses through the cytoplasm to receptors located on vesicles of the endoplasmic reticulum, causing the opening of a channel and release of Ca^{2+} into the cytoplasm. Ca^{2+} and other second messengers can produce the gating of ion channels but can also regulate a large variety of other proteins, important for transduction but also for modulation of sensory cascades, for example during adaptation to maintained stimulation.

The response of a sensory cell must ultimately lead to an opening or closing of an ion channel. Many metabotropic sensory cells utilize channels gated by second messengers. Among the most interesting are the cyclic-nucleotide-gated channels, which show a structural similarity to voltage-gated K^+ channels but are permeable to a wider variety of cations, including Ca^{2+}. Each channel monomer contains a cyclic-nucleotide binding site, and since the channels are tetramers, the complete channel has four binding sites which must all be occupied by cyclic nucleotide for maximal activation of the channel. These channels can be gated when receptor activation modulates the activity of either a cyclic-nucleotide cyclase or a phosphodiesterase, producing an increase or decrease in cAMP or cGMP. Cyclic-nucleotide cascades are important examples of signal processing pathways used by metabotropic sensory receptor cells.

CHAPTER 5

Mechanoreceptors and touch

The single-celled organism *Paramecium*, familiar to most biology students (Figure 5.1A), has a remarkably well-developed sensitivity to a variety of stimuli. As H. S. Jennings showed over a hundred years ago (Jennings, 1906), *Paramecium* and many other protozoans can respond to touch, certain chemicals, osmotic pressure, temperature, strong light, and even gravity. These stimuli alter the movement of the cilia that surround the body, changing the direction of motion. When, for example, a *Paramecium* encounters an immovable barrier such as a small twig in pond water, it initiates a complicated behavior that Jennings called an *avoiding reaction* (Figure 5.1B). The *Paramecium* abruptly ceases to beat its cilia, then reverses the direction of beat for several seconds to withdraw from the vicinity of the obstacle. It next pivots about its posterior and finally heads off in some new direction. If it hits the barrier again, it initiates another avoiding reaction until it successfully swims free.

Mechanoreception in *Paramecium*

Jennings was the first to realize that the *Paramecium* must have a sense of touch that is different for different parts of the cell. Mechanical stimulation to the anterior end of the cell causes the animal to stop and reverse its cilia, but stimulation to the posterior end makes the animal swim faster as if to escape whatever is prodding it. Touch to the middle causes no response at all. Much later, Naitoh and Eckert (1969) showed that touch to the body of a *Paramecium* produces an electrical response (Figure 5.2). Stimulation of the anterior part of the cell generates a depolarization probably as the result of a selective increase in permeability to cations including Ca^{2+}, whereas touch to the posterior end produces a hyperpolarization due to a selective increase in permeability to K^+ (Naitoh and Eckert, 1973). As expected, mechanical stimulation to the middle of the animal causes no change in membrane potential.

The changes in membrane potential are directly responsible for producing the changes in the behavior of the organism. Depolarization produced by touching the anterior pole elicits a Ca^{2+}-dependent action potential (see Eckert and Brehm, 1979), causing a significant entry of Ca^{2+} from the pond water into the cytoplasm. This increase in Ca^{2+} in some way causes the reversal of direction of ciliary beat. This phenomenon was clearly demonstrated by the clever experiments of Naitoh and Kaneko (1972), who exposed *Paramecia* to a low concentration of the detergent Triton X-100. This procedure disrupted the plasma membrane of the cell, so that ions could flow freely between the external medium and the cytoplasm. The detergent did not, however, disrupt the mechanism of ciliary beat. Naitoh and Kaneko then demonstrated that the direction of beat could be manipulated by changing the free-Ca^{2+} concentration in the bathing solution, which in Triton-extracted animals also changed the Ca^{2+} concentration inside the cell. When the external concentration was less than 10^{-6} M (i.e. 1 μM), the cilia beat in their normal direction. Increasing the Ca^{2+} concentration above this value caused the cilia to reverse and beat backwards. The mechanism of this effect is unknown.

Sensory Transduction. Second Edition. Gordon L. Fain, Oxford University Press (2020). © Gordon L. Fain 2020.
DOI: 10.1093/oso/9780198835028.001.0001

(A)

(B)

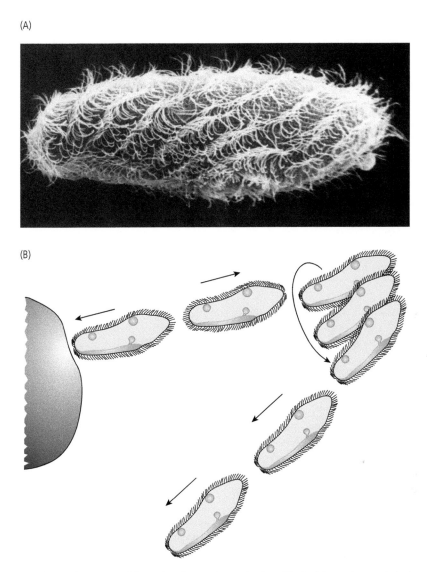

Figure 5.1 *Paramecium*: the avoiding reaction. (A) Scanning electron micrograph of a rapidly fixed *Paramecium* caught in the act of forward swimming, showing the typical "metachronal" wave pattern of beating cilia. Anterior of *Paramecium* is at left. (B) Avoiding reaction. (A from Lieberman et al., 1988; B after Jennings, 1906.)

The hyperpolarization produced by touching the posterior of the *Paramecium* increases the rate of ciliary beat and speeds the animal forward. This response is somehow also produced directly by the change in membrane potential, because the effect of touch can be mimicked simply by the injection of hyperpolarizing current (Brehm and Eckert, 1978). The greater the hyperpolarization, the greater the rate of forward beat. The mechanism is again unknown. A protozoan would seem to be an ideal organism for understanding the molecular apparatus responsible for mechanoreception, because the channels for signaling touch as well as the molecules that produce the behavioral response are all present within a single cell. Unfortunately, little further progress has been made.

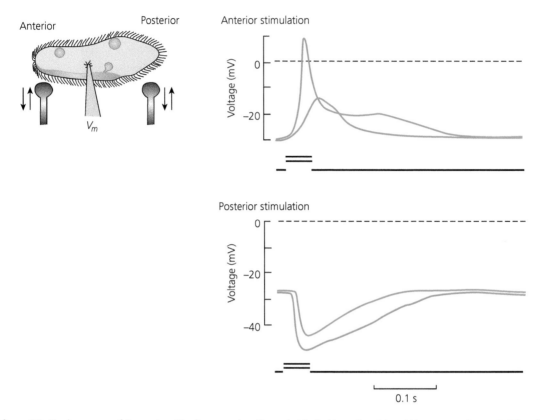

Figure 5.2 Touch responses of *Paramecium*. Stimuli were produced by an electrically driven micro-stylus which was pressed up against the cell. The timing and relative amplitude of the stimuli are shown in the traces below each of the electrical recordings. Two amplitudes of pressure were applied at each location. (After Eckert, 1972.)

Several mechanoreceptive channels have been identified in bacteria, but these channels seem not to mediate touch sensitivity but rather to protect the bacterium against osmotic shock (Cox et al., 2018). When a bacterium is exposed to hypo-osmotic shock, the bacterium swells, producing a change in membrane tension which gates the opening of large mechanosensitive channels. The two most extensively studied are called $M_{sc}S$ and $M_{sc}L$, mechanosensitive channels of "small" and "large" conductance (Martinac et al., 1987; Kung et al., 2010). Both have very large pores, with single-channel conductances fifty to a hundred times larger than voltage-gated Na^+ and K^+ channels. These mechanosensitive channels are so large than much of the small-molecular-weight molecules in the bacterial cytoplasm can pass through them to the extracellular medium, re-establishing osmotic equilibrium and preventing lysis of the cell. Channels related to $M_{sc}S$ and $M_{sc}L$ have

been discovered in fungi and higher plants: the genome of the plant *Arabidopsis* has ten proteins related to $M_{sc}S$. But there have been no reports of similar proteins in animals.

Transduction of touch in the round worm *C. elegans*

Among the better understood of the touch receptors in the animal kingdom are those of the worm *Caenorhabditis elegans*. This animal has become a favored species for biological research because, though a metazoan, it has fewer than a thousand cells and only 302 neurons. Extensive investigation of the genetics, anatomy, and behavior of this animal has led to the identification of several groups of cells that contribute to touch reception. These include six microtubule cells, so-called because they contain a characteristic, dark-staining bundle of microtubules (Figure 5.3 and

Chalfie and Sulston, 1981; Chalfie and Thomson, 1982). These cells mediate the response of the animal to light touch. They are responsible for a behavior remarkably similar to the one just described for *Paramecium*: when the anterior part of the animal is lightly stroked, the animal moves backward; and when the posterior region is stimulated, it moves forward. Once again, gentle touch in the middle has no effect.

Since *C. elegans* is transparent, the microtubule cells can be identified in the living animal. When all of the microtubule cells are selectively killed with strong laser light, the response to light touch of the anterior and posterior body is eliminated but not the response to much stronger noxious mechanical stimulation. Ablation of the microtubule cells also has no effect on touch of the nose, which seems to be mediated by a different mechanoreceptive system (Christensen and Corey, 2007; Geffeney et al., 2011).

In an important series of experiments, O'Hagen and colleagues (2005) succeeded in making patch-clamp recordings from microtubule cells (called PLM cells) near the worm's posterior (see Figure 5.3A). When a glass probe was gently pushed against the worm's body, the PLM cell responded with transient depolarizations at both the beginning and end of stimulation. Stimulation of a voltage-clamped cell produced a rapidly adapting increase in inward current again at both the beginning and end of stimulation (Figure 5.4A). The current is produced largely by an increase in permeability to Na^+ (Figure 5.4B) and can be blocked by amiloride (Figure 5.4C), a compound commonly used to block epithelial Na^+ channels of the DEG/ENaC family like those in the human kidney. Channels from this family have also been implicated in mammalian salt detection (see Chapter 8).

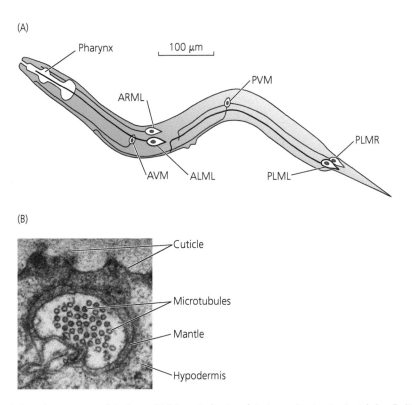

Figure 5.3 Microtubule mechanoreceptors of *C. elegans*. (A) Schematic drawing of *C. elegans* showing six microtubule cells. (B) Electron micrograph of a transverse section of the sensory process of one of the microtubule cells. Magnification 36,000×. (A after Chalfie and Sulston, 1981; B from Chalfie and Sulston, 1981.)

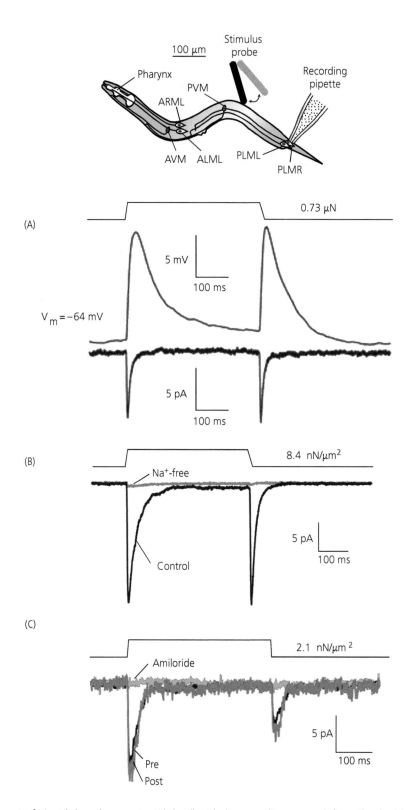

Figure 5.4 Responses of microtubule mechanoreceptors. Whole-cell patch-clamp recordings were made from either the right or left PLM microtubule neuron. Pressure was applied with a stimulus probe. (A) Responses to an applied force of 0.73 μN. Upper trace shows stimulus timing, middle trace gives the change in membrane potential in an unclamped cell from a resting membrane potential (V_m) of −64 mV, and lower trace shows the change in current from the same cell, recorded under voltage clamp at a holding potential equal to V_m. (B) Mechanoreceptive current in control solution (black) and Na+-free external saline (blue). (C) Mechanoreceptive current before (Pre, black), during (Amiloride, blue), and after (Post, gray) exposure to 200 μM amiloride. (From O'Hagan et al., 2005.)

To understand the molecular components of mechanoreception in *C. elegans*, a large number of mutant strains were isolated that specifically lack the response to light touch (Chalfie and Sulston, 1981). These screens led to the identification of *mechanosensory* or *mec* genes and their encoded products, the MEC proteins. All of these proteins seem to be essential to the function of the microtubule cells. Some are developmental and contribute to formation of the cells (Chalfie and Au, 1989), and others may alter the folding or assembly of the proteins responsible for sensation (Chen et al., 2016). The majority seem, however, to have some specific role in transduction.

Among the most interesting are MEC-4 and MEC-10, which have been shown to be pore-forming channel proteins (Goodman et al., 2002; O'Hagan et al., 2005; Arnadottir et al., 2011). They are quite different in structure and amino acid sequence from voltage-gated channels like the *Shaker* K+ channel or the piezo proteins. MEC-4 and MEC-10 have only two membrane-spanning regions connected by a large extracellular domain (Hong et al., 2000) and have a structure closely resembling channels of the DEG/ENaC family (Eastwood and Goodman, 2012). No current can be recorded when MEC-4 and/or MEC-10 are expressed in a *Xenopus* oocyte (Goodman et al., 2002), but there are two altered forms of the channels called MEC-4d and MEC-10d, which together form a channel that is continuously open. The MEC-4d/MEC-10d current is inhibited by amiloride, much like the current recorded from PLM cells in Figure 5.4C. This current can be increased several-fold if two additional proteins, called MEC-2 and MEC-6, are also expressed in the same oocyte. The channel itself seems, however, to be composed only of MEC-4 and MEC-10, probably as a heterotrimer formed from two MEC-4s and one MEC-10 (Chen et al., 2015).

In addition to these proteins, the response to light touch in *C. elegans* requires MEC-7 and MEC-12. These proteins code for α and β tubulin, structural components of the microtubules that run through the long sensory process formed by each of the cells (Figure 5.3B). Mechanosensitivity also requires MEC-1, MEC-5, and MEC-9, which appear to be extracellular, secreted into the space around the sensory process. Both the intracellular microtubules and extracellular proteins are essential components

of the transduction mechanism, whose exact function is still debated. They may tether the channels to the rest of the cell or provide a substrate to increase the sensitivity of the channels to changes in membrane tension (Cueva et al., 2007). Another cytoplasmic protein called β-spectrin forms part of the cytoskeleton of the cell and seems also to assist in the conversion of tension to an electrical response (Krieg et al., 2014).

From the remarkable advances in the genetics and molecular biology of mechanoreception in *C. elegans* we have learned two very important lessons. First, there may be many different kinds of channels used by mechanoreceptors to sense membrane stretch. The MEC-4 and MEC-10 proteins are completely unlike the stretch-sensitive channels used in osmotic regulation of bacteria or the piezo proteins. Second, mechanoreception may be produced in *C. elegans* and in higher animals by complex structures formed by both intracellular and extracellular proteins in addition to the channels. These proteins may be absolutely essential to the sensitive detection of pressure pushing against the membrane of the cell.

Crayfish stretch receptor

Although *C. elegans* is ideal for genetics and molecular biology and has given us considerable insight into the molecular mechanism of transduction, it is a much less advantageous preparation for the detailed investigation of function. The stretch receptor of the crayfish has provided no cloned genes or sequenced proteins, but electrical recording is relatively easy, and we know a lot about how this mechanoreceptor works (see Rydqvist et al., 2007).

Between each of the segmented plates of the abdomen of the crayfish and other crustaceans, there are muscle receptor organs, which in crayfish are called MRO_1 and MRO_2. These organs sense flexion and extension of the abdomen (see Bullock and Horridge, 1965; Cattaert and Le Ray, 2001). They are made up of muscle fibers invested by the fine dendritic terminals of a single large mechanoreceptor (Figure 5.5). The dendrites are full of microtubules and surrounded by extracellular connective tissue (Tao Cheng et al., 1981; Tskhovrebova et al., 1991), much like the mechanoreceptors of

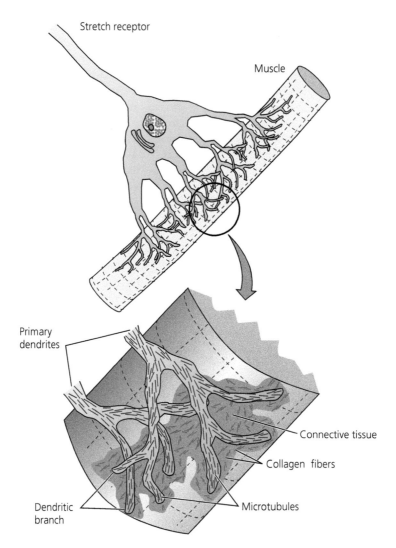

Stretch receptor

Muscle

Primary
dendrites

Connective tissue

Collagen fibers

Microtubules

Dendritic
branch

Figure 5.5 Crayfish stretch receptor. The fine dendrites of stretch receptor neurons are surrounded by collagen fibers and connective tissue and are closely apposed to accessory muscle fibers. (After Wiersma, 1967.)

C. elegans. When the abdomen is flexed, the muscle fibers of the receptor organs are stretched and the dendrites of the mechanoreceptors are distended, and this movement causes the receptors to depolarize and generate action potentials.

Intracellular recordings were first made from these stretch receptors many years ago by Eyzaguirre and Kuffler (1955), who showed that MRO$_1$ and MRO$_2$ both have resting potentials of the order of –70 to –80 mV and depolarize to stretch. The two kinds of mechanoreceptors produce somewhat different responses (Figure 5.6). For MRO$_1$, stretching produces a continuous depolarization and a continuous firing of action potentials for as long as the stretch is maintained. For MRO$_2$, on the other hand, stimulation produces a similar depolarizing receptor potential but only a short burst of action potentials which ceases before the stretch is released (Figure 5.6B). For this

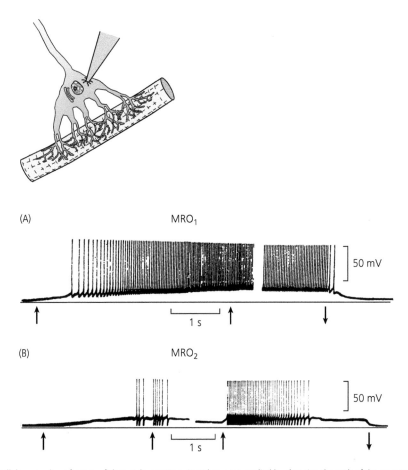

(A) MRO_1

50 mV

1 s

(B) MRO_2

50 mV

1 s

Figure 5.6 Intracellular recordings from crayfish stretch receptors. Stretches were applied by clamping the ends of the accessory muscle with tweezers attached to sliding supports. The supports were then gently moved with micrometer screws to pull the ends of the muscle apart. (A) MRO_1, a slowly adapting receptor. At the first arrow, the stretch was applied and then gradually increased up to the second arrow. The stretch was then maintained until the third (downward) arrow. The gap in the record contained several seconds of rhythmic spiking. (B) MRO_2, a rapidly adapting receptor. Slow stretch at the first arrow produced three action potentials. An additional stretch at the second arrow was maintained for 4–5 s, though spiking ceased after less than 0.5 s. The cell was then stimulated with a rapid stretch at the third arrow, which was maintained until the fourth (downward) arrow. (Data re-labeled and displayed with contrast reversed from Eyzaguirre and Kuffler, 1955.)

reason, the MRO_1 if often called *slowly adapting* and MRO_2, *rapidly adapting*.

The difference in adaptation of the two muscle receptor organs is at least in part the result of a difference in the voltage-gated channels in the receptor membrane, that is in the Na^+ and K^+ channels that produce the action potentials. Part of the evidence for this conclusion comes from experiments in which depolarizing current is injected into the receptor cells in the absence of any stretch. For the slowly adapting receptors, depolarizing current pro-

duces a maintained burst of action potentials; but for the rapidly adapting, the action potentials cease after a second or two (Nakajima and Onodera, 1969a; Rydqvist and Purali, 1993). If injected current can produce such a pronounced difference in firing even without muscle stretch, then the transduction machinery of the mechanoreceptor would seem to play little role, at least in this form of adaptation of the response.

The modulation of the rate of action potential firing is not, however, the only sort of adaptation these receptors show. A key experiment in favor of an additional mechanism of adaptation is shown in Figure 5.7 (from Rydqvist and Purali, 1993). Here the two kinds of receptor cells have been voltage clamped and the current of the receptor measured directly. Because the voltage of the cell was kept at a constant value by the voltage-clamp circuit (see Chapter 3), voltage-gated Na^+ and K^+ channels could have made no contribution to the current waveform, and indeed the receptor current is little affected by agents like TEA and 4-AP that block

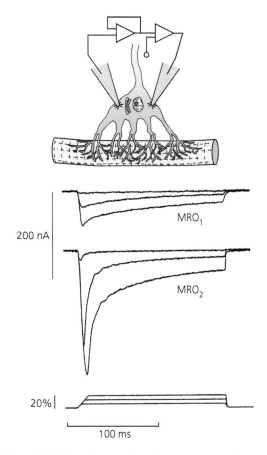

voltage-gated K^+ channels. Nevertheless, the receptor current of both the rapidly and slowly adapting mechanoreceptors reaches a peak and then slowly declines during a maintained stretch, indicative of some process of adaptation that has nothing to do with action potentials and that is greater for the rapidly adapting (MRO_2) receptors than for the slowly adapting (MRO_1) cells.

This adaptation may be produced by the mechanical properties of the muscles to which the receptors are attached. When the muscle fibers are stretched and held at a given length, the stretch of the muscle initially exerts a strong tension on the dendrites of the stretch organ, but this tension gradually relaxes. The reason may be purely physical: the viscosity of the muscle causes the tension to dissipate. In an ingenious series of experiments, Nakajima and Onodera (1969b) produced stretches of constant tension instead of constant length. To do this experiment, they measured the tension exerted by the muscle during the applied stretch, and they increased the length of the stretch as the tension began to relax to keep tension constant, producing a kind of tension clamp. In this way they discovered that stretches of constant tension show much less adaptation than stretches of constant length. These results suggest that the difference in the adaptation of the voltage-clamp currents for MRO_1 and MRO_2 receptors may result primarily from a difference in the mechanical properties of their associated muscle fibers (Rydqvist et al., 1994).

To identify the stretch-sensitive channels of the crayfish stretch organ, Erxleben (1989) made on-cell and isolated patch recordings from the cell body and dendrites of the mechanoreceptors. He found two kinds of channels (Figure 5.8), one he called SA, for stretch-activated, and the other RSA, for rectifying stretch-activated. Both responded to membrane tension and both were permeable to monovalent and divalent cations. They differed in their voltage dependence. The probability of gating of the RSA channels during the application of pressure was a strong function of membrane voltage, decreasing as the voltage depolarized. The SA channels, on the other hand, were largely independent of voltage. The RSA channels were most often recorded from patches from the cell body, and the density of channels was usually rather small. SA channels, in contrast, could

Figure 5.7 Voltage-clamp recordings of generator currents from slowly and rapidly adapting crayfish stretch receptors. MRO_1 and MRO_2 receptors were voltage clamped with two microelectrodes and held at a command potential set equal to the resting membrane potential (−70 to −80 mV). Responses are shown to extensions of 6 percent, 12 percent, and 18 percent of the muscle length, whose amplitude and timing are indicated by the lowermost traces. Currents were negative (inward), gated by the stretch and produced by an increase in permeability mostly to Na^+. (From Rydqvist and Purali, 1993.)

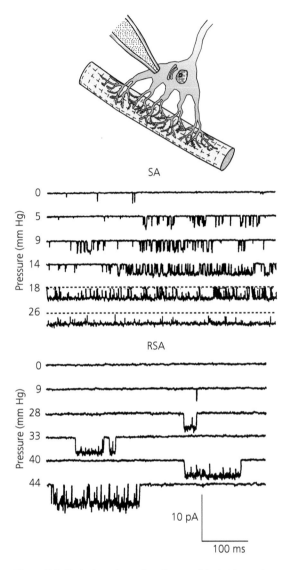

Figure 5.8 Single-channel recordings from crayfish stretch receptors. On-cell patch recordings made from the cell body and main dendrites of stretch receptors. Channel opening was produced as in Figure 3.5 by the direct application to the patch pipette of suction from a calibrated pressure transducer. The amplitude of suction is given to the left in units of millimeters of mercury (Hg). Pressure was applied continuously for the duration of each record. Patches were voltage clamped at the resting membrane potential (for SA) and 50 mV negative of the resting membrane potential (for RSA). (From Erxleben, 1989.)

be recorded from the dendrites, and each patch typically contained a large number of channels.

Erxleben could not record from the finest dendritic terminals, the ones closely apposed to the muscle fibers and buried deep within the extracel-

lular matrix. For this reason, it is difficult to be sure that the channels he identified are responsible for mechanotransduction. Nevertheless, the SA channels have all the properties one would expect. They are activated by stretch. Their conductance is largely voltage independent (as is the conductance gated by stretch for the receptor current recorded with voltage clamp), and the SA channels are abundant in dendritic membrane. It would be easy to imagine these channels arranged much like those of *C. elegans*, in some way dependent upon the numerous microtubules within the cytoplasm and extracellular matrix as in Figure 5.5. Further clarification may be possible when methods of investigating gene function in crustaceans become more readily available.

Insect mechanoreceptors

Insects contain several different kinds of touch-sensitive cells usually separated into two types (Bullock and Horridge, 1965; McIver, 1985; Keil, 1997). The most numerous and prominent are the type I mechanoreceptors, which are the cells that signal detection in an insect bristle or sensillum and are associated with complex extracellular structures. A typical example is the hair plate sensillum (Figure 5.9A). A single mechanoreceptor surrounded by several kinds of supporting cells sends a dendritic process often called an outer segment up into the bristle shaft. Bending of the bristle signals detection. The mechanosensory part of the outer segment is thought to be the distal area with a dense conglomeration of microtubules, called the tubular body or distal tip (Figure 5.9B). This part of the cell is surrounded by a prominent region of extracellular matrix called the cap.

Another commonly found type I receptor is the campaniform sensillum (Figure 5.9C), with a tubular body full of microtubules and a cap similar to the hair plate sensillum; the dendrite, however, terminates at the base of an extracellular dome rather than within a bristle. Campaniform sensilla sense compression of the cuticle surface, caused, for example, by leg or wing movement. A final important class of type I mechanoreceptor is contained within the chordotonal organ and is composed of one or more scolopidia (Figure 5.9D), each containing a ciliary dendrite surrounded by a

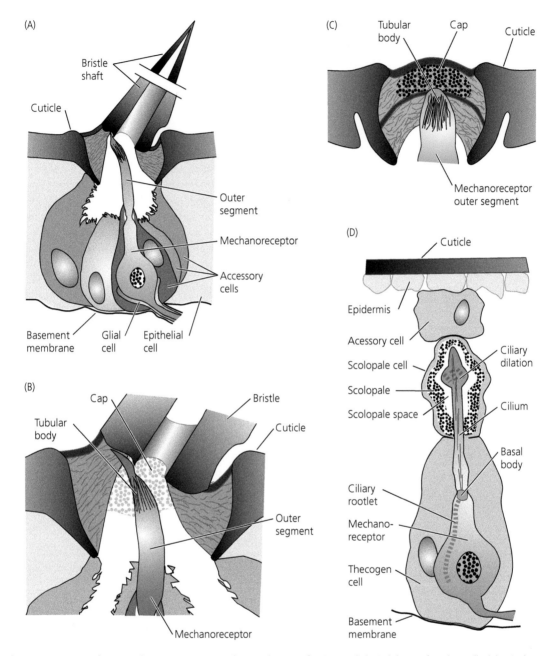

Figure 5.9 Anatomy of insect mechanoreceptive organs. Schematic drawings of major morphological classes of touch sensilla. (A) Hair plate (bristle) sensillum. (B) Magnified view of distal part of hair plate sensillum. (C) Campaniform sensillum. (D) Scolopidial organ. The thecogen cell is a type of supporting cell. (After Thurm, 1964; Bullock and Horridge, 1965; Keil, 1997.)

supporting cell called a scolopale cell. This cell secretes an elaborate extracellular structure analogous to the cap of the other two receptor types. The receptor neurons of chordotonal organs are not associated with bristles or domes, but instead the

scolopale and distal tip of the receptor are covered by an accessory cell which abuts the epidermis and inside surface of the cuticle. It is presumably the accessory cell which transmits vibrations from the surface of the cuticle to the receptor dendrite.

Chordotonal organs detect pressure induced either internally or externally against the body cuticle, providing proprioceptive information about body and joint position. They are also common in tympanal organs and in Johnston's organ, which are used in insects to detect external vibration and sound (see Chapter 6).

These three kinds of type I mechanoreceptors share common features (McIver, 1985). The receptor neurons are bipolar, that is they have two processes—an axon and a sensory dendrite—emerging in opposite directions from the nucleus. All have a prominent cilium located at the base of the outer segment, much like a vertebrate rod or cone. The dendrites in all type I receptors are associated with a prominent extracellular structure such as a cap or scolopale and are bathed in a medium high in K^+ secreted by accessory cells (Thurm and Küppers, 1980).

Insects also have numerous type II mechanoreceptors of several different types (Grueber et al., 2002). Many are multipolar neurons with widely branching dendrites. In an insect larva such as a grub or caterpillar, the dendrites lie just below the soft cuticle, and they (instead of type I cells) are the principal receptors for touch. Other neurons with an even more complex branching pattern are responsible for noxious touch in insect larvae. In adult insects, type II receptors are present and often found associated with internal organs where they sense distension, for example in a newly mated female butterfly when the bursa copulatrix fills with the spermatophore of the male. They are also found investing muscle, where they probably act as stretch receptors much like those of crayfish.

Many years of patient investigation, particularly of the hair plate sensilla of Figure 5.9A, have given us a fairly clear notion of how insect mechanoreceptors function. It is particularly easy to record from these cells, because it is possible to cut the end of the bristle and stick a pipette over the whole of the shaft. The pipette can be moved to stimulate the bristle, and the same pipette can be used to record the extracellular currents that are generated as the bristle responds to the movement (Figure 5.10A). The receptors are excited when the bristles are deflected. Excitation is directionally sensitive: the bristles are only excited

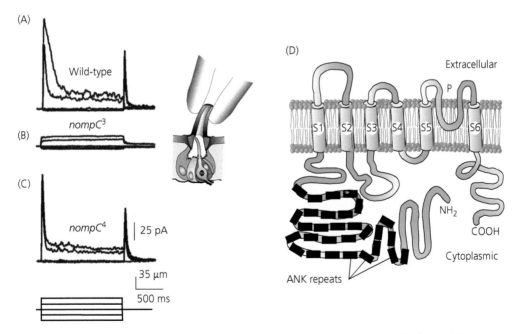

Figure 5.10 Mechanoreceptive responses of *Drosophila* sensilla. (A–C) Recordings from touch-sensitive bristles of *Drosophila*. A pipette was used both to apply the stimulus and to record extracellular current. Recordings give the responses to bristle movement for (from top to bottom) wild-type animals (A), *nompC³* showing almost no response (B), and *nompC⁴* showing a response that declines (adapts) more rapidly than wild-type (C). Bottom traces give timing and relative magnitudes of stimuli. (D) Membrane topology of the NOMPC TRP-channel protein. Ankyrin (ANK) repeats are represented as black boxes. (A–C from Walker et al., 2000.)

when bent toward the body of the animal (Thurm, 1965b), and movement in the opposite direction may produce inhibition (Thurm, 1983; Walker et al., 2000). The response is produced with very short latency, of the order of 100 μs or less (Thurm, 1983), and it is extraordinarily sensitive. In bristles with long, filiform hairs, a deflection of only a few tenths of a degree of angle can produce maximal excitation of the mechanoreceptor. The excitation rapidly adapts (Thurm, 1965a; Walker et al., 2000), but the respective contributions of mechanical elements and voltage-gated channels to adaptation have not yet been elucidated as clearly as for crayfish stretch receptors.

The first information about the molecular mechanism of mechanoreception came from experiments isolating mutant fruit flies showing abnormal behavior, consisting of moderate to severe clumsiness and lack of coordination (Kernan et al., 1994). Several of the isolated mutations were characterized as producing no mechanoreceptor potential and were called *nomp* genes. One gene in particular, *nompC*, had a particularly interesting phenomenology (Walker et al., 2000). For lines *nompC¹–nompC³*, electrical recordings from the mechanoreceptor organs of the fly showed almost no response (Figure 5.10B). For the mutant line called *nompC⁴*, the response was present but decayed abnormally rapidly (Figure 5.10C). This result suggested that the NOMPC protein had some specific role in transduction.

The *nompC* mutations were then shown to be caused by amino acid alterations in a novel protein called NOMPC (or alternatively TRPN1). The sequence of the NOMPC protein indicated that it is almost certainly an ion channel similar in structure to the *Shaker* protein, containing six transmembrane domains, a putative P loop between domains 5 and 6, and cytoplasmic amino and carboxyl termini (Figure 5.10D). This protein is again different from the other proposed stretch-sensitive channels we have described and is distantly related to the TRP family of ion channels. Some members of the TRP-channel family have cytoplasmic ankyrin repeats, which we encountered in Chapter 3 for the heat-sensitive TRPV1 channel (see Figure 3.6A). What is remarkable about NOMPC is that it has twenty-nine ankyrin repeats (black boxes in Figure 5.10D), the largest of any known TRP protein.

Although there was some initial skepticism about the role of the NOMPC channel in insect mechan-oreception (Christensen and Corey, 2007), there is now abundant evidence that this channel is indeed a stretch-sensitive channel in many insect sensilla. An antibody to NOMPC has been shown to be localized to the tubular body in the vicinity of the microtubules (Cheng et al., 2010; Liang et al., 2013). Single mechanosensitive channels can be recorded after expression of NOMPC in an insect cell line, showing that the NOMPC protein by itself—even in excised patches—can respond to changes in membrane tension (Yan et al., 2013). Moreover, expression of NOMPC in cells not normally sensitive to touch makes these cells mechanosensitive (Yan et al., 2013).

Attention has focused on the ankyrin repeats, which are now thought to have an essential role in the mechanism of transduction (Liang et al., 2013; Zhang et al., 2015; Jin et al., 2017). Deletion of all of the repeats (Δ1–29ARs) or even just twelve of the repeats (Δ1–12ARs) completely eliminates the response of the receptor to mechanical stimulation (Figure 5.11A). Doubling the number of ankyrin repeats (29+29ARs) significantly reduces the response but does not eliminate it. This experiment is an important control, because the simple deletion of the ankyrin repeats can affect expression or transport of the NOMPC protein to the tubular body (Cheng et al., 2010). If doubling the number of repeats also affects the response, we can be more confident that the ankyrin repeats are having some specific role in producing the response.

In addition to this evidence, Zhang and colleagues (2015) have shown that a channel not normally sensitive to membrane stretch can be converted to a stretch-sensitive channel merely by the addition of ankyrin repeats. In Figure 5.11B, recordings were made from isolated patches of membrane pulled from an insect cell line in which various forms of the voltage-gated $K_V1.2$ potassium channel had been expressed. Recordings labeled S1268-G160 were made from $K_V1.2$ channels to which the twenty-nine ankyrin repeat units of NOMPC had been added. When the membrane patch was stretched by application of a decrease in pressure of 50 mm of mercury (Hg), there was a large outward current when Na^+ was present in the external solution but none if the Na^+ was replaced with Cs^+ (gray trace). Wild-type channels (WT

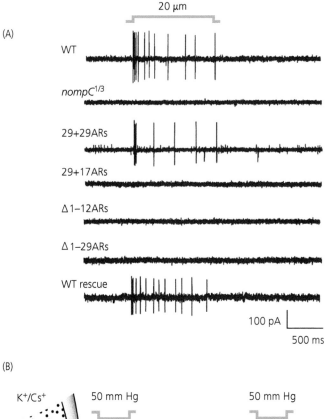

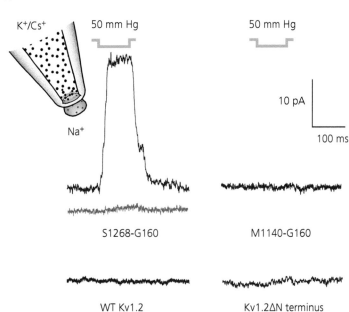

Figure 5.11 Ankyrin repeats are essential for NOMPC-mediated mechanoreception. (A) Extracellular recordings of action potentials from mechanoreceptors in wild-type (WT) and mutant *Drosophila* larvae. *NompC^1/3* are null mutants for NOMPC (see Figure 5.10B). Other mutants express NOMPC with modified ankyrin sequences on a null background. 29+29 mutants have two full ankyrin-repeat sequences, 29+17 have one complete and one partial sequence, and the other mutants have a single sequence lacking some or all of the ankyrin repeats. WT rescue expresses full-length ankyrin repeat. (B) Recordings from chimeric channel S1268-G160-Kv1.2 made from $K_v1.2$ voltage-gated K$^+$ channel to which the 29 ankyrin repeat units of NOMPC had been added. Channels were expressed in a cell line, and recordings were made from outside-out patches held at +60 mV. Deformation of the membrane by negative pressure produced a large current in Na$^+$ solution but none in Cs$^+$ solution (gray trace). Full-length Kv1.2 (WT Kv1.2), M1120-G160-Kv1.2 chimeric channels with ankyrin repeats but without a NOMPC linker region between S1 and the repeats, and Kv1.2 with a truncated N terminus (Kv1.2ΔN terminus) were not responsive to the same stimulus. (From Zhang et al., 2015.)

$K_V 1.2$), or channels completely lacking an amino terminus ($K_V 1.2 \Delta N$ terminus), or channels with the ankyrin repeats but without a NOMPC linker region between the repeats and the rest of the pro-tein (M1140-G160) all failed to show any sensitivity to membrane stretch.

Cryo-EM of NOMPC (Figure 5.12A) indicates that the assembled NOMPC protein is a tetramer

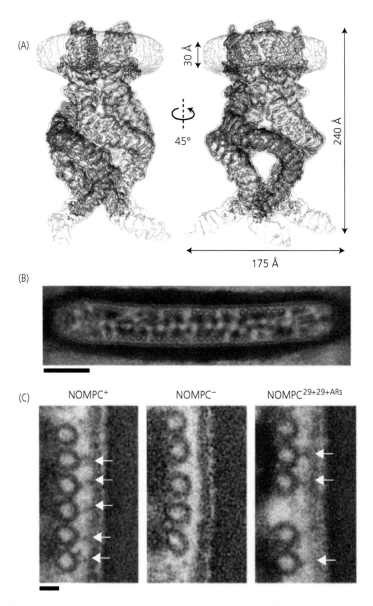

Figure 5.12 Structure of NOMPC: role of ankyrin repeats. (A) Cryo-EM electron density maps of NOMPC shown from the side at two different orientations. Each of the four subunits is represented with a different color. Spirals extending downward from the plasma membrane (gray torus at top) are ankyrin repeat sequences. (B) Electron-microscopic picture of structure of the dendritic tip of a campaniform sensillum (scale bar, 200 nm). Circular profiles are rows of microtubules just below the plasma membrane. (C) Cross-sections through the mechanosensitive dendritic tips of campaniform mechanoreceptors, depicting the extracellular sheath, cell membrane, microtubules, and membrane-microtubule connectors (arrows). Connectors are present in WT flies (NOMPC+, left) but not in *nompC*[1] null mutants (NOMPC−, middle). Mutants with double the number of ankyrin repeats (NOMPC[29+29ARs]) are shown on the right; connectors are somewhat longer. Scale bar, 20 nm. (A from Jin et al., 2017; B and C from Zhang et al., 2015.)

like *Shaker* and other voltage-gated potassium channels. The four ankyrin repeat sequences of each of the NOMPC monomers coil around one another on the cytoplasmic side of the protein, forming a structure resembling a spring (Jin et al., 2017). It is quite likely that this "spring" connects the channels to microtubules and is responsible for the sensitivity of the channel to membrane deformation.

The ankyrin repeats associate with microtubules and span the gap between the plasma membrane and the microtubules of the tubular body (Liang et al., 2013; Zhang et al., 2015). Figure 5.12B is a micrograph of the structure of the tubular body (or dendritic tip) of a campaniform sensillum like that in Figure 5.9C. The circular profiles are rows of microtubules just below the plasma membrane on either side. Figure 5.12C shows one of the microtubule rows at higher power, first from a wild-type fly (NOMPC+), then from a fly lacking the NOMPC channels (NOMPC−), and finally from a fly with channels having twice the number of ankyrin repeats (NOMPC$^{29+29ARs}$). In a wild-type fly, there are clearly visible filaments stretching from the plasma membrane to the microtubules, marked by the white arrows. These filaments are lacking in animals without the NOMPC channels and are present and significantly longer in channels with twice the number of ankyrin subunits. These filaments are most likely formed by the ankyrin "springs" of NOMPC protein.

In addition to the *nompC* gene, another gene called *nompA* was isolated in initial screening (Kernan et al., 1994). The protein product of this gene is also essential to the production of the response and seems to form part of the extracellular cap produced by supporting cells (Chung et al., 2001; Cheng et al., 2010). In mutant animals, the cap forms but is abnormal in its organization, and the tip of the dendritic process of the sensory neuron does not properly insert into the base of the bristle. The response of a type I receptor in an insect sensillum seems therefore to require both tethering of the channel to cytoplasmic microtubules and connections of some sort to extracellular cap structures. Although the NOMPC protein can respond to touch without any of these specializations in isolated membrane patches (Yan et al., 2013), the sensitivity of the receptor may be much greater in the intact sensillum.

In type II receptors in *Drosophila* larvae, NOMPC seems also to be the stretch-sensitive channel mediating light touch (Yan et al., 2013). It is of some interest, however, that stronger, noxious touch is mediated differently, apparently by a combination of a *Drosophila* piezo protein and a member of the DEG/ENaC family of channels called *pickpocket* or *ppk* (Zhong et al., 2010; Kim et al., 2012). Whether these channels are also linked in some way to intracellular and extracellular proteins is not presently known.

Mechanoreceptors and touch in mammals

The neurons that form touch-sensitive endings in the trunk and appendages of the mammalian body have their nuclei in the dorsal root ganglia (DRGs), which are adjacent to each of the segments of the spinal cord. The DRG cells send myelinated axons branching in two directions, inward into the dorsal horn of the spinal cord and outward to the skin. The branches that enter the skin lose their myelin and form fine dendrites usually (though not always) invested by extracellular matrix and accessory cells, forming some sort of external specialization. Not all of these terminations mediate our sensitivity to touch: some are pain receptors and thermoreceptors in skin, and others are mechanoreceptors that innervate muscle spindles and Golgi tendon organs.

A considerable amount is known about the anatomy and physiology of touch-sensitive skin receptors (see Darian-Smith, 1984a; Willis and Coggeshall, 1991; Hamann, 1995; Abraira and Ginty, 2013; Owens and Lumpkin, 2014; Zimmerman et al., 2014). Figure 5.13 indicates the position and shape of the four principal touch cells in a typical bit of primate finger. The skin of the finger is usually referred to as glabrous, from the Latin *glaber* meaning without hair. As we shall see, hairy skin has a different pattern of innervation, with the hairs themselves making an important contribution to touch sensitivity.

In the glabrous skin of the finger, there are four structures that are thought to be primarily responsible for our sense of touch. The most important in sensing tactile form and coarse texture are undoubtedly the Merkel cells, clustered on the apices of dermal ridges just below the epidermis. I describe these cells and their role in transduction in some detail later in the chapter. Adjacent to the Merkel cells are the Meissner corpuscles, enveloped within dermal ridges just below the epidermis. In humans, where

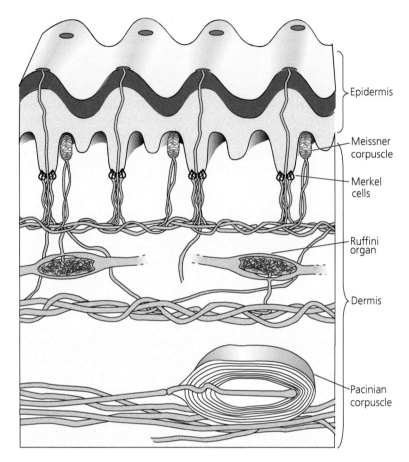

Epidermis

Meissner corpuscle

Merkel cells

Ruffini organ

Dermis

Pacinian corpuscle

Figure 5.13 Touch receptors in the glabrous skin of a primate finger pad. Glabrous (hairless) skin of a primate contains four types of touch receptors: Meissner corpuscles, Merkel cells, Ruffini organs, and Pacinian corpuscles. (After Darian-Smith, 1984a.)

these corpuscles are especially large and prominent, two to six nerve fibers lose their myelin sheath and enter the corpuscle. The nerve fibers then become interleaved with layers of Schwann cell lamellae and thick collagen fibers, which appear to be continuous with collagen in the skin. We do not know in detail what these external specializations do, but comparison with the Pacinian corpuscle (Figure 2.9) suggests they are likely to play some role in adaptation of the receptor response.

Deeper within the dermis of the skin are the Ruffini organs or corpuscles. A single nerve terminal loses its myelin sheath and splits into numerous fine processes that meander within a dense protuberance of collagen. The collagen undoubtedly con-tributes in some way to the response of the neuron, though its function is again not well understood. The fourth receptor type is the Pacinian corpuscle, whose anatomy and physiology are described in Chapter 2. The Pacinian corpuscle is the largest of the mechanoreceptive endings in skin, but it is also the least common. The apical end of the human finger contains only about 200 of them (Figure 5.14).

Recordings of the action potentials of DRG axons coming from glabrous skin show four different kinds of touch responses, apparently corresponding to the four anatomical types of receptor specializations. These spike discharges are usually characterized as slowly adapting types I and II (SAI and SAII), rapidly adapting (RA), and most rapidly adapting or

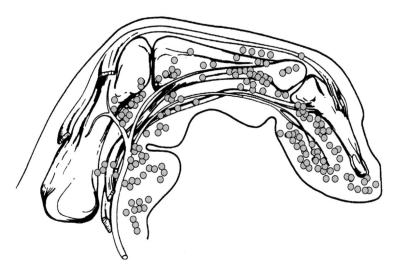

Figure 5.14 Distribution of Pacinian corpuscles in a human finger. Pacinian corpuscles (blue) in the radial half of the index finger of a 7-month human fetus. (After Darian-Smith, 1984a.)

Pacinian corpuscle (PC). To discover which mechanoreceptor type produces which response, spike discharges have been recorded from areas of skin known to contain endings of a specific type (for example Iggo and Muir, 1969), or the skin has been marked in some way to locate the receptor organ. Pacinian corpuscles were mechanically isolated from the skin and subjected to carefully controlled stimulation (as in Figure 2.9). In this way, it was possible to show that isolated Pacinian corpuscles are very rapidly adapting and respond to quite high frequencies of mechanical buzzing or tactile flutter, in the same way as the PC response recorded from DRG axons.

The different mechanoreceptive organs provide different information about the world around us. Slowly adapting receptors are primarily responsible for our sense of form, that is for our ability to tell what we have in our hand without looking. The SAI fibers (the Merkel cells) are particularly important for this sense, and it is these receptors that are primarily used, for example, to distinguish the patterns of dots in Braille characters (see Darian-Smith, 1984a; Johnson and Hsiao, 1992). SAII fibers are most sensitive to skin stretch and are probably Ruffini endings (Chambers et al., 1972), though this conclusion remains controversial. The rapidly adapting RA receptors (Meissner corpuscles) sense low-frequency flutter and vibration, but they probably also contribute to our sense of tactual form and texture. We are better at making fine distinctions of shape when we move objects about in our hand and manipulate them, than if we hold them in a fixed position. These receptors may also contribute to grip control. Pacinian corpuscles respond to high-frequency flutter and are very sensitive to any sudden indentation of the skin.

Hairy skin

The follicles or roots of simple hairs like those found on the trunk or back of the hand are surrounded by nerve endings of several DRG cell axon terminals, which form many elongate lanceolate endings closely apposed to the hair follicle (Figure 5.15). The inner face of each cylindrical ending typically abuts the basal lamina of the follicle basement membrane and is surrounded by myelin and collagen fibers. Movement of the hair pushes against the membrane of the lanceolate endings, producing membrane stretch. The nerve population innervating the hairs can be quite variable (Abraira and Ginty, 2013), consisting of rapidly adapting myelinated A fibers with a variety of conduction speeds, as well as unmyelinated C fibers responding to slow movement across the skin, sometimes called "caress detectors."

A single hair can receive innervation from more than one type of axon, with (for example) C fibers and A fibers interdigitating together around the follicle (Li et al., 2011). In addition to the lanceolate endings, the largest hairs are surrounded by Merkel cells, which provide a slowly adapting response like that of the SAI fibers of glabrous skin. Hairs are also encased by circumferential lanceolate endings, which wrap around the longitudinal endings (not shown in Figure 5.15). The circumferential endings are also sensory and seem to signal gentle skin

stroking over a large area of the skin, as well as high-threshold pulling of the hair perhaps signaling pain (Bai et al., 2015; Ghitani et al., 2017). This complex pattern of innervation of the hairs can provide a varied repertoire of sensory input when hair is moved.

Merkel cells

Because of their importance to our sense of fine touch, Merkel cells have been the focus of extensive investigation. These cells have a peculiar anatomy (see Figure 5.16A and Iggo and Muir, 1969). They lie below the epidermis in both glabrous and hairy skin and are often found in clusters of as many as several hundred cells in elevated structures called *touch domes*. An axon coming from the DRG loses its myelin, and its fine dendrites approach and invest the Merkel cells, with one nerve terminal often innervating many cells (Lesniak et al., 2014). The Merkel cell is approximately spherical but is flattened in the region juxtaposed to the nerve terminal. Merkel cells contain numerous dense-core vesicles like those containing catecholamines (norepinephrine and dopamine). Moreover, these cells express many of the proteins required for synaptic transmitter release (Haeberle et al., 2004; Hoffman et al., 2018).

These observations would seem to suggest that the Merkel cells are secondary sense cells, responding to pressure and transmitting this response at synapses onto nerve fibers, much like photoreceptors (Figure 2.12A) or hair cells of the inner ear (Figure 3.11). In agreement with this notion, molecular ablation of Merkel cells produces a profound loss of SAI responses and behavioral touch discrimination (Maricich et al., 2009, 2012). Moreover, Merkel cells can respond to membrane pressure, both within the skin (Ikeda et al., 2014) and as isolated single cells (Maksimovic et al., 2014; Woo et al., 2014). The recordings in Figure 5.16B were made with patch-clamp electrodes from areas of rat skin that had been carefully dissected to remove superficial layers of epidermis. Large, rapidly adapting inward currents can be recorded from a Merkel cell (see also Maksimovic et al., 2014), whether the cell is stimulated directly with a blunt probe or indirectly in adjacent areas of the skin. The

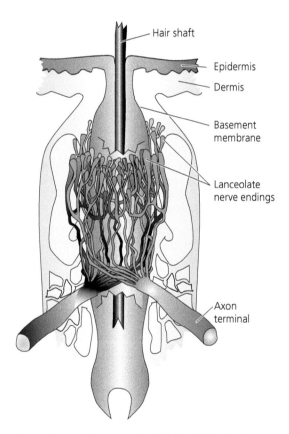

Hair shaft

Epidermis

Dermis

Basement membrane

Lanceolate nerve endings

Axon terminal

Figure 5.15 Touch receptors in hair follicles. Follicles of simple hairs are innervated by a few DRG axons that lose their myelin, split into several branches, and form longitudinal lanceolate or palisade endings that surround the hair. Endings from different axons can innervate the same hair and intermix. The follicle may also be surrounded by circumferential lanceolate endings and Merkel cells (not shown). (After Halata, 1975; Garcia-Anoveros et al., 2001.)

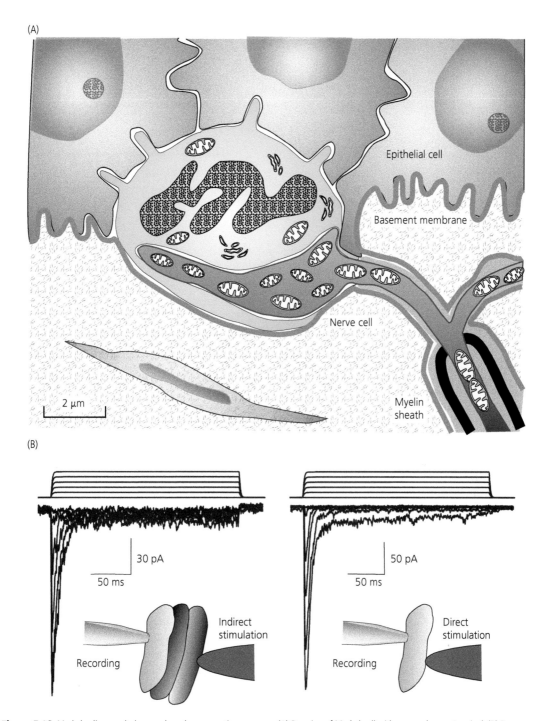

Figure 5.16 Merkel cells: morphology and mechanoreceptive response. (A) Drawing of Merkel cell with apposed nerve terminal. (B) Responses of rat Merkel cells (light blue) recorded with whole-cell patch clamp. Cells were stimulated by displacements in 1-μm steps of a fire-polished blunt glass probe (dark blue). Relative amplitude and duration of stimulus is indicated by traces above the recordings. Stimuli were applied either indirectly by pressure to the tissue (records to left) or directly onto the Merkel cell itself (records to right). (A after Iggo and Muir, 1969; B from Ikeda et al., 2014.)

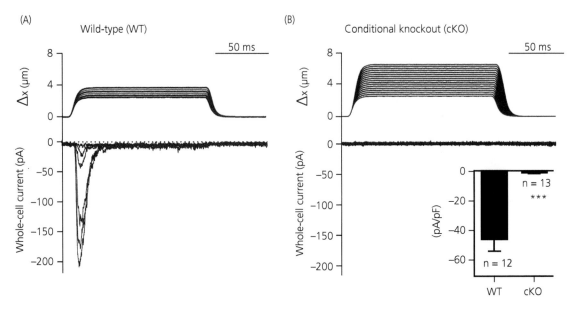

Figure 5.17 Merkel cell response requires Piezo2. Merkel cells from mouse skin were genetically labeled with GFP and isolated by fluorescence-cell sorting after tissue dissociation. Responses were recorded with whole-cell patch clamp during stimulation as in Figure 1.11A with a blunt glass probe. Upper traces indicate amplitude and duration of stimuli; lower traces give whole-cell current responses. (A) Wild-type mice. (B) Mice with conditional knockout of gene for Piezo2 (cKO) in Merkel cells. Inset: histogram showing mean values ± s.e.m. for WT and cKO mice. Currents have been divided by cell capacitance (pF) to correct for difference in cell dimension (total cell capacitance is proportional to cell membrane area). *** indicates significance at $p < 0.001$. (From Woo et al., 2014.)

response is produced by channels that are non-selective cationic and permeable to Ca^{2+}, much like those of the hair cells of the inner ear. These experiments show that Merkel cells play a primary role in producing the response to light touch.

The stretch-sensitive channel of Merkel cells is apparently our friend from the first chapter, the piezo channel. Figure 5.17 shows patch-clamp recordings from Merkel cells isolated from mouse skin (Woo et al., 2014). Green fluorescent protein (GFP) was expressed in Merkel cells by linking the GFP gene to *Atoh1*, a transcription factor expressed by all of the Merkel cells. Merkel cells could then be isolated from the skin and identified. Patch-clamp recordings from an isolated Merkel cell produced rapidly adapting currents to gentle poking with a blunt probe (Figure 5.17A), much like those from Merkel cells still intact in the skin (Figure 5.16B) and quite similar also to the inward currents produced by Piezo2 channels expressed in a cell line (Figure 1.11C). When, however, the gene for Piezo2 was knocked out specifically from the Merkel cells (Woo et al., 2014), no current could be recorded (Figure 5.17B).

Since the piezo channels are non-selectively permeable to cations (see Chapter 1), the opening of the channels during membrane stretch will produce a depolarization. This depolarization must then be communicated in some way to the nerve terminals that invest the touch domes (Figure 5.16A). Merkel cells are known to express many of the proteins required for synaptic transmitter release, as I have said, and nerve-terminal firing can be largely silenced when vesicle release is inhibited. The transmitter may be norepinephrine (Hoffman et al., 2018). Norepinephrine can excite the nerve terminals coming from the touch domes, and touch responses can be inhibited by ICI 118,551, which selectively blocks β_2-adrenergic receptors (Figure 5.18A). The selective deletion of β_2-adrenergic receptors also largely eliminates the response of the sensory terminals to touch (Figure 5.18B). The binding of norepinephrine to the receptors must then alter the membrane potential of the nerve terminals by a mechanism not presently understood.

Piezo2 may also be the stretch-sensitive channel used by other touch organs in the skin. When

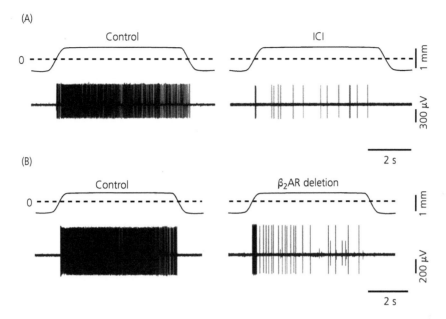

Figure 5.18 Merkel cells may release norepinephrine onto β_2-adrenergic receptors. Recordings were made from nerve terminals coming from touch domes in the skin of the mouse hindlimb. The skin was indented with a mechanical stimulator under computer control. The traces above the action-potential recordings show the magnitude and time course of the stimulus, with the dotted line indicating the point of contact of the stimulator with the skin. (A) Responses recorded after 30-s exposure to normal solution (left) and a solution containing 50 µM ICI 118,551 (right), which selectively blocks β_2-adrenergic receptors. (B) Responses recorded from a litter-mate control (left) and a mouse that has a deletion of β_2-adrenergic receptors selectively in cells derived from the neural crest, including somatosensory neurons. (From Hoffman et al., 2018.)

Piezo2 expression was labeled with GFP, dye could be observed not only in Merkel cells but also in Meissner corpuscles and in the longitudinal and circumferential lanceolate endings of hair follicles (Ranade et al., 2014). Deletion of Piezo2 produces a considerable decrease in the responsiveness of DRG neurons to touch, particularly evident in the reduction of low-threshold touch-sensitive fibers. Mice without Piezo2 show behaviorally a reduced sensitivity to light touch. Humans with nonsense mutations in Piezo2 also show marked defects in touch perception and proprioception (Chesler et al., 2016). Of particular interest is the loss of sensation to high-frequency flutter, indicating that the functioning of Pacinian corpuscles may require Piezo2. As in the experiments with mice, however, not all sensitivity to touch was eliminated, particularly in hairy skin, suggesting that Piezo2 is not the only stretch-sensitive channel used in human touch sensation.

Summary

Mechanosensitive channels are ionotropic. A mechanical force applied to the surface of the plasma membrane gates channel opening with a latency of the order of a few hundred microseconds. It is inconceivable that metabotropic cascades could gate channels in so short a time. The experiments I have described from *Paramecia* to mammals now give a fairly clear idea how mechanoreception occurs. At least three very different sorts of channels are used: some are related to ENaC, such as MEC-4 and MEC-10; some are members of the TRP family of channel proteins, such as NOMPC; and some have a completely different structure, like the bacterial $M_{sc}S$ and $M_{sc}L$ channels and Piezo1 and Piezo2. Regardless of their molecular conformation, these channels nearly all share a non-selective permeability to cations, such that stimulation produces a depolarization. The only known exception is *Paramecium*, for which stimulation in the posterior of the cell produces a hyperpolarization caused by a selective increase in K^+ permeability.

Touch receptors often require some form of extracellular specialization, like the MEC-1, MEC-5, and MEC-9 proteins of *C. elegans*, the caps of insect sensilla, and the myelin and collagen of mammalian corpuscles. Accessory cells may form external specializations such as the lamellae of Meissner or Pacinian corpuscles, which play an essential role in some aspect of the receptor response. Intracellular organelles such as microtubules are also often found near the sensory membrane. These intracellular and extracellular components are very likely necessary either to direct the force of the mechanical disturbance to the channel, to modify the probability of gating of the channel pore to increase sensitivity, or to modulate the sensitivity of the response during adaptation.

Multicellular animals generally require several different kinds of touch receptor cells. Even the lowly *C. elegans* has several types, such as the microtubule cells and touch-sensitive neurons in the nose, and these cells may use different channels with different mechanisms of channel gating (see for example Kang et al., 2010). Insects have type I cells and sensilla but also multipolar type II receptors, and mammals have a variety of different sorts of touch-sensitive organs, each with somewhat different response characteristics. One important distinguishing feature of these different receptor organs is their rate of adaptation. There are two sorts of crayfish stretch receptors, slowly and rapidly adapting, which differ in two ways. Their voltage-gated channels have different properties, so that one receptor stops firing spikes after a few seconds of maintained depolarization, whereas the other keeps generating action potentials for as long as the depolarization is continued.

The accessory muscle cells seem also to have different mechanical properties, one relaxing more rapidly than the other to a stretch of constant length. Similarly, in mammalian skin there are Merkel cell organs which adapt very slowly and sense tactile form, and Pacinian corpuscles which adapt rapidly and respond to vibration and flutter. Accessory structures seem again to play an important role in the speed of adaptation.

Despite the remarkable advances in *C. elegans*, *Drosophila*, and mouse, there is still much we do not understand about the mechanism of mechanotransduction. For *C. elegans* we have excellent evidence of the importance of extracellular proteins and microtubules for producing the mechanical response, but no clear idea how these various structures are working in concert. Our understanding of mechanosensation in *Drosophila* is severely limited by the lack of voltage-clamp recordings from the receptor cells, due to the difficulty of approach of the electrode through the chitin of the sensillum. The recent demonstration of the role of Piezo2 in the stretch-sensitive response of the Merkel cell is impressive, but how does the rapidly adapting response of the channel produce a slowly adapting SAI spike discharge? If the transmitter released by Merkel cells is norepinephrine which binds to β-adrenergic receptors on sensory-nerve terminals, how is depolarization of the terminals produced? Is there a G-protein cascade, and if so, how are the channels gated? Can the axon below the Merkel cell also respond to stretch, as some experiments seem to indicate (Maksimovic et al., 2014)? The good news is that our repertoire of techniques for approaching these problems continues to expand, and we are likely to learn much more of interest about our sensation of touch in the coming years.

CHAPTER 6

Hearing and hair cells

When animals emerged from the sea onto dry land, they entered a place initially without insect chirping or bird song. The detection of vibrations in air would nevertheless have been advantageous for finding prey and escaping from predators, and both insects and vertebrates adapted primitive ears and existing mechanoreceptors to form more advanced organs of hearing. Insects modified sensilla like those used in arthropod proprioception, whereas vertebrates adapted hair cells of the inner ear and of lateral line organs, which in fish are used to detect the motion of water. Once terrestrial animals could hear, they could detect other approaching animals, and they could hear the sounds that they themselves were making to find and attract mates.

In insects, hearing evolved independently in at least nine different orders but always from scolopidia and chordotonal organs (Figure 5.9D). And because chordotonal organs are located all over the insect body, insect ears can be found almost anywhere—on the foreleg of a cricket (Figure 6.1A), but in a moth along the body wall where the thorax joins the abdomen (Hoy and Robert, 1996; Yack, 2004; Gopfert and Hennig, 2016; Ter Hofstede and Ratcliffe, 2016). There are two broad categories of insect auditory organs—the tympanal organs, for example in beetles and locusts, and Johnston's organs of flies, honey bees, and mosquitoes. These two kinds of ears have a different anatomy and function rather differently.

Insect hearing: tympanal organs

The majority of insect ears are tympanal organs (Hoy and Robert, 1996). Chordotonal scolopidia like those in Figure 5.9D terminate in a specialized region of cuticle, which is so thin that it forms a tympanum and can vibrate to sound much like a vertebrate (and human) ear drum (Yack, 2004). The internal surface of the tympanum often abuts air sacs of trachea, forming an enclosed chamber like the inside of a kettle drum. In most cases the ear is tuned to a particular range of frequencies, in part from the acoustical properties of the tympanum and associated air sacs, and in part from tuning by the scolopidia and sensory neurons. In Mediterranean cicadas, the ear responds to the song of the males peaking at about 10 kHz (Sueur et al., 2006). The ears of South American rainforest katydids are tuned to song peaking at 23 kHz, partially as a result of the properties of the tympanum and partially from specialized structures called the crista acustica and acoustic vesicle, which may permit discrimination of different frequencies of sound somewhat like the vertebrate ear (Montealegre et al., 2012). Many moths can detect ultrasonic frequencies to hear the echolocating calls of bats. As Roeder and colleagues first showed (Roeder, 1967), bat sounds heard by moth ears can induce a variety of avoidance maneuvers. Some moths begin to flutter or fly erratically, and others fold their wings and drop to the ground. Tiger moths can produce competing clicks either to jam the bat echolocation signals or to warn bats away (Ter Hofstede and Ratcliffe, 2016).

The neurons of insect tympanal organs respond to sound much like the mechanoreceptors from which they evolved. Figure 6.1B shows some of the first intracellular recordings, in this work from an auditory receptor of a locust. The first record in the

Sensory Transduction. Second Edition. Gordon L. Fain, Oxford University Press (2020). © Gordon L. Fain 2020.
DOI: 10.1093/oso/9780198835028.001.0001

(A)

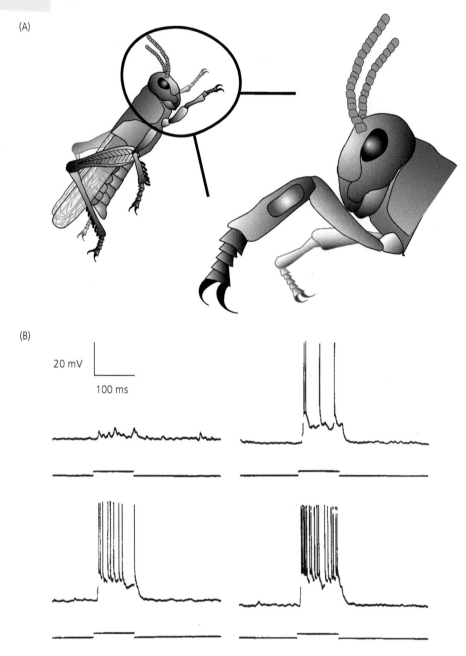

(B)

Figure 6.1 Tympanal organs. (A) Tympanal organ of a cricket on the foreleg (oval indicated in blue). (B) Intracellular recordings from a locust tympanal receptor neuron. The amplitude of the sound was increased in 10-decibel steps (about a factor of three) from upper left to upper right, and then from lower left to lower right. Increasing the sound amplitude increased the amplitude of the depolarization and frequency of action potentials. (B from Hill, 1983.)

upper left is to a weak sound near the most sensitive frequency of the receptor, typically between 6 and 9 kHz. In each of the three succeeding records, the frequency has been kept constant, but the amplitude of the sound has been increased by 10 decibels (dB), about a factor of three. The cell responded with a graded depolarization and action potentials, whose frequency of firing increased with increasing sound intensity. Tympanal ears can have just a few receptor cells (moths have one to four) or a great many (locusts can have seventy to eighty). In some species different receptors have different sensitivities to broaden the range of sound intensities over which the ear can respond (Ter Hofstede and Ratcliffe, 2016).

Insect hearing: Johnston's organ

Ears formed from Johnston's organs are less common than tympanal ears and can sense air vibration only within a limited range of frequencies and at short distances. Even so, they have been the subject of many investigations because *Drosophila* have them. The powerful genetic tools available for the fruit fly can be employed to investigate the physiology and molecular mechanism of transduction, and much is now known about the function of the fly ear.

Male fruit flies court females by flapping their wings. The frequency of wing flapping is several hundred per second, producing a "buzz" like the one a housefly makes against a kitchen window. The frequency and pattern of wing flapping is different in each species of fruit fly, producing a characteristic courtship song (Ewing and Bennet-Clark, 1968). The female can detect the sound produced by the wings only if she is within a few millimeters of the male. Other males also hear the sound and are attracted to it, perhaps in the hope of themselves finding a female.

Both males and females detect the air vibrations of courtship songs with scolopedia mechanoreceptors located in the fruit-fly antenna (Figure 6.2A and Albert and Göpfert, 2015; Kamikouchi and Ishikawa, 2016). As air moves back and forth, it sets in vibration a feather-like arista attached to the funiculus (see also Figure 7.5A). In a clever series of experiments, Göpfert and Robert (2001, 2002) used laser Doppler vibrometry to measure the movement of the different parts of the antenna and showed that sounds of a few hundred Hertz cause the arista and funiculus to rotate as a unit in phase with the stimulus. The top of the funiculus has a slender stalk inside, which inserts into a cavity of the pedicel (Gopfert and Robert, 2002; Kamikouchi et al., 2006). This stalk forms a hook, which is surrounded by the distal processes of about 500 sensory neurons (Figure 6.2B). As the arista and funiculus rotate back and forth, the hook vibrates against one group and then another group of neurons, producing the sensory response.

The sensory neurons can be divided into five groups called A–E, based on their location in the pedicel and their central projections (Kamikouchi et al., 2006). The neurons in groups A and B are most sensitive to vibration and sound, whereas those in C and E respond best to static movement of the arista. The cells in D are intermediate in their properties. Intracellular or patch recordings from the neurons are difficult to perform, because the pedicel would need to be opened to expose the cell bodies, and the dissection would most likely destroy the functioning of the organ. The cells, however, lie just below the cuticle and can be specifically labeled with a genetically encoded Ca^{2+} sensor such as one of the GCaMPs (Grienberger and Konnerth, 2012; Chen et al., 2013). These dyes can be expressed specifically in the neurons of Johnston's organ and will fluoresce when the neurons are excited and Ca^{2+} enters the cell, either through stretch-sensitive channels or through voltage-gated Ca^{2+} channels.

A typical experiment is shown in Figure 6.3A (Yorozu et al., 2009). Fluorescence from the Ca^{2+} dye genetically expressed only in the neurons of Johnston's organ was imaged within the fly CNS, where the axonal terminations of the different groups of sensory neurons are segregated and more easily distinguished. The pictures below show approximate demarcations of groups A and E; the dotted lines indicate the area of group C. When the fly was stimulated with a pulse of sound at 400 Hz, the cells in subgroup A were stimulated, but no response could be detected in groups C or E (Figure 6.3A, left). When, on the other hand, a static current of air was directed against the front of the head of the animal to simulate a breeze, the cells in group E

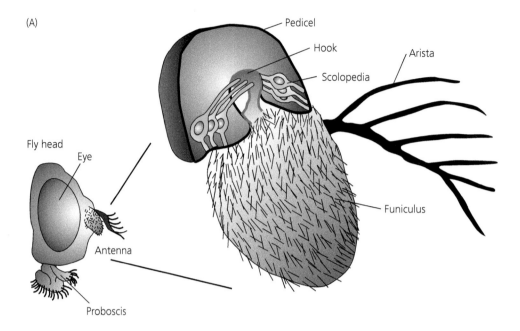

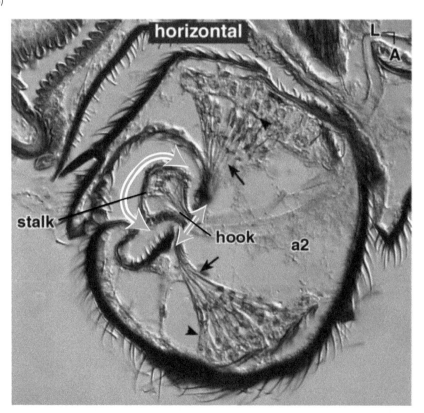

Figure 6.2 Johnston's organ: anatomy. (A) Antenna of *Drosophila*. Vibration of feather-like arista together with funiculus causes movement of hook in pedicel and stimulates scolopedia neurons. (B) Light micrograph of horizontal section of *Drosophila* antenna showing pedicel (also called the second antennal segment or a2). Orientation is upside-down (vertically reversed) from diagram (A). Blue arrows show rotation of the funiculus stalk causing the hook to move back and forth, alternately stimulating the ciliary processes of the two groups of neurons on the left and right (black arrows). Arrowhead at bottom indicates neuronal cell body. Letters in upper right indicate orientation of section: A (anterior) is downward, L (lateral) is to left. (A after Gopfert and Robert, 2002; B from Kamikouchi et al., 2006.)

responded (Figure 6.3A, right). The cells in group C were not excited by either of these stimuli but did respond if a static current of air was directed against the back of the animal rather than the front. Similar experiments have shown that groups C and E respond to any static movement of the aristae, including gravity (Kamikouchi et al., 2009). Responses to sound seem to be mediated exclusively by cells in groups A and B.

An alternative method of investigating the activity of Johnston's organ is to record from second-order neurons in the fly CNS that receive input from the sensory cells. Identifiable giant-fiber neurons have dendrites near the terminations of the sensory cells. They can be recorded with whole-cell patch clamp and receive electrical synapses (gap junctions) only from the sound-receptive cells in groups A and B (Lehnert et al., 2013). In the experiments of Figure 6.3B, the fly was stimulated with a 100-Hz tone in the presence of tetrodotoxin (TTX), to block voltage-gated action potentials. The resulting currents were unaffected when chemical synaptic transmission was blocked with Cd^{2+} but were eliminated by genetic deletion of *Drosophila* gap junctions. These observations argue that the currents were generated by the opening and closing of

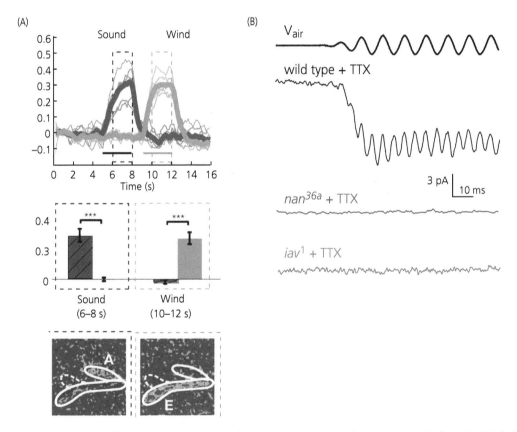

Figure 6.3 Johnston's organ: physiology. (A) Responses to sound and wind. Neurons were made to express a genetically encoded Ca^{2+} indicator dye, and fluorescence was measured from the axonal terminations of cells in the fly CNS (lowermost diagrams). The antenna was stimulated at different times either with sound (courtship song, 400 Hz at 90 dB) or with wind (velocity of 0.9 m s^{-1}). Neurons from A group responded to sound but not to wind; neurons from E group responded to wind but not to sound. Histograms give means ± s.e.m. *** indicates significance at p < 0.001. (B) Whole-cell voltage-clamp recordings from the giant fiber neuron in the *Drosophila* CNS, which has been shown to receive gap-junctional input from neurons of Johnston's organ. Recordings were made in the presence of tetrodotoxin (TTX) to block action potentials. Traces are averages of 100–500 trials to a 100-Hz tone. No response could be recorded in mutants of the TRP channels *nanchung* (*nan*^{36a}) and *inactive* (*iav*^{1}). (A from Yorozu et al., 2009; B from Lehnert et al., 2013.)

the mechanoreceptive channels of group A and B sensory neurons spreading through electrical synapses onto the giant-fiber neurons.

The lower two traces show the effects of knocking out of the genes of one or the other of two TRP channels, called *Nanchung* (*nan*) and *Inactive* (*iav*). In either case, the response of the giant-fiber neuron was eliminated, confirming previous observations with less-sensitive methods of recording (Kim et al., 2003; Gong et al., 2004). These experiments seem to indicate that the proteins of the *Nanchung* and *Inactive* genes serve as mechanoreceptive channels in Johnston's organ. It is odd, however, that these putative channel proteins have been localized to the proximal part of the sensory cell far away from the dendritic tip where the neuron presses up against the hook of the pedicel (see for example Gong et al., 2004). The scolopedia of Johnston's organ also express NOMPC, the stretch-sensitive channel of insect touch (see Chapter 5). When NOMPC is knocked out, the responses of sensory neurons in Johnston's organ are either reduced (Lehnert et al., 2013) or completely eliminated (Effertz et al., 2011; Effertz et al., 2012). It remains unclear whether *Nanchung*, *Inactive*, *nompC*, or some combination of the protein products of these genes can form the channels in the *Drosophila* ear (Albert and Göpfert, 2015).

The gene for NOMPA is also expressed in Johnston's organ as elsewhere in fly mechanoreceptors (see Chapter 5), and it is localized to the extracellular dendritic cap of the sensory cells (Chung et al., 2001) where it seems to have a similar role of transmitting mechanical stimuli to the stretch-sensitive membrane of the sensory cells. An extensive screening has turned up a large number of genes for additional proteins expressed in the fly ear, which may also have some role in mechanoreception. These proteins included some surprises, such as four fly visual pigments and the fly photoreceptor light-gated channels (Senthilan et al., 2012). Hearing in *Drosophila* is unaffected by light, and we have no idea what these photoreceptor proteins are doing in this location.

Hair cells

Hearing in vertebrates is achieved by a mechanoreceptor which is in a class all to itself, both because of its peculiar and elaborate mechanism of transduction and because of its importance for our vestibular and auditory function. This cell is called the hair cell, which we have already encountered in Chapter 3 (Figure 3.11). Hair cells are found exclusively in vertebrates in three places: the lateral line organs of fish, which sense movement of water; the vestibular system of the inner ear, which detects position and movement of the head; and the cochlea, the vertebrate organ of hearing.

Hair cells in all of these structures and in all vertebrate species which have been investigated share a great many common characteristics (Hudspeth, 1989; Fettiplace and Kim, 2014; Hudspeth, 2014; Fettiplace, 2017). They have a similar anatomy, with a characteristic bundle of microvilli projecting from the apical surface of the cell and principally responsible for mechanotransduction (Figures 3.11 and 6.4). The little that we know about different vertebrate classes suggests that hair cells in the lateral line organs of fish use the same kinds of transduction proteins as those in the cochlea of a mammal. It is striking, moreover, that many of the mutations that cause deafness in humans also produce deafness in mice and in zebrafish, and that mutations producing deafness very often also affect vestibular function.

All hair cells seem to produce electrical responses in a similar way (see Chapter 3). Movement of the hair bundle is always directional, giving a depolarization for motion toward the tallest of the microvilli and hyperpolarization for motion in the opposite direction. This direction of polarity was first discovered in lower vertebrates (Loewenstein and Wersäll, 1959; Flock, 1965; Hudspeth and Corey, 1977) but is just as true for the mammalian ear (see for example Kros et al., 2002). Movement of the hair bundle gates the opening and closing of ion channels, producing a change in membrane potential (Figure 3.11). Transduction is extremely rapid and incredibly sensitive: a detectable voltage response can be recorded for angular rotation of the hair bundle by only 0.01 of a degree or about 1 nm (Crawford and Fettiplace, 1985).

The 20–300 microvilli that form the hair bundles can take different shapes. In the bullfrog sacculus (Figure 6.4A), there is a single large, true cilium called a kinocilium, which expands into a ball at its top. All hair cells initially have a kinocilium, which

(A)

(B)

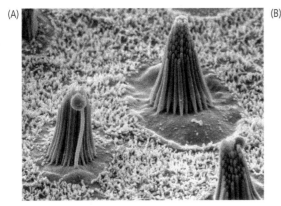

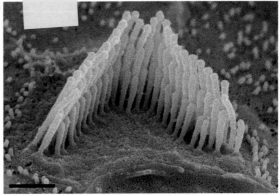

Figure 6.4 Scanning electron micrographs of hair bundles. (A) Bullfrog sacculus. The otolithic membrane has been removed. Note the prominent kinocilium with ball at top. (B) Stereocilium bundle of an outer hair cell from the mouse cochlea. Scale bar, 1 μm. (A courtesy of J. A. Assad and D. P. Corey; B courtesy of Jeffrey Holt.)

has an essential role in establishing the polarity of the hair bundle during development (Schwander et al., 2010; Kindt et al., 2012); but in a few organs (in particular the mammalian cochlea), the kinocilium eventually degenerates and is not present in the adult. Next to the kinocilium are the microvilli, which are usually called stereocilia, even though they have none of the attributes of cilia. The name is unfortunate, but I shall use it nevertheless because it is the name used nearly universally by scientists who work on these cells. The stereocilia are arranged in order of increasing height across the apical surface of the cell. In the mammalian ear (Figure 6.4B), the hair bundle is often in the shape of a V or W, again with stereocilia lined up in order of increasing height.

As for microvilli in other sense cells (see Chapter 2), the stereocilia are full of parallel actin fibers and also contain cross-linking proteins such as plastin-1, fascin-2, and espin. At the base of the stereocilium, the actin bundle narrows to form a structure called a rootlet, which inserts into a dense aggregation of actin called the cuticular plate (Figure 6.5A). The cuticular plate is in turn surrounded by microtubules. This arrangement of stereocilia and cuticular plate is of considerable importance in the functioning of the hair bundle because, when the stereocilia are pushed, they move nearly as rigid rods pivoting about their insertion point (Crawford and Fettiplace, 1985; Howard and Ashmore, 1986; Karavitaki and Corey, 2010). The cross-linked actin bundle gives

the stereocilium the necessary rigidity, and the narrowing of the bundle near the cuticular plate permits the stereocilia to rotate back and forth without flexing.

Hair cells are secondary receptors, producing changes in membrane potential usually without action potentials (see Figure 3.11B). They communicate their responses to the CNS by making synaptic connections onto ten to twenty afferent terminals of the eighth cranial nerve at the base of the cell (Figure 6.5B). These synapses have a peculiar morphology (see Fettiplace, 2017). Adjacent to the synapses are prominent, electron-dense presynaptic bodies surrounded by a halo of synaptic vesicles, analogous to the T bodies of fly photoreceptors (Figure 2.12A). In the hair cells of many species, these presynaptic structures are nearly spherical (Figure 6.5C), but in mammals they more nearly resemble a bar or ribbon. The membrane of the hair cell just below the presynaptic body is also electron dense and can be seen in freeze-fracture micrographs to contain a regular array of intramembranous particles (Figure 6.5D). These particles delineate the active zone of the synapse, that is the area where synaptic vesicle release occurs. The synaptic transmitter released by hair cells is glutamate. There are also efferent synapses onto the hair cells made by axons coming from the superior olivary complex of the CNS, which can synapse either onto the afferent synaptic terminals (as shown in Figure 6.5B) or directly into the body of the hair cell.

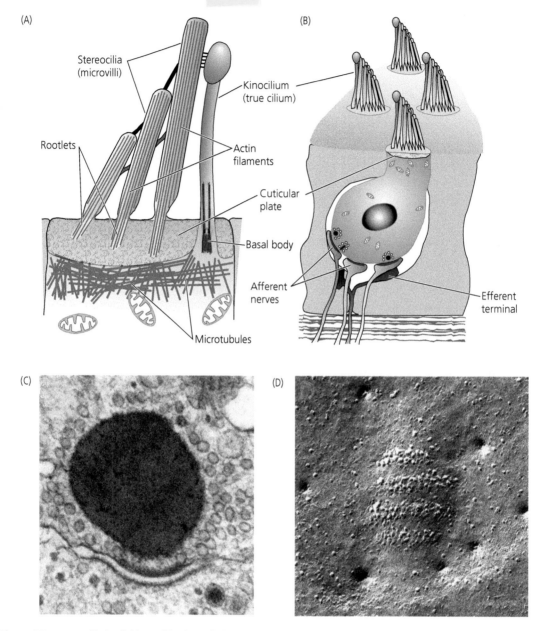

Figure 6.5 Anatomy of hair cell. (A) Top of frog hair cell, showing cuticular plate, stereocilia, and kinocilium. (B) Hair cell showing synapses onto afferent nerves and from efferent nerves. Efferent nerves can synapse onto endings of afferent nerve terminals (as shown) or directly onto the hair cells. (C) Electron micrograph of a hair-cell synapse from a frog sacculus. (D) Freeze-fracture micrograph of presynaptic membrane from a frog saccular hair cell. Magnification for C and D 90,000×. (A and B after Pickles, 1988; C and D from Roberts et al., 1990.)

Tip links

The latency of the vertebrate hair cell response is exceedingly short, less than 40 μs at room temperature (or 10 μs at 37°C; see Corey and Hudspeth, 1979b, 1983; Crawford et al., 1989), just as short as that of touch receptors (see Chapter 5) and auditory receptor cells in insects (Gopfert and Hennig, 2016). This latency is much too short to be produced by a metabotropic cascade. The mechanism of transduction is therefore almost certainly ionotropic, with channels like those in touch receptors sensitive to tension on the channel protein itself. This finding poses the following question. If the stereocilia rotate as rigid rods, where are the channels located, and how is tension conveyed to them? One conceivable answer is that the channels are located near the base of the stereocilia on the cuticular plate, so that pivoting of the stereocilia would pull the membrane back and forth (Ohmori, 1988); but other measurements show that they are probably positioned near the tips of the stereocilia (Hudspeth, 1982; Denk et al., 1995).

Recall (from Chapter 3) that the channels of hair cells are permeable to cations and rather non-selective, admitting Na^+, Li^+, and K^+, but also divalent cations like Ca^{2+}. This characteristic of the channels made possible the following experiment (Denk et al., 1995). A single hair cell was recorded with whole-cell patch clamp by sealing onto the cell with an electrode filled with a fluorescent Ca^{2+} indicator dye called Calcium Green-1™. This compound is a member of a large family of dyes whose fluorescence changes as the free-Ca^{2+} concentration changes, like the GCaMPs used in the experiments of Figure 6.3A. But unlike the GCaMPs, the dye in this experiment was not expressed genetically but was introduced into the cell by diffusion from the patch pipette to fill the cytoplasm, including the inside of the stereocilia. The stereocilia were then exposed to a medium containing an elevated extracellular Ca^{2+} concentration and were stimulated by pushing on the hair bundle with a rapid jet of solution. Since the channels are permeable to Ca^{2+}, stimulation in a direction to depolarize the hair cell caused an influx of Ca^{2+} into the cell, which could be recorded by measuring dye fluorescence.

Denk and collaborators used the technique of two-photon microscopy (see Helmchen and Denk, 2005; Svoboda and Yasuda, 2006; Grienberger and Konnerth, 2012). This method made it possible for them to localize with considerable accuracy the source of the fluorescence within the cell. The results of their experiments indicate that the change in Ca^{2+} concentration is greatest near the tips of the stereocilia, not at their bases (Figure 6.6). These results (see also Lumpkin and Hudspeth, 1995) confirmed others using somewhat different techniques (Hudspeth, 1982; Jaramillo and Hudspeth, 1991), which

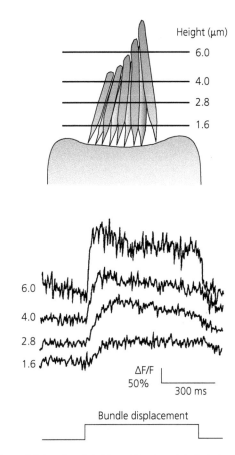

Figure 6.6 Location of stretch-sensitive channels in the hair cell. Change in indicator-dye fluorescence produced by the entry of Ca^{2+} through mechanoreceptive channels in different regions of a hair bundle, as measured with two-photon confocal microscopy. Cell had been filled with the fluorescent Ca^{2+} indicator dye Calcium Green-1™. The plane of section of the measurement is indicated in the drawing above (note numbers next to traces below). The amplitude of fluorescence is given as a ratio of fluorescence change (ΔF) to resting fluorescence (F). (From Denk et al., 1995.)

have all shown that the channels are located near the stereocilia tips. This result has the consequence that the channels cannot be gated by tension applied to the membrane of the cuticular plate. How, then, are they gated?

One interesting possibility was first suggested by Jim Pickles and colleagues (1984). They noticed that the stereocilia are extensively connected by an extracellular network of fibrous cross-links running along the axis of excitation of the hair bundle, up the ranks of stereocilia (Figures 6.7A and B). They called these connecting fibrils *tip links*. We now know that tip links are composed of two extracellular fibrous proteins called cadherin-23 and protocadherin-15 (Kazmierczak et al., 2007). Cadherin-23 is at the upper end of the tip link and protocadherin-15 is at the lower end. Both are present as homodimers, with the dimer of one protein joined to the dimer of the other at the dimer amino termini. Both insert their carboxyl termini into regions of membrane density at the lateral wall or tip of the stereocilium (Figure 6.7C). Mutation of the genes of either cadherin-23 or protocadherin-15 can eliminate the hair-

cell response (Kazmierczak et al., 2007; Richardson et al., 2011). In addition to the tip links, there are numerous shorter, connecting links at the top of the stereocilium, which may be composed of a protein called stereocilin (Richardson et al., 2011) and may aid in the cohesion of the hair bundle.

Pickles and collaborators speculated that movement of the hair bundle toward the kinocilium and the largest of the stereocilia stretches the tip links and pulls the channels open, whereas motion in the opposite direction causes the tip links to relax and

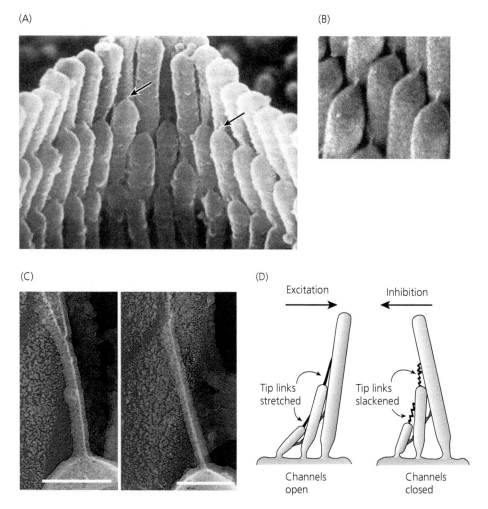

Figure 6.7 Tip links. (A) Scanning electron micrograph showing tip links (arrows) from outer hair cell of the guinea pig cochlea. (B) Field-emission scanning electron micrograph of a chicken auditory receptor. Magnification 27,500×. (C) Freeze-etch image of upper insertions of tip links of hair cells from guinea pig cochlea. Scale bars, 100 nm. (D) Proposed role of tip links in channel gating. (A from Pickles, 1988; B and C from Kachar et al., 2000.)

allows the channels to close (Figure 6.7D). This notion is consistent with the position of the channels near the tips of the stereocilia. A role for the tip links in mechanotransduction was corroborated by the remarkable discovery of Assad et al. (1991). They showed that exposing hair cells to very low Ca^{2+} solution, which had long been known to abolish transduction (Sand, 1975), breaks the tip links but has very little if any effect on the other crosslinks between the stereocilia. The tip links break because the association of their component cadherins is highly Ca^{2+} dependent. As the tip links regenerate, transduction returns with approximately the same time course (Zhao et al., 1996).

Hair cell transduction proteins

Imagine now the tip links pulling against the membrane of the stereocilium. How does this stretching of the membrane gate the opening of the channels? Experiments similar to those of Denk and collaborators but with high-speed calcium imaging indicate that the calcium signals are much smaller and slower at the upper insertion points of the tip links than at the lower insertion points (Beurg et al., 2009). This observation strongly argues that the channels are only present at the bottom of the tip links. Studies of touch receptors in insects and mammals would lead us to suspect that channels in this location would be associated with intracellular cytoskeletal and extracellular matrix proteins. What are these molecules? Are the channels connected to actin, or are there other proteins linking the channels to the cytoskeleton? And finally, what protein or proteins can form the channel?

To answer these questions it would be useful to have more information about the molecules responsible for hair cell function. Biochemical isolation or RNA expression studies are more difficult for cells in the ear than for sensory neurons in the eye, because a single human cochlea has only about 16,000 hair cells compared to over 100 million photoreceptors. A more fruitful approach might be to search for mutations isolated from animals showing specific deficits in sensory detection, but which animal should we use? *Caenorhabditis elegans* and *Drosophila* don't have hair cells—only vertebrates have them. The most useful species turns out to be

our own, because one of every 1000–2000 children is born with genetic deafness. A large and active research effort has helped identify many genes that are essential for human hearing (see Richardson et al., 2011; Fettiplace, 2017). Many of these proteins are necessary for the development of the hair cells or are expressed in accessory cells or extracellular structures, but others are now known to be localized to the stereocilia and have an essential role in the mechanism of transduction.

Among these essential proteins of the stereocilium are those associated with the actin structure of the microvillus, such as actin itself and actin-binding proteins like plastin-1 and fascin-2. Other essential proteins are unconventional myosins, which include myosin IIIA (Walsh et al., 2002), myosin VI (Avraham et al., 1995), myosin VIIA (Gibson et al., 1995; Self et al., 1998), and myosin XV (Probst et al., 1998). These proteins do not participate in muscle contraction but are thought to function in other ways, for example in membrane trafficking, protein targeting, endocytosis, cytoplasmic vesicle movement, and actin polymerization. Many genes have been discovered whose mutations produce Usher syndrome, which can cause deafness or both deafness and blindness (Richardson et al., 2011). These genes code for Usher proteins, which include cadherin-23 and protocadherin-15, the components of the tip links; and sans, harmonin, and myosin VIIA, which have all been localized to the top of the tip links where cadherin-23 is inserted into the stereocilium. There are also other proteins known to cause deafness which, in some cases, are localized to the stereocilium or otherwise implicated in some aspect of hair cell function (see Corey et al., 2018). The role of all of these proteins in hearing is a topic of much active research.

The channels

This rich harvest of molecules essential for hair cell function resembles in many ways the group of MEC proteins necessary for touch perception by the *C. elegans* microtubule cells (see Chapter 5). In both groups of molecules there are intracellular cytoskeletal proteins, extracellular matrix proteins, and proteins that probably interconnect the matrix and cytoskeleton. There are also genes for channels, including *transmembrane channel-like protein 1 (TMC1)*. TMC1 has long been thought to be a candidate for the channel producing the hair-cell response (see Fettiplace, 2016; Qiu and Muller, 2018). It is of particular interest because thirty-five different mutations of the *TMC1* gene are known to produce genetically inherited hearing loss. It is expressed by vestibular and cochlear hair cells and localized to the tips of the stereocilia (Kawashima et al., 2011). In mouse, the channels of hair cells expressing a single-site mutation in *Tmc1* show a reduced Ca^{2+} permeability (Pan et al., 2013).

TMC1 and a close homolog, TMC2, have ten predicted transmembrane domains. Biochemical and ultrastructural experiments indicate that they assemble as dimers (Pan et al., 2018). In this regard TMC1 is like the TMEM16 family of ion channels (see Pedemonte and Galietta, 2014), which include the Ca^{2+}-activated Cl^- channels of olfactory receptor cells and photoreceptors which we will encounter in Chapters 7 and 9. TMEM16 channels, unlike most other ion channels, do not have a central pore but instead have two pores, one for each of the dimers. Because the structure of TMEM16 was already known (Brunner et al., 2014; Dang et al., 2017), Pan and colleagues (2018) could locate the putative pore of the hair-cell protein TMC1 and design mutations near this pore region. They then expressed mutant versions of TMC1 in mouse hair cells with cysteines placed near the pore. When they exposed the expressed channel to a bulky reagent that reacts with the cysteine, the reagent covalently attached to the channel and altered the channel permeability. Nearly all of the mutations they tested had some effect on the stretch-sensitive current of the hair cells. This is strong evidence that TMC1 and TMC2 are at least part of the channel mediating hearing by hair cells.

It has not been possible to show that TMC1 is stretch sensitive by expressing it in a cell line, as has been done for example for Piezo2 (Coste et al., 2010; Syeda et al., 2016) and NOMPC (Zhang et al., 2015). TMC1 may only respond to stretch in association with the other intracellular and extracellular components of the stereocilium. Hair cells also express Piezo2 but at the apical membrane on the cuticular plate near the rootlet (Beurg and Fettiplace, 2017; Wu et al., 2017). Deletion of the gene for Piezo2 produces a relatively minor loss in the sensitivity of hearing by a mechanism not yet fully elucidated.

As we saw in Chapter 3, the mechanoreceptive channels of hair cells are rather non-selective but cationic, much like the channels of many other mechanoreceptors, and they are permeable to Ca^{2+}. If the extracellular Ca^{2+} concentration is increased from micromolar to millimolar, the increased Ca^{2+} blocks the channels, and hair-cell responses can drop by half (see Fettiplace and Kim, 2014). The Ca^{2+} may have to bind to a site within the channel pore in order to permeate, and if too much Ca^{2+} is present in the extracellular solution, the Ca^{2+} may occupy this binding site sufficiently often to impede the entrance of Na^+ or K^+ into the channel. The block of the channel by Ca^{2+} is of considerable practical importance, since the Ca^{2+} concentration in the cerebrospinal fluid bathing the cells of the brain and most of the hair cell is between 2 and 3 millimolar; but as we shall see later in the chapter, the Ca^{2+} concentration bathing the part of the hair cell containing the stereocilia is much lower, more like 20–30 *micro*molar.

If the extracellular Ca^{2+} concentration is made extremely low, the tip links break, as we have seen. If, however, exposure to the low-Ca^{2+} solution is brief, a small number of tip links can remain intact. This remarkable effect was exploited by Crawford et al. (1991), who exposed hair cells to a very low Ca^{2+} concentration for a short time and then returned the cells to normal solution. With this treatment, many of the tip links were severed, but a few remained intact. In the most favorable circumstance, only a single channel continued to be gated by the movement of the hair bundle, and this circumstance made it possible to record the activity of single mechanoreceptive channels.

An example of such a recording is given in Figure 6.8 (from Ricci et al., 2003). The tracing at the top of the figure shows the timing of the stimulus given to the hair bundle, and recordings of the hair cell currents are shown below. These recordings are voltage-clamp records obtained with whole-cell patch clamp and are not responses from isolated patches of membrane. For this reason, the records are a little noisier than isolated-patch recordings, because the noise of the whole of the cell is collected instead of just the noise of a small piece of membrane. Fortunately, the single-channel conductance is large, and there are clear channel openings gated by the movement of the hair bundle. When many such records

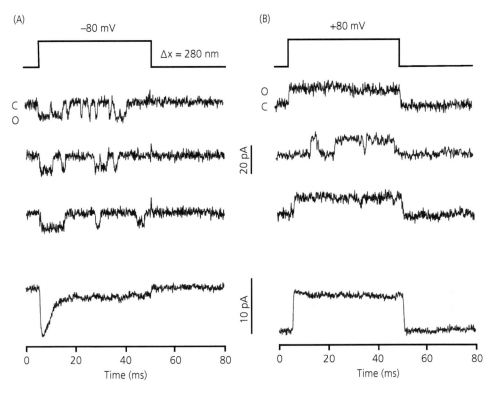

Figure 6.8 Single-channel currents from hair cells of the turtle cochlea. Cells were exposed briefly to a free-Ca^{2+} concentration of 200 nM and then returned to a solution with 2.8 mM Ca^{2+}. Whole-cell voltage-clamp recordings were made at holding potentials of −80 mV in (A) and +80 mV in (B). Hair bundles were rapidly deflected a distance of 280 nm (0.28 μm) with a rigid glass pipette polished to a tip diameter of about 1 μm. The timing of mechanical stimulation is indicated by the uppermost traces. The three traces in both columns below are successive recordings from the same cell in response to the mechanical stimulus (C, current with channels closed; O, current with single channels open). The lowermost traces give averages of fifty responses and show adaptation of the response at −80mV but not at +80mV. Currents change sign as a result of the change in holding potential and driving force. (From Ricci et al., 2003.)

are averaged, as in the lowermost traces of Figure 6.8, the gating of channel opening by mechanical stimulation can be clearly observed.

The conductance of a single channel in these recordings is about 100 pS and can be as large as 300 pS in some hair cells. Similar values have been obtained by stimulating single stereocilia in intact hair cells (Pan et al., 2013) and from measurements of noise analysis (Pan et al., 2018). This value is rather large, at least five times greater than the conductance of a voltage-gated Na^+ channel in an axon. From this number we can estimate the approximate number of channels per hair cell. We proceed in the following way. The total conductance of the whole of a cell gated by the movement of an intact hair bundle can be obtained by measuring the peak amplitude of the largest current response recorded under voltage clamp. We then divide this number

by the driving force to give the maximum value of the change in conductance—see Eq. (3.8) in Chapter 3. We assume that when an intact hair cell produces its largest current response, all of the mechanoreceptive channels are opened. The number of channels per hair cell is then estimated by dividing the total change in conductance of the cell by the conductance of a single channel. This number can further be divided by the number of stereocilia to give the number of channels per tip link.

The result is surprising: there may be as few as two channels per tip link (Fettiplace and Kim, 2014; Corey et al., 2018). If channels are present only at the lower end of the tip link, there could be one channel per protocadherin-15 monomer, fitting very well with the idea that both protocadherin-15 and TMC1 are dimers. For comparison, a squid giant axon has several hundred voltage-gated Na^+

channels per square micrometer of membrane and many thousands along the entire length of the cell. The hair cell may have relatively few mechanically gated channels, so that the force applied to a hair bundle by sound or head movement is divided among only a small number of tip links, increasing the probability of channel opening. Each channel is, however, provided with a large conductance, which can produce a detectable signal from each channel opening and increase the sensitivity of transduction.

Gating and bundle stiffness

When force is applied to the hair bundle, the bundle moves like a spring according to Hooke's law:

$$F = kx \qquad (6.1)$$

where F is the force applied to the hair bundle, x is the distance the hair bundle moves, and k is the spring constant of the hair bundle. Equation (6.1) says that the stiffer the spring and the larger the spring constant k, the smaller the movement of the hair bundle to any given applied force.

The spring constant of the hair bundle is composed of three elements. First, there is the stiffness of the stereocilium rotating about the cuticular plate. Second, and in parallel, there are the elastic elements that pull on the channels. These are called gating springs and may be the tip links or proteins in the stereocilium associated with the channels. Finally, of considerable interest for understanding the mechanism of hair cell transduction, there are the channels themselves, which may also contribute to the spring constant of the hair bundle.

The contribution of the channels to the stiffness of the hair bundle has been called *gating compliance* (see Hudspeth, 2014). It was first measured in the following way (Howard and Hudspeth, 1988). A flexible glass fiber was prepared by heating and pulling a thin piece of glass. The fine end of the fiber was placed against the kinocilium of a hair cell from the bullfrog sacculus, where the glass of the fiber adhered spontaneously and quite tightly (Figure 6.9). The other end of the pipette was moved under electronic control with a piezoelectric device. A voltage pulse given to the piezoelectrical device

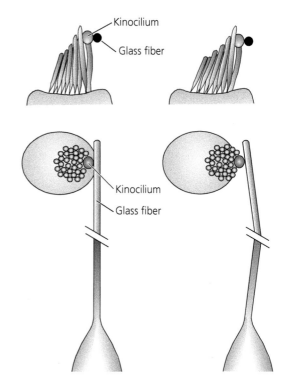

Figure 6.9 Measuring the gating spring. The hair bundle is moved with a fine glass fiber adhering to the kinocilium. See text. (After Howard and Hudspeth, 1988.)

produced a rapid movement of the fiber, pulling or pushing on the hair bundle.

When the glass fiber was moved, a force was applied to the hair bundle. The value of this force was determined by measuring the spring constant of the glass fiber. This measurement was done at the beginning of the experiment before attaching the fiber to the hair bundle, either by placing small weights on the end of the fiber and measuring how far it moved under the microscope, or by measuring the Brownian motion of the tip of the fiber. The force applied to the hair cell could then be calculated from Eq. (6.1) by multiplying the spring constant of the fiber by the distance the fiber deflected.

Once the value of the applied force F was known, the spring constant of the hair bundle could be calculated, again with Eq. (6.1) but this time taking the variable x to be the displacement of the hair bundle instead of the distance of motion of the flexible fiber. Dividing F by the hair bundle displacement x gave the stiffness k of the hair bundle. The stiffness of the bundle was not constant but depended on the distance the bundle moved. Howard and Hudspeth therefore evaluated k as a function of x from the slope of the force/distance relationship, that is they calculated

$$k(x) = \frac{\Delta F}{\Delta x} \qquad (6.2)$$

where ΔF is the change in force applied to the hair bundle, Δx is the change in hair bundle position, and $k(x)$ is the spring constant of the bundle as a function of the distance the bundle moved.

The results of these calculations for the hair cells of the bullfrog sacculus are given in Figure 6.10. The black symbols show the amplitude of the receptor potential as a function of bundle displacement. This curve is nearly the same as the one in Figure 3.11C from earlier work on this preparation. This curve shows that when the hair bundle was moved toward the kinocilium, the membrane potential depolarized, reaching a maximum for displacements of about 100 nm. Movements in the opposite direction caused the membrane potential to hyperpolarize.

The blue symbols in Figure 6.10 give bundle stiffness as a function of bundle displacement from Eq. (6.2). They show that the bundle is most stiff when the channels are either entirely open or entirely closed, reaching a minimum when the probability of channel opening is about 0.5. Similar results have been obtained for hair cells from the cochlea of turtles (Ricci et al., 2002) and mammals (Russell et al., 1992; Kennedy et al., 2005), as well as for mechano-receptors in the *Drosophila* ear (Albert et al., 2007; Albert and Göpfert, 2015). This dependence of bundle stiffness on the distance of displacement is likely to result from the gating of the channels, since it disappears (in vertebrates) if the channels are blocked with an aminoglycoside antibiotic (Howard and Hudspeth, 1988; Russell et al., 1992).

Why does stiffness reach a minimum when half the channels are open? Imagine pulling on a bundle with many elastic cords, each attached to the door knob of a large number of doors (see Hudspeth, 1989). Suppose you were to measure the stiffness of the cord bundle by pulling and measuring the distance the cord moved. The cord would seem equally stiff when all the doors were closed or all completely open. It would seem less stiff if some of the

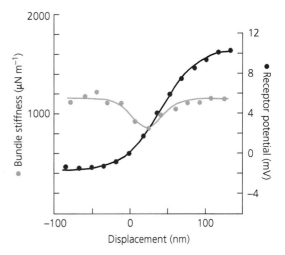

Figure 6.10 Stiffness of hair bundle and receptor potential as functions of bundle displacement. Amplitude of receptor potential (black symbols and curve) and bundle stiffness k(x) (blue symbols and curve) as functions of hair bundle displacement. See text. (After Howard and Hudspeth, 1988.)

doors were free to open, because the cord attached to those door knobs would relax as the door moved. The cord would seem least stiff when half the doors were free to open.

Suppose, while you were pulling on the cord, some of the doors suddenly opened. The opening of the doors would cause the cord to move toward you, in the same direction you were pulling to open them. For the hair cell, the opening of channels by an applied force would also cause a kind of positive feedback, producing increased movement in a direction that opens more channels, that is toward the kinocilium or tallest of the stereocilia (see for example Kennedy et al., 2005). Such a movement would produce a non-linear dependence of channel opening on stimulus intensity, because the movement would be greatest when the channels were least stiff, when about half the channels were open (Figure 6.10).

This non-linear dependence can function as a kind of amplification, increasing the response of the hair cell to an incoming movement or sound (Hudspeth, 2014). As channels open to a constant applied force, the decrease in stiffness of the bundle would cause an even larger displacement of the stereocilium. This greater displacement would in turn produce more channel opening and a larger response. I return to mechanisms of signal amplification later in the chapter, after I describe the organization of the mammalian cochlea.

Adaptation of hair cells

For hair cells, as for other mechanoreceptors, the response to constant stimulation is not steady but declines in amplitude as the cell adapts to the stimulus. For crayfish stretch receptors and Pacinian corpuscles, much of adaptation is the result of external specializations rather than of the transduction mechanism per se and is produced by the viscosity of accessory muscle fibers or membrane lamellae. For the hair cell, on the other hand, adaptation can occur even when the cell is completely isolated from the rest of the tissue (see for example Assad et al., 1989; Assad and Corey, 1992), and even for single channels—the decay of the response to a constant stimulus is clearly visible in the lowermost trace of Figure 6.8A. Much of adaptation seems therefore to be intrinsic to the way mechanical force is coupled to the channel proteins.

As the membrane current of the hair cell declines during adaptation, the whole operating range of the cell is shifted (Eatock et al., 1987; Assad et al., 1989; Assad and Corey, 1992). This alteration has the important effect of adjusting the region of bundle displacement to the value where the hair cell is most sensitive. The way this effect happens is shown in Figure 6.11 (from Assad and Corey, 1992). A single hair cell isolated from the bullfrog sacculus was voltage-clamped with whole-cell patch clamp, and the current of the cell was recorded as the hair bundle was displaced. In part A of the figure, the upper traces show the current responses and the lower traces show the timing of bundle displacement. Many current responses have been superimposed. Those shown furthest to the left were recorded from a resting hair bundle, whereas the two groups of responses to the right of the figure were recorded after the tip of the hair bundle was moved 450 nm. Positive movements of the hair bundle (displacements toward the kinocilium) produced negative (inward) currents, as channels opened and increased the permeability of the cell to cations.

The peak amplitude of each response was measured and used to construct a plot of response versus displacement. This plot is shown in part B as the dashed curve for the resting hair bundle. This curve should be familiar—we have encountered similar plots in Figures 3.11C and 6.10. After the hair bundle was deflected by 450 nm to its new reference position, two new series of responses were recorded. Keep in mind that to do this experiment the hair bundle had first to be deflected by a steady amount. The bundle was then deflected again a further distance from this steady deflected position to measure the value of the current produced by this incremental movement. The amplitudes of the responses to the incremental movements are plotted as the solid blue and black curves in Figure 6.11B.

Steady deflection of the hair bundle in the bullfrog sacculus produced a shift of the operating range of the hair cell, so that it became most sensitive over a different region of hair bundle deflection. This shift had the useful consequence that the sensitivity of the cell in the region of large bundle deflection was greatly increased. A hair cell at rest does not distinguish very well displacements of the bundle to 400 or to 500 nm. This difficulty can be appreciated by comparing the intersections of the two vertical dashed lines in Figure 6.11B with the dashed curve. Such large deflections almost saturate the response, opening nearly all of the mechanoreceptive channels. If, however, the bundle is first deflected by 450 nm and the bundle then moved from this new point of departure, the operating range of the hair cell can be seen to have shifted along the displacement axis. Responses to deflections to positions 400 and 500 nm from the new resting position of the bundle are now easier for the hair cell to detect, since the difference in current amplitudes for these two positions is greater. This difference can be seen by comparing the intersections between the two vertical dashed lines and the solid blue and black curves, on the one hand, with those between the vertical dashed lines and the dashed curve, on the other.

Steady deflection of the hair bundle resets the dynamic range of the response, and this alteration happens rather rapidly (Eatock et al., 1987; Assad and Corey, 1992). This process is likely to make an important contribution to sensory detection. For the sacculus, it may allow hair cells to detect transient acceleration of the head in the presence of the static

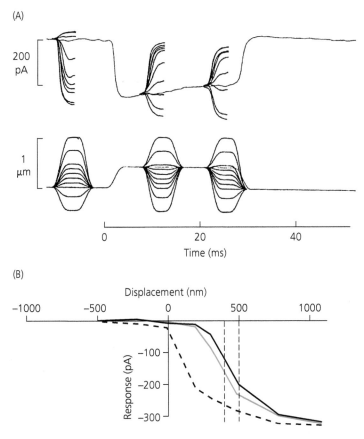

(A)

200 pA

1 μm

0   20   40

Time (ms)

(B)

Displacement (nm)

−1000   −500   0   500   1000

Response (pA)

−100

−200

−300

Figure 6.11 Adaptation of hair cell. Movement of hair bundle to a new steady position resets the operating range of the response. (A) The upper traces show current responses to a series of bundle displacements whose timing and amplitude are given in the lower traces. The results of many separate trials have been superimposed. (B) Response as a function of bundle displacement. Dashed curve is for the resting hair bundle, and solid curves are for deflections after the hair bundle had been moved to a new reference position 450 nm toward the kinocilium. Blue curve is for the first series of deflections, black for the second—see part (A) of figure. (From Assad and Corey, 1992.)

acceleration of gravity. For the cochlea, it may aid in accommodation to steady noise, though important contributions to this process probably also come from synaptic integration and processing in the CNS.

The mechanism responsible for adaptation is still unclear. In the experiments of Figure 6.8, the decline in single-channel current seen in part A at a membrane potential of −80 mV is much smaller or even absent in part B at a membrane potential of +80 mV. The usual interpretation of this result is that at +80 mV the driving force for entry of Ca^{2+} into the cell is greatly reduced or eliminated; and since adaptation is also reduced or eliminated, adaptation must depend upon the entry of Ca^{2+}. Although this conclusion

seems secure for reptiles and amphibians, there is some doubt about the Ca^{2+} dependence of adaptation in mammals (Peng et al., 2013; but see also Corns et al., 2014). There is also little agreement about what, if anything, Ca^{2+} is supposed to be doing. The Ca^{2+} could act directly on the channel, affecting the probability of channel opening. Mechanoreceptive channels even in the absence of accessory proteins (see for example Coste et al., 2010) and even in a pure lipid membrane (Syeda et al., 2016) can adapt to a maintained stimulus. But the Ca^{2+} could equally well be acting on one of the other proteins near the channel at the lower end of the tip link. Because several unconventional myosins are thought to be localized to the region of the tip links, it is conceivable that Ca^{2+} can activate a myosin and produce the sliding of the tip links or some other structural change in the vicinity of the channel (see for example Holt et al., 2002). Myosin movement seems an unlikely explanation for the rapid adaptation of Figure 6.8A but could be the basis of a slower adjustment in sensitivity (see Hudspeth, 2014; Fettiplace, 2017). A more detailed explanation of adaptation may have to await identification of the complex of molecules forming the channel, clarification of the roles of the many proteins localized to the vicinity of the tip links, and a better understanding of the way stereocilium movement is coupled to channel gating.

Organs of the lateral line

Having described the anatomy and physiology of hair cells, I now turn to the organs that use them. The most primitive are the lateral line organs of fish and amphibians, which detect movement at short distance in the surrounding water, produced for example by other organisms or by turbulence generated by tidal currents and obstructions like logs or rocks in the middle of a stream. This ability to detect water vibration is of considerable importance in a variety of behaviors (Dijkgraaf, 1967). Many fish wave their fins during sexual display, producing patterns of water vibration recognized by a member of the opposite sex. Some species are able to school even when blinded, and naturally blind cave fish use their lateral line organs to navigate without colliding into barriers or other organisms.

In its simplest form, a mechanoreceptive lateral line organ occurs superficially on the skin, as in most amphibians and small fish. Figure 6.12A shows them on the tail of a minnow (from Dijkgraaf, 1967). Each organ can be identified from the small protuberance that emerges from the skin. A single organ consists of a number of hair cells (Figure 6.12B), whose kinocilia and stereocilia are embedded in a clear gelatinous structure called the cupula. In many adult teleost fishes, the majority of the lateral line organs do not lie exposed to the surface but are buried in pits or canals, sometimes containing only a single organ but often containing several. There are systems of canals around the head as well as down the trunk of the animal.

The gelatinous cupula is firmly attached to the kinocilium and stereocilia but free to slide across the surface of the epithelium. The stereocilia are arranged in order of increasing height and directionally sensitive (Flock, 1965, 1967), in just the same way as in the sacculus or cochlea. They are oriented to produce maximal stimulation of the hair bundle along the long axis of the body for superficially located organs, as in the minnow of Figure 6.12A; or, for organs in canals, along the long axis of the canal. A single organ typically contains two populations of hair cells, oriented in opposing directions. Thus half of the cells in the organ signal movement of fluid in one direction through the canal or along the surface of the body, and the other half signal movement in the opposite direction (Figure 6.12B). Since the kinocilium and stereocilia are embedded in the cupula, the mechanical properties of the hair bundle and the opening and closing of the mechanosensitive channels can make a substantial contribution to the mechanics of displacement of the organ (van Netten and Khanna, 1994).

Recordings from hair cells (Harris et al., 1970; Kroese and van Netten, 1989; Olt et al., 2016) show that they respond much like those in the bullfrog sacculus (Figure 3.11), although many of the detailed features of the response have not yet been examined. Vibration of the cupula produces a change in membrane potential, presumably by much the same sort of mechanism used by hair cells in other organ systems and species. It seems likely that adaptation plays an important role in the response of the hair cells, perhaps in part produced by the mechanics of fluid movement within the canal and by flexing of the cupula (summarized in Kroese and van Netten, 1989).

The vestibular system

The organs of the vestibular system are located within the inner ear adjacent to the cochlea (Figure 6.13). There are two different types (Wersäll and Bagger-Sjöbäck, 1974; Wilson and Melvill Jones, 1979; Guth et al., 1998), called semicircular canals and otolith organs. Each of the two sides of the head has three semicircular canals representing the three axes in space, each approximately orthogonal to the other two. In mammals there are two otolith organs,

(A)

HC
structure,
organs,
channel

(B)

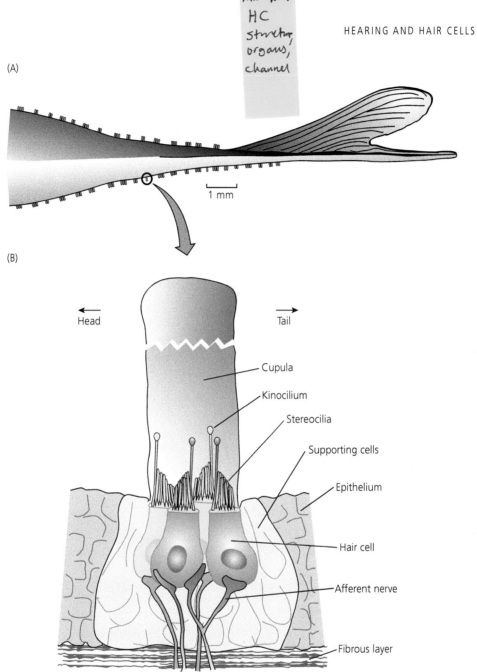

Head

Tail

Cupula

Kinocilium

Stereocilia

Supporting cells

Epithelium

Hair cell

Afferent nerve

Fibrous layer

1 mm

Figure 6.12 Lateral line organs. (A) Distribution of lateral line organs on the tail of the European minnow *Phoxinus phoxinus*. (B) Anatomy of a single organ. Note the two populations of cells with hair bundles oriented in opposing directions. (After Dijkgraaf, 1967; Flock, 1967.)

called the sacculus and the utriculus (Figure 6.13B); but many vertebrates also have a third, called the lagena.

The hair cells of the semicircular canals are contained only in ampullae, which are swellings adjacent to the sacculus and utriculus (Figure 6.13B). The epithelium of the ampulla forms a ridge called the crista ampullaris, which contains a uniform population of hair cells whose stereocilia and kinocilia are embedded in a gelatinous structure again called a

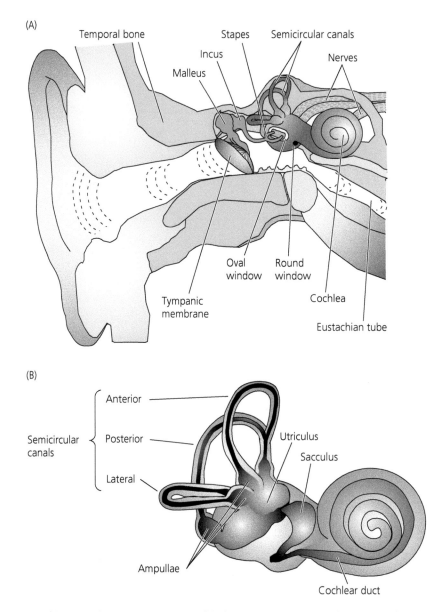

Figure 6.13 Anatomy of the mammalian inner ear. (A) Structure of the human ear. (B) Enlargement of components of the mammalian inner ear, showing semicircular canals, otolith organs, and cochlea. See the text.

cupula (Figure 6.14). Angular acceleration of the head causes the fluid of the semicircular canal to push against or pull away from the cupula, exciting or inhibiting the hair cells.

Consider the pair of canals on the two sides of the head that are oriented horizontally (Figure 6.15). If the head moves to the right (clockwise looking

down from above), the fluid in the semicircular canals on the two sides of the body will tend from inertia to lag behind, that is to move counterclockwise relative to the head. Since the stereocilia of the hair cells in the crista ampullaris are all oriented with their kinocilia pointing in the same direction on either side of the head (smaller arrows in the middle

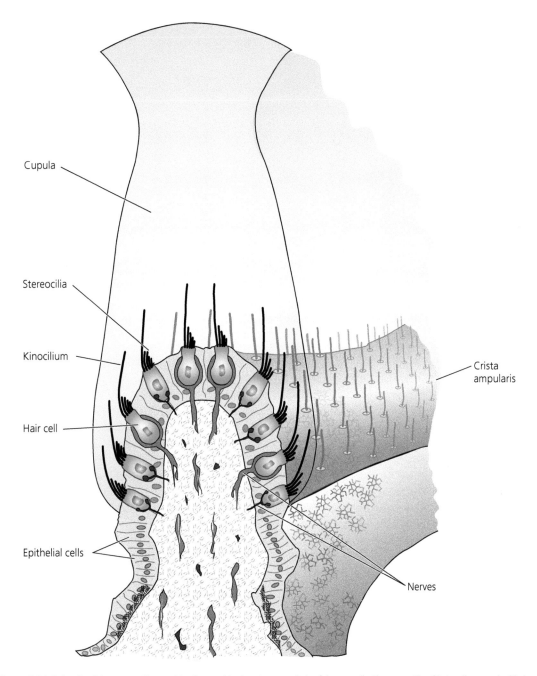

Cupula

Stereocilia

Kinocilium

Hair cell

Epithelial cells

Crista
ampularis

Nerves

Figure 6.14 Hair cells of the mammalian semicircular canal in the crista ampularis of the ampulla. The stereocilia of hair cells are embedded in a gelatinous cupula and are stimulated when the cupula is moved by the movement of fluid in the semicircular canal. (After Wersäll and Bagger-Sjöbäck, 1974.)

of Figure 6.15), counterclockwise flow of fluid (large arrows) will depolarize the hair cells on the right side and hyperpolarize those on the left. These signals are then communicated at synapses to the ter-

minals of the eighth cranial nerve, which provide information to the CNS about angular acceleration.

The otolith organs are organized rather differently. The sensory epithelium containing the hair

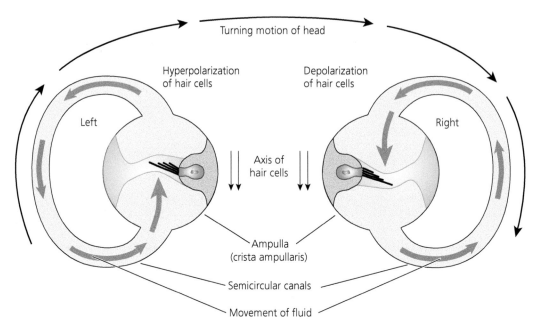

Figure 6.15 Stimulation of hair cells of semicircular canals by head movement. Turning the head laterally to the right produces inertial fluid movement in the horizontally oriented semicircular canals on either side of the head, so that hair cells on the left are hyperpolarized and those on the right are depolarized.

cells is called the macula, Latin for "spot." It is oriented approximately in a horizontal plane for the utriculus and in a vertical plane for the sacculus. The hair cells of the macula are covered by an otolithic membrane, interrupted by a series of tubular apertures through which the hair bundles penetrate (see Figure 3.11A). Above the otolithic membrane and on top of the hair bundles, there is a dense network of crystals of calcium carbonate called otoconia, emmeshed in a fine, fibrous web. When the head moves, these crystals tend to lag behind like an inertial mass, causing the otolithic membrane to slide relative to the macula, bending the hair bundle and stimulating the hair cells. This process is not so different from the mechanism of stimulation of mechanoreceptors in the statocyst of a lobster (see Figure 2.10).

Unlike the hair cells of the semicircular canals, the stereocilia of the cells in the utriculus and sacculus are not all pointed in the same direction. They are instead arranged so that movement can be detected in all possible directions within the plane of orientation of the organ. For the utriculus, for example, all the hair cells have their stereocilia

perpendicular to the horizontal plane, but the direction of orientation is systematically varied so that any horizontal movement maximally excites some population of cells. Movement in the vertical plane is detected by the hair cells of the sacculus. Together, these two organs signal both the static position of the head by sensing the acceleration of the force of gravity, and linear acceleration, caused by sudden sideward or upward movement of the head. It is possible that the otolith organs sense both position and movement with different populations of hair cells (Baird, 1994), some showing very little adaptation and best suited to the detection of static position, and others adapting rapidly and more suited to signaling movement (see also Geleoc et al., 1997; Holt et al., 1997).

Sudden motion of the head can occasionally dislodge one of the otoconia, sending it out of the utriculus so that it becomes lodged within one of the semicircular canals. The calcium carbonate crystal may then rest against the cupula, producing violent bursts of dizziness and nausea whenever the head is moved. This condition, called benign positional vertigo, is not uncommon especially among gymnasts

Amp!

and ballet dancers, or men and (statistically more often) women over the age of 50 who spend a lot of time at the neighborhood gym. It is easily cured by a systematic series of movements of the head called the Epley and Semont maneuvers. These movements methodically rotate the position of the head, so that the crystal is moved away from the cupula to a place where it can do no further harm.

The cochlea

Although many vertebrates including amphibians and reptiles have ears of some description (see for example Figure 6.21), the most elaborate organ of hearing in nature is the mammalian cochlea, located just inside the ear next to the organs of the vestibular system (Figure 6.13). Changes in air pressure which are created by sound can impinge upon the ear drum (tympanic membrane), where they are coupled by the three bones of the middle ear (the malleus, incus, and stapes). These bones cause the oval window to vibrate. The inside of the cochlea is filled with fluid, and vibration of the oval window sets this fluid in motion.

Hair cells in the mammalian cochlea reside within the organ of Corti (Figure 6.16A). The motion of the oval window pushes against fluid in the scala vestibuli and scala media (Figure 6.16B), causing a wave of pressure that can traverse the entire length of the cochlea through to the apex and round to the scala tympani, finally pushing out against the round window. The organ of Corti rests upon the basilar membrane, which divides the cochlea approximately in two. The oscillating wave of fluid pressure moving down the scala media can push against the basilar membrane and cause it to oscillate in phase with the sound.

The basilar membrane can be caused to vibrate in this way only if it is mechanically compliant at the frequency of vibration of the sound. If the basilar membrane were made of stiff metal, it would vibrate only at quite high frequencies. If it were made of soft rubber, low frequencies would set it in motion. The key to understanding frequency discrimination in the cochlea was the discovery that the mechanical compliance of the basilar membrane changes from base to apex: it is narrow and rather stiff at its base near the oval window, but wider and much more compliant near the apex. This change in stiffness has a dramatic effect on its motion. As von Békésy first showed (von Békésy, 1960), sound causes the basilar membrane to oscillate with an amplitude that changes down the length of the cochlea. For high-frequency tones the amplitude of oscillation peaks near the base, where the basilar membrane is stiffest. For low-frequency tones, the amplitude peaks near the apex where the membrane is most compliant.

The organ of Corti sits directly on top of the basilar membrane and moves when the basilar membrane moves. The organ of Corti is a complicated structure (Figure 6.17A), containing several sorts of supporting cells and (in humans) about 16,000 hair cells in four rows: one of inner hair cells and three to four of outer hair cells. Although less numerous, the inner hair cells are the primary sensory cells of the cochlea, providing synaptic input to 90–95 percent of the afferent nerve fibers of the cranial eighth nerve at numerous synapses on the base and along the side of each inner hair cell (Figure 6.5B). It is the signals of these afferent fibers that provide the major input to auditory centers in the CNS. The outer hair cells receive mostly efferent input and produce a remarkable amplification of the vibration of the basilar membrane, described later in the chapter.

When the basilar membrane moves, it moves the hair cells (Figure 6.17B). The stereocilia of the inner hair cells are closely apposed to a gelatinous and fibrous extracellular matrix called the tectorial membrane. The outer hair cell stereocilia are inserted into small pits in the tectorial membrane and are even more closely associated with it. The tectorial membrane is made of collagen and glycoproteins and varies in stiffness from apex to base much like the basilar membrane. Mutations in one of its principal constituents, called α-tectorin, can produce loss of hearing or complete deafness (Richardson et al., 2011). The motion of the basilar membrane causes the stereocilia to be deflected by the tectorial membrane, producing an electrical response. Because the amplitude of motion of the basilar membrane varies with position along the cochlea, individual hair cells are more sensitive to some frequencies than to others. They are most sensitive to a best or characteristic frequency (see Figure 1.4), which

(A)

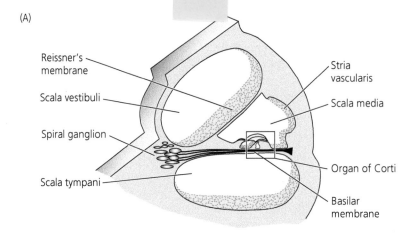

Reissner's membrane

Scala vestibuli

Spiral ganglion

Scala tympani

Stria vascularis

Scala media

Organ of Corti

Basilar membrane

(B)

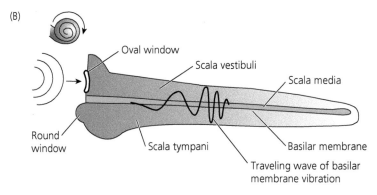

Oval window

Scala vestibuli

Scala media

Round window

Scala tympani

Basilar membrane

Traveling wave of basilar membrane vibration

Figure 6.16 The mammalian cochlea. (A) Cross-section of the cochlea. (B) The cochlea imagined as uncoiled to illustrate the motion of the basilar membrane. Pressure on the oval window is communicated through the fluids of the scala vestibuli and scala tympani to the round window, producing a traveling wave of basilar membrane oscillation. (A after Pickles, 1988; B after Nobili et al., 1998.)

varies in an orderly fashion from high frequencies at the base of the cochlea to low frequencies at the apex. In this way, the cochlea acts as a frequency analyzer, producing responses in a population of nerve fibers which reflect the distribution of frequencies of the incoming sound.

Endolymph and endocochlear potential

In many of the experiments I have described so far (including those in Chapter 3), the stereocilia of the hair cell were bathed in an artificial saline containing high Na^+ with a composition similar to that of the blood or cerebrospinal fluid. Within the living animal, however, the solution bathing the hair cell

stereocilia has quite a different composition. That is true not only of the cochlea but also of the semicircular canals and otolith organs, in the ears of lower vertebrates as well as mammals. This solution bathing the stereocilia is high in K^+ and low in Na^+ and is called endolymph (Figure 6.18).

We know most about the physiology of endolymph for the mammalian cochlea. Here the apical ends of the hair cells near the hair bundle are connected to one another (and to supporting cells) by tight junctions to form a structure called the reticular lamina (Figure 6.17A). The reticular lamina, basilar membrane, and another structure, Reissner's membrane (Figure 6.16A), separate the cochlea into three different compartments. Two of these,

(A)

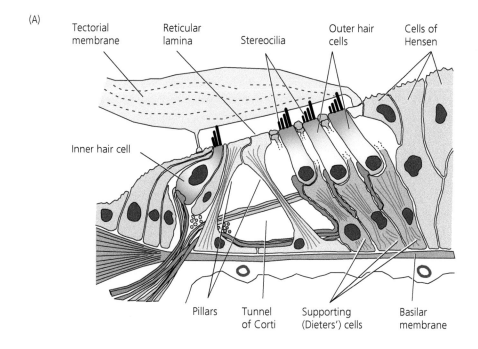

Tectorial membrane

Reticular lamina

Stereocilia

Outer hair cells

Cells of Hensen

Inner hair cell

Pillars

Tunnel of Corti

Supporting (Dieters') cells

Basilar membrane

(B)

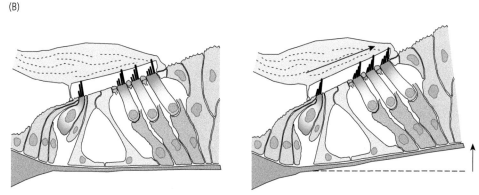

Figure 6.17 The organ of Corti. (A) Schematic cross-section. (B) Movement of the basilar membrane causes bending of the stereocilia of the hair cells. (After Ryan and Dallos, 1984; Pickles, 1988.)

the scala vestibuli and the scala tympani, are filled with a fluid called perilymph similar in composition to blood or cerebrospinal fluid, high in Na^+ and low in K^+. The third, the scala media, is filled with endolymph, which bathes the stereocilia and is high in K^+ and low in Na^+.

The endolymph of the cochlea is secreted into the scala media by the cells of the stria vascularis (Figure 6.16A). These cells have a very active Na^+/K^+ ATPase, as well as other channels and transport

proteins that move K^+ into the scala media. Disruption of this K^+ transport produces degeneration of the hair cells and deafness (see Nin et al., 2016; Fettiplace, 2017). In mammals, the cells of the stria vascularis also generate a large potential difference of approximately +80 mV between the scala media and the other compartments of the cochlea. This potential, called the endolymphatic or endocochlear potential, has the effect of greatly increasing the driving force for current flow into the stereocilia,

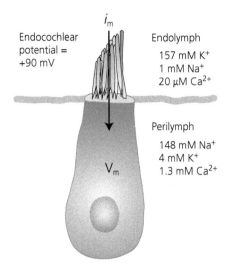

Endocochlear
potential =
+90 mV

Endolymph

157 mM K⁺
1 mM Na⁺
20 µM Ca²⁺

Perilymph

148 mM Na⁺
4 mM K⁺
1.3 mM Ca²⁺

i_m

V_m

Figure 6.18 The endolymph and endocochlear potential. See text. (After Fettiplace, 2017.)

since the difference in potential between the intracellular and extracellular space is much larger than if the endocochlear potential were not present. Mutations that reduce this potential can produce hearing loss and deafness (Nin et al., 2016).

To understand the role of the endolymph and endocochlear potential, we remember from Chapter 3 that the change in current flow produced by a change in membrane conductance is given by Eq. (3.8),

$$\Delta i_m = \Delta g\left(V_m - E_{rev}\right)$$

The stretch-sensitive channels of hair cells are cationic and rather non-selective (see Chapter 3), and because the stereocilia are exposed to endolymph with high K⁺, most of the current through the channels will be carried by K⁺. The cytoplasm of the hair cell also has a high concentration of K⁺, so the K⁺ concentration on the two sides of the stereocilium membrane is approximately equal and E_{rev} is close to zero. If all of the hair cell were exposed to endolymph, V_m would also be zero. But because most of the membrane of the cell including the synaptic terminal is exposed to perilymph with high Na⁺ and low K⁺ (Figure 6.18), and because the cytoplasm of a hair cell has an ion composition similar to that of a neuron in the CNS, the resting membrane potential of a hair cell is between –40 and –60 mV (see for example Johnson et al., 2011).

If there were no endocochlear potential, ions would flow from a region of zero potential (the extracellular space) to a region of –40 to –60 mV (the inside of the cell). The value of ($V_m - E_{rev}$) would therefore be –40 to –60 mV, and the increase in K⁺ permeability would produce a depolarization from the resting membrane potential toward the reversal potential (which is near zero). The endocochlear potential has the further effect of increasing the potential difference across the membrane of the stereocilia to something like –120 to –140 mV, because the extracellular potential adjacent to the part of the hair cell containing the stereocilia and stretch-sensitive channels is not actually zero but rather of the order of +80 to +120 mV. The endocochlear potential can more than double the driving force for current movement.

The endocochlear potential increases the driving force producing a change in membrane potential when the stretch-sensitive channels are open, but it is less clear why hair bundles in the cochlea and vestibular system need to be surrounded by an elevated concentration of K⁺. The reason usually given is that if the current were carried by Na⁺, the Na⁺ would have to be pumped out of the cell to prevent it from accumulating. The K⁺ entering the cell from the endolymph, on the other hand, can flow passively out of the cell down its concentration gradient from a high concentration inside the cell to a low concentration in the perilymph. There is therefore no need to pump it out of the cell with a Na⁺/K⁺ ATPase, thus sparing the hair cell a considerable metabolic load. It may be advantageous to shift this load away from the hair cell to a separate and more numerous cell type. In addition, the stria vacularis may be more easily fed by the circulatory system and may transport K⁺ more efficiently than would the hair cell.

Outer hair cells and tuning

An initial solution to the frequency selectivity of mammalian hair cells was given by the demonstration of von Békésy of a traveling wave passing down the basilar membrane, peaking at different points for different sound frequencies. Von Békésy's measurements were, however, made from cadavers and reflected only passive basilar membrane motion. As recordings began to be made from afferent nerve fibers with metal microelectrodes (Figure 1.4), it became rapidly apparent that single axons in mammals show much sharper frequency tuning than can be explained purely by passive basilar membrane motion. Later observations from the living cochlea showed a correspondingly sharper turning in the vibration of the basilar membrane (see Ashmore, 2008; Fettiplace, 2017), nearly as great as that of the afferent nerve fibers. Clearly, some mechanism in the living cochlea must actively magnify motion of the basilar membrane and increase the frequency selectivity of the hair cells. This mechanism is usually referred to as the *cochlear amplifier*.

We have already seen that the gating compliance of the channels can function as an amplifier, producing positive feedback and increasing the response of the hair cell to an incoming sound. This mechanism, along with the electrical resonance described in the following section, may be the most important signal boosters in lower vertebrates. In the cochlea of mammals, however, there is a different kind of amplifier of remarkable properties produced by the three rows of outer hair cells. These cells make little contribution to the sensory signal conveyed to the CNS—this function seems to be served almost entirely by the inner hair cells. The outer hair cells nevertheless have stereocilia and respond to sound much like the inner hair cells (see for example Russell and Sellick, 1983). They have in addition an extraordinary ability to change shape in response to a change in membrane potential. Isolated outer hair cells become shorter when they are depolarized and longer when hyperpolarized (Brownell et al., 1985), and this motion can amount to as much as 4 percent of the length of the cell. It is extremely rapid and doesn't require a direct contribution from ATP (Ashmore, 1987).

Early experiments showed that the motor must be a network of many independent units, because the two halves of a cell could be made to move in opposite directions (Dallos et al., 1991). Even isolated membrane patches can show voltage-dependent motion (Ashmore, 1992; Kalinec et al., 1992). The movement seems to be produced by a membrane protein, which undergoes a voltage-dependent change in conformation that alters the shape of the cell. The change in shape is accompanied by a voltage-dependent movement of charge within the membrane (Santos-Sacchi, 1991; Ashmore, 1992), probably produced by a change in protein conformation.

After considerable effort, the protein responsible for outer hair cell movement was identified by carefully isolating a large number of inner and outer hair cells and looking for transmembrane proteins selectively expressed in one cell type but not in the other (Zheng et al., 2000). A large, integral membrane protein was found in the outer hair cells and given the name prestin. When the gene for prestin was transfected into a cell line, it conferred all of the properties of outer hair cell movement. Changes in membrane potential in the transfected cell generated voltage-dependent movement of charge similar to that seen in the outer hair cells (Figure 6.19), and sinusoidal changes in voltage produced sinusoidal movements of the cell not seen in control, untransfected cells from the same cell line (Zheng et al., 2000; Ludwig et al., 2001). This result shows that prestin by itself is able to generate changes in cell shape.

Prestin has been found so far only in the outer hair cells of mammals (see Dallos, 2008) and in the hair cells of birds (Beurg et al., 2013). It is thought to have fourteen hydrophobic sequences long enough to span the membrane and is probably an oligomer (see Ashmore, 2018). It is a member of the SLC26 family of proteins, which can code for transporters that move anions like sulfate across the plasma membrane. There is no evidence that prestin in mammals actually transports anions, but Cl⁻ plays an important role in its function, because removal of intracellular Cl⁻ abolishes cell movement (Oliver et al., 2001). The Cl⁻ may attach to a binding site within the prestin molecule much as if prestin were actually transporting Cl⁻ across the membrane. The binding of Cl⁻ is inhibited by salicylate, and this may account for hearing loss that can be induced by high doses of aspirin.

The evidence is now compelling that prestin and the outer hair cells function as the principal cochlear amplifier in the following way. As sound deflects the stereocilia of the outer hair cells, stretch-sensitive channels open and produce a depolarization of the outer hair cells. This depolarization acts on the prestin molecule to generate a decrease in the height of the row of outer hair cells, producing a kind of

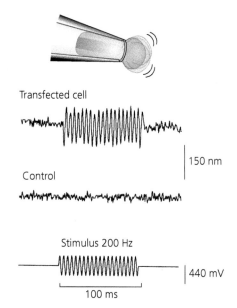

Stimulus 200 Hz

Figure 6.19 Prestin produces voltage-dependent cell movement. A single cell from a cell line expressing prestin was drawn up into a micropipette and stimulated with a sinusoidal voltage difference between the inside and outside of the pipette. Cell movement was measured by imaging the tip of the cell onto a photodiode and recording the change in current as more or less of the light falling on the diode was occluded. (From Zheng et al., 2000.)

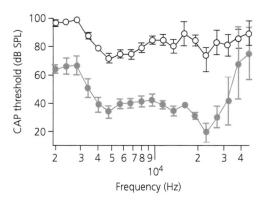

Figure 6.20 Effect of prestin deletion on audition. Extracellular recording of massed action potentials of the auditory nerve (compound action potential or CAP), used to estimate auditory threshold. CAP threshold is plotted on the ordinate, with sound amplitude given in decibels of sound pressure level (db SPL). The frequency of sound is given on the abscissa. Black, wild-type mice; blue, 499 KI mice with mutant and largely non-functional prestin. For mutant mice, much louder sounds were required to produce the same threshold of action-potential firing over most of the range of hearing. (From Dallos et al., 2008.)

pincer motion that pulls the outer hair cells toward the tectorial membrane and moves the whole organ of Corti upward to accentuate movement of the basilar membrane (see Figure 6.17B). When the prestin molecule is knocked out (Liberman et al., 2002) or replaced by a mutant protein no longer able to mediate the voltage-dependent change in shape of the outer hair cells (Dallos et al., 2008), the sensitivity of hearing is greatly reduced (Figure 6.20).

One possible difficulty with this hypothesis is that outer hair cells would have to change shape at frequencies up to at least 20 kHz in humans and at even higher frequencies in mice and bats. The membrane potential of the outer hair cells and the change in prestin conformation would both have to be synchronized with the oscillations of basilar membrane motion (see Ashmore, 2008). The maximum rate at which the membrane potential of the outer hair cells would be able to follow the oscillations of the sound would, however, be limited by the electrical time constant of the cell membrane, which in most neurons is typically 1 ms. If the time constant of outer hair cells had this value, prestin would not be

able to respond to frequencies much above 1000 Hz. The data in Figure 6.20 show, however, that deletion of functional prestin depresses the sensitivity of the mouse ear at much higher frequencies.

The solution to this puzzle seems to have been discovered at least in part by Johnson and coworkers (2011), who showed that the time constants of outer hair cells vary with the position of the cell from apex to base along the cochlea and can be much shorter than originally supposed. The time constant of a compact neuron like an outer hair cell is given by

$$\tau = RC \qquad (6.3)$$

where τ is the time constant in seconds, R is the resistance of the cell in ohms, and C is the cell capacitance in farads. What Johnson and colleagues showed is that τ becomes systematically shorter from apex (low frequencies) to base (high frequencies) along the length of the basilar membrane, because both R and C decrease in cells responding to higher frequencies.

The resistance R is the inverse of the membrane conductance G. The value of G increases (and R decreases) with any augmentation in the number or single-channel conductance of open channels. The principal channel types in the outer hair cells are the stretch-sensitive channels in the stereocilia whose probability of opening (Johnson et al., 2011) and single-channel conductance (Beurg et al., 2014), both increase from apex to base; and voltage-gated potassium channels in the soma, whose conductance also increases in the same direction (Mammano and Ashmore, 1996; Johnson et al., 2011). These two conductances have quite different reversal potentials, one near zero and the other near –80 or –90 mV; but because both increase in parallel, the membrane potential of the outer hair cells can remain almost the same in the different parts of the cochlea. And because conductance increases, R decreases.

The membrane capacitance of a cell is nearly proportional to the amount of its plasma membrane, since capacitance per unit membrane is very nearly a constant. The outer hair cells become progressively smaller from apex to base (see Fettiplace, 2017), so cell capacitance also decreases. The net result is that both R and C decrease, and the time constant becomes progressively shorter from low to high frequencies. It is probably brief enough to allow the outer hair cell membrane potential to change in phase with the oscillation of the basilar membrane over most of the range of audible sound. And because prestin can change conformation very rapidly (see Ashmore, 2008; but see also Santos-Sacchi and Tan, 2018), prestin and the outer hair cells are likely to be principally responsible for the cochlear amplifier of mammalian hearing.

Outer hair cells receive a large efferent input from axons coming from the superior olivary complex of the CNS. This input is cholinergic but inhibitory and hyperpolarizes the outer hair cells, decreasing the amplitude of their response to sound (see Fuchs, 2014; Fettiplace, 2017). As a consequence, electrical stimulation of this input produces a decrease in cochlear amplification, effectively elevating the threshold of hearing. The function of this pathway may be to improve the detection of sound in background noise or to protect the hair cells against damage from loud noise (see for example Maison et al., 2013).

Electrical resonance

Although some sort of frequency tuning occurs in the ears of nearly all vertebrate species, the way this tuning is produced can be quite variable and, in many lower vertebrates, seems not to occur in the way as in mammals (see Fettiplace and Kim,). For many non-mammalian species, the peak amplitude of motion of the basilar membrane may not change with distance in the same way as in mammals, nor do these species have outer hair cells. They nevertheless show an orderly progression of frequency selectivity among their hair cells, often produced by a systematic change in the electrical and/or mechanical properties of the sensory receptors themselves.

One particularly interesting mechanism of tuning was first discovered in the turtle (Fettiplace and Crawford, 1978). The turtle ear is formed as in mammals with an eardrum or tympanum. Its vibrations are transmitted by a small bone called the columella, which functions like the bones of the mammalian inner ear (see Figure 6.21A). These vibrations are transferred to fluid-filled cavities adjacent to the basilar papilla, where the auditory hair cells are located (see Baird, 1960; Grassé, 1970). The hair cells rest on a basilar membrane and their stereocilia abut a tectorial membrane much as in mammals (Miller, 1978), and vibrations of the basilar membrane bend the stereocilia and excite the hair cells as in Figure 6.17B.

Although it remains possible that the mechanical properties of the basilar membrane change somewhat from one end of the basilar papilla to the other, the tuning of the hair cell seems to be largely dependent on an electrical resonance whose frequency varies systematically down the length of the papilla. When steady current is injected into a cell with an intracellular microelectrode, the membrane potential oscillates (Figure 6.22A). This is an *electrical* not a mechanical oscillation, and different hair cells oscillate at different frequencies close to the characteristic frequency of the cell for sound (Crawford and Fettiplace, 1981). As a consequence, the sinusoidal change in potential in response to sound is greatly amplified when the frequency of the sound is the same as that of the electrical resonance. This mechanism is similar to the way tuning works in a radio: as the dial is moved, the electrical resonance of the receiver is changed to different frequencies to amplify the incoming signals of different radio stations.

The electrical resonance in a hair cell is produced largely by ion channels whose properties change systematically from one end of the cochlea to the

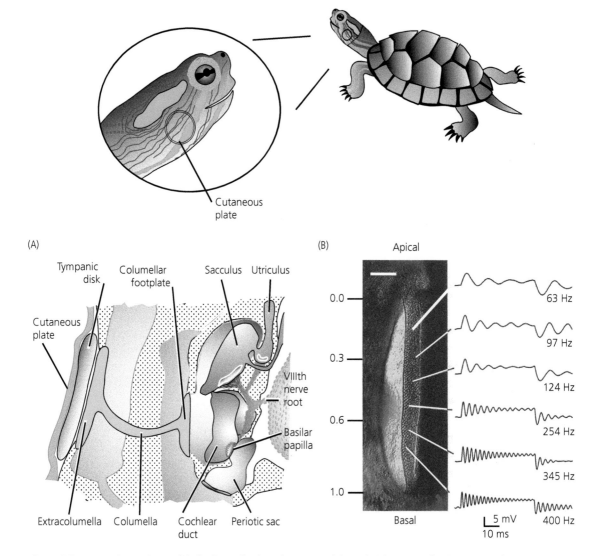

Figure 6.21 Tonotopic organization of the basilar papilla, the auditory organ of the turtle. (A) Anatomy of turtle inner ear. Changes in pressure on cutaneous plate are transferred into the inner ear by the columella and detected by the hair cells of the basilar papilla. (B) Basilar papilla. On the left is the surface of the excised papilla; scale bar, 100 µm. On the right are selected examples of electrical resonance recorded from hair cells at different positions along the papilla. Resonant frequencies are given beside each of the traces and increase from apex to base. Each record is the voltage response to a small depolarizing current step. (A after Baird, 1960; Grassé, 1970; Miller, 1978; B from Ricci et al., 2000.)

other. In addition to mechanoreceptive channels in the stereocilia, turtle hair cells have a large population of voltage-gated Ca^{2+} channels, as well as K^+ channels activated by intracellular Ca^{2+} (K_{Ca} channels), both probably located in close proximity to one another near the synapses of the cell (Figure 6.22B). When the stereocilia move and the

mechanoreceptive channels open, the hair cell depolarizes. This depolarization opens the voltage-gated Ca^{2+} channels, producing further depolarization. As the Ca^{2+} enters the cell through the Ca^{2+} channels, the increase in intracellular Ca^{2+} causes the K_{Ca} channels to open, producing a delayed hyperpolarization. This hyperpolarization closes the

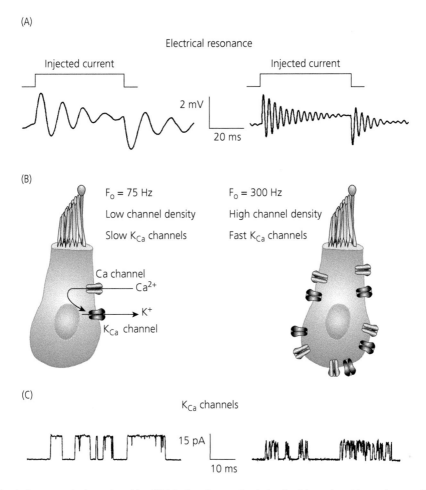

Figure 6.22 Electrical resonance in the turtle cochlea. (A) Injection of current into hair cells of the turtle cochlea produces an electrical resonance at a frequency close to the characteristic frequency (F_o) of the cell. (B) Electrical resonance is produced by voltage-gated Ca^{2+} channels and Ca^{2+}-activated K^+ channels (K_{Ca}), whose properties change systematically down the length of the cochlea. (C) K_{Ca} channels open and close more rapidly in hair cells for which the electrical resonance occurs at a higher frequency. (A from Art and Fettiplace, 1987; B after Fettiplace and Fuchs, 1999; C from Art et al., 1995.)

voltage-gated Ca^{2+} channels, reducing the entry of Ca^{2+} and eventually also closing the K_{Ca} channels. The cell is then ready to undergo the next cycle of depolarization and hyperpolarization, producing the electrical resonance that amplifies the sound response.

The frequency of the resonance depends upon the density of the channels and the properties of the K_{Ca} channels, which both vary systematically from one end of the cochlea to the other (Art et al., 1995). For cells responding to higher frequencies, the opening and closing of the K_{Ca} channels occurs much more rapidly than for those responding to lower frequencies (Figure 6.22C). This change in channel

kinetics quickens the cycle of depolarization and hyperpolarization. Hair cells appear to have a large number of different kinds of K_{Ca} channels, produced by alternative splicing of a single gene (Navaratnam et al., 1997; Rosenblatt et al., 1997). The distribution of these different splice variants varies systematically from cell to cell down the cochlea (Ricci et al., 2000), producing an orderly progression of electrical resonance in the basilar papilla, which is the turtle's auditory organ (Figure 6.21B). This has the consequence that different hair cells of the papilla respond best to different frequencies of sound, and the difference in resonant frequency is thought to be largely responsible for the tuning of the hair cell. The resonant frequency is limited by the kinetics of the Ca^{2+} and K^+ channels as well as the membrane time constant of the hair cells, which probably explains why turtles are most sensitive to frequencies at about 500 Hz and cannot hear sound above about 800–1000 Hz. So if you want to sing to your pet turtle, forget the Queen of the Night aria. Try Deep River instead.

Summary

Insects have two principal kinds of ears, both adapted from arthropod chordotonal organs. In tympanal ears, scolopedia are pressed up against a tympanum, which is formed from a thin layer of cuticle. As the tympanum vibrates to changing air pressure, the sensory neurons are depolarized and fire action potentials. In most species, the tympanum and sensory apparatus are tuned to one principal frequency, often that of the male mating call.

The other kind of insect ear is called Johnston's organ, which is used by flies, bees, and mosquitoes to hear the species-specific buzzing of male wing beat. These organs have been most thoroughly investigated in *Drosophila*, where the oscillations of air produce an in-phase rotation of a feather-like arista coupled to the funiculus of the fruit-fly antenna. As the funiculus rotates, a hook-like structure moves back and forth against several hundred scolopedia within the organ, producing signals in distinct populations of sensory cells responding to sound and to static changes in wind or gravity. Present evidence indicates that depolarizations in membrane potential in the sensory cells are produced by TRP channels, perhaps including NOMPC and the protein products of the *Inactive* and *Nanchung* genes.

In vertebrates, sound and motion are detected by a specialized mechanoreceptor called the hair cell. What distinguishes hair cells from other touch and stretch receptors is the hair bundle at the distal end of the cell. Consisting in most cases of 30–300 microvilli (called stereocilia) and one true cilium (called the kinocilium), the hair bundle has evolved to be a highly sensitive detector of movement. It contains everything required to transduce the mechanical signal: the mechanoreceptive channels, as well as all of the specializations necessary to focus the energy of bundle deflection on channel gating and to modulate the sensitivity of the response during maintained stimulation.

Each stereocilium contains only a small number of channels. An essential role in channel gating is played by specialized extracellular structures called tip links, which connect adjacent stereocilia and couple bundle movement to the force that gates the channels. When the hair bundle moves, the stereocilia pivot as rigid rods under tension, causing the tip links to pull against the membrane of the stereocilia. The movement of the bundle to an applied force is also affected by channel opening and closing, and this gating compliance of the channels may boost the amplitude of the response to sound.

As for other mechanoreceptors, the channels appear to be associated with proteins both within the stereocilium and in the extracellular matrix. Some of these proteins have been identified, either from an analysis of genetically inherited hearing disorders or from targeted gene mutation. These include cadherin-23 and protocadherin-15, which form the tip links of the stereocilia; proteins of the actin bundle; several forms of unconventional myosin; Usher proteins such as sans and harmonin, which are localized to the top of the tip links; α-tectorin, a principal component of the tectorial membrane; and TMC1 and TMC2, which form at least in part the stretch-sensitive channel that is part of the transduction complex.

Hair cells adapt to maintained stimulation. When the hair bundle is moved from its resting position to a new steady position, the change in tension causes the whole operating range of the hair cell to be transferred to this new position. Transduction then becomes more sensitive to movement around this position. In some and perhaps all species, adaptation is dependent on Ca^{2+}, though what Ca^{2+} does is still uncertain.

Hair cells are found only in vertebrates within the lateral line organs of fish and amphibia, in the vestibular system, and in the cochlea of the ear. The mechanosensitive lateral line organs are located either on the surface of the body or within shallow canals about the head or down the long axis of the animal. The stereocilia of lateral line hair cells are embedded in a gelatinous structure called the cupula, which is deflected by currents in the surrounding water. The pressure against the cupula pushes against the hair bundle, producing a change in receptor membrane potential.

The vestibular system within the inner ear contains two different kinds of organs. The three semicircular canals located approximately orthogonal to one another sense movement of the head. One end of each canal has a structure called the crista ampullaris, which contains a uniform population of hair cells whose stereocilia are again embedded in a gelatinous cupula. Movement of the head causes the fluid of the semicircular canal to push against the cupula, exciting or inhibiting the hair cells. The mammalian inner ear also contains the two otolith organs—the sacculus and utriculus, which both detect the steady position of the head and linear acceleration. The hair bundles in these organs are associated with a layer of crystals of calcium carbonate, called otoconia, which act as an inertial mass to deflect the stereocilia and produce changes in membrane potential.

The most elaborate of the organs that utilize hair cells is the cochlea of the mammalian ear. The hair cells are supported by a membrane, called the basilar membrane, which is put into motion by pressure waves falling on the ear drum. Sound moves the fluid of the cochlea, setting up a wave of oscillation that travels down the basilar membrane. The nature of basilar membrane oscillation depends upon the frequency of the sound. For high frequencies, the peak amplitude of the wave occurs near the base of the cochlea, where the basilar membrane is least compliant; for low frequencies, the peak occurs nears the apex. As the basilar membrane moves, it pushes the stereocilia of the hair cells against a gelatinous matrix called the tectorial membrane, deflecting the stereocilia and producing the change in conductance that generates the receptor potential.

There are two kinds of hair cells in mammals including humans, both responding to movement of the basilar membrane but with quite different functions. The single row of inner hair cells makes the very great majority of synapses onto afferent nerve fibers and is primarily responsible for the sensory signal sent to the CNS. The more numerous outer hair cells respond to the sound but produce a mechanical feedback that acts as a cochlear amplifier, effectively boosting the movement of the basilar membrane. The outer hair cells of mammals contain a protein called prestin, which undergoes a rapid change in conformation as the membrane potential changes. As a consequence, the length of the outer hair cell can oscillate up and down in phase with the sound, boosting the movement of the basilar membrane and improving frequency selectivity. Tuning in lower vertebrates can employ a variety of additional mechanisms, and one of the most interesting is an electrical resonance produced by voltage-gated Ca^{2+} channels and Ca^{2+}-activated K^+ channels. The density and properties of the channels change systematically down the length of the cochlea, producing an orderly change in electrical resonance and best frequency of the hair cells.

CHAPTER 7

Chemoreception and the sense of smell

Chemoreception emerged very early in the evolution of living forms, principally as a means of finding food but also for avoiding potentially harmful substances in the environment. It has common mechanisms nearly everywhere it occurs: chemicals bind to receptor proteins, either opening an ion channel directly or activating a transduction cascade that changes the concentration of a second messenger. In its simplest form, chemoreception is responsible for chemotaxis, the remarkable ability of a bacterium to move toward attractants like sugars and amino acids and away from repellants like acids and alcohols. At its most sophisticated, it provides the cues that enable a dog to recognize its keeper by odor alone, or that guide a salmon as it travels sometimes a great distance upriver to its spawning ground.

Chemotaxis

Consider *Escherichia coli*, a common bacterium found in the human intestine. This organism has from four to eight flagella (Figure 7.1), 5–10 μm long and composed of many copies of the protein flagellin, attached to a hook-like structure that inserts into the cell wall and plasma membrane (Chaban et al., 2015). The movement of the flagellum is driven by proteins that use the energy of the transmembrane proton gradient to turn the flagellum at rates of up to several hundred revolutions per second. When all the flagella turn counterclockwise, they come together and rotate as a bundle, pushing against the viscosity of the fluid and propelling the bacterium forward at a speed of about 20 μm/s. When the flagella turn clockwise, they flail apart, producing an event called a tumble, which brings the bacterium to a halt (Berg, 1988). The bacterium has so little inertia moving through the viscosity of water that it is able to stop within a very short distance (less than 10^{-10} m or 1 Å). After about 0.1 s, the bacterium recommences swimming in a new direction taken at random. The motion of the organism, then, consists of a series of runs interrupted by abrupt changes in direction, producing a random walk through the medium.

The first systematic study of bacterial taxis was made by Engelmann and Pfeffer in the latter part of the nineteenth century (see Berg, 1975). Their observations were refined by Adler and collaborators (Adler, 1966, 1975; see Hazelbauer, 2012), who more thoroughly characterized the class of compounds that attract and repel. *Escherichia coli* has proved to be a favorable organism for chemotaxis research, because so much is known about its genetics and molecular biology. Many different chemotaxis (Che) mutants have been isolated, some for receptor proteins and others for members of the transduction cascade that modulate movement of the flagella. This has made it possible to isolate the molecules responsible for chemotaxis and to study them in some detail.

We now have a rather complete picture how chemotaxis occurs (see Parkinson et al., 2015).

Sensory Transduction. Second Edition. Gordon L. Fain, Oxford University Press (2020). © Gordon L. Fain 2020.
DOI: 10.1093/oso/9780198835028.001.0001

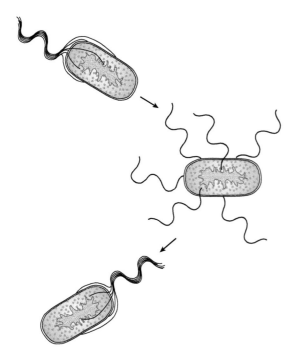

Figure 7.1 A tumble changes the direction of motion of the bacterium *E. coli*. Forward motion of the bacterium is produced by counterclockwise rotation of the flagella, interrupted by a sudden rotation of the flagella in a clockwise direction causing a tumble. After about 0.1 s, the bacterium starts off again in a new direction that is taken at random. (After Berry and Armitage, 1999.)

Attractants bind to receptors in the bacterial plasma membrane, either by themselves or in combination with periplasmic binding proteins present in the space between the outer membrane of the bacterial cell wall and the plasma membrane (Figure 7.2). Binding to the receptor protein initiates a sensory cascade that reduces the number of tumbles. This means that, in the presence of increasing concentration of attractant, the bacterium makes fewer reversals of direction and swims more expeditiously to the source of the attractant. Repellents do the opposite: the bug turns more often and so finds safer waters.

Escherichia coli has five receptor proteins called *methyl-accepting chemotaxis proteins (MCPs)* (see Ortega et al., 2017): Trg (sensing for example ribose and galactose), Tar (aspartate and maltose), Tsr (serine), Tap (dipeptides and pyrimidines), and Aer (O_2). Although shown in Figure 7.2 as a monomer, the receptor proteins are known to exist in arrays of trimers of homodimers. The intracellular surface of the receptor complex is associated with a histidine protein kinase called CheA and an obligatory adaptor protein called CheW, which couples CheA activity to receptor binding. CheA phosphorylates another protein called CheY to produce CheY•P, which acts as the second messenger of the transduction cascade. CheY•P binds to the C-ring at the base of the flagellum to produce a tumble. The binding of attractants causes a subtle change in the conformation of the cytoplasmic domains of the receptor, which reduces the activity of the CheA kinase, reduces phosphorylation of CheY, and decreases the probability of a tumble. This allows the bacterium to swim for a longer period between reversals of direction. Repellants do just the opposite. CheY•P is short lived and rapidly dephosphorylated by another protein called CheZ.

One interesting feature of bacterial chemoreception is that it adapts. An *E. coli* bacterium senses change in the concentration of an attractant or repellant in order to swim toward an increasing concentration of an attractant or away from a decreasing concentration of repellent. An increase in concentration of attractant produces a reduction in the number of tumbles and longer runs, permitting the bacterium to approach the source of the goodie. Once the bacterium gets to the attractant, it is important that it doesn't just swim away but stays in the vicinity. Adaptation in the presence of a constant concentration of attractant brings the frequency of tumbles back to its basal rate, and the bug then does a random walk around the nutrient. If the bacterium begins to move away, a decrease in attractant concentration causes the tumble rate to accelerate, eventually bringing the bacterium back where it belongs.

Adaptation is produced by two further proteins, a methyl transferase called CheR, and a methyl esterase called CheB (Figure 7.3). These two proteins put on and take off methyl groups from the receptor proteins, giving the MCPs their name. The amount of methylation at any point in time depends upon the relative rates of these two enzymes. The greater the methylation of the receptor proteins, the more efficiently the receptors activate the CheA protein to phosphorylate the second messenger CheY, and the greater the frequency of tumbles.

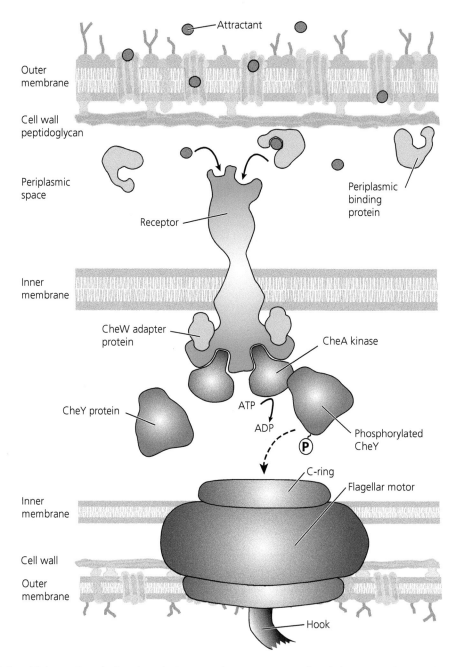

Figure 7.2 Bacterial chemotaxis: mechanism of transduction. In *E. coli* an attractant passes through the outer membrane, binds in some cases together with a periplasmic binding protein to a receptor, and changes the conformation of the receptor to initiate a transduction cascade. See text. Circle with P is phosphate group. (After Grebe and Stock, 1998; Mowbray and Sandgren, 1998; Armitage, 1999.)

Methylation is regulated by phosphorylation of the esterase (see Figure 7.3), which accelerates the rate of removal of methyl groups and reduces net methylation of the receptor proteins. Phosphorylation of the esterase is produced by CheA, the same protein that phosphorylates CheY.

Adaptation for an attractant therefore works in the following way (Figure 7.4). Binding of attractant, which *reduces* the protein kinase activity of CheA and *decreases* the production of CheY•P, also *reduces* the rate of activity of the methyl esterase. This leads to a reduction in the rate of removal of methyl groups from the receptor proteins and a gradual increase in their number. The greater the methylation, the more efficient the coupling between the receptor protein and CheA, and this increased coupling brings the rate of the kinase and the production of CheY•P back nearly to its basal rate.

Adaptation is produced by delayed activation of a secondary, compensating pathway. The binding of attractants initially inhibits the rate of phosphorylation by CheA, decreasing the amount of phosphorylated second messenger and reducing the frequency of the tumbles. But it also reduces phosphorylation of CheB (the methyl esterase), producing a delayed decrease in esterase rate and an increase in the net amount of methylation. An increase in methylation increases the rate of phosphorylation of CheY by CheA and gradually returns the rate of tumbles nearly to its resting value. As long as the bacterium swims up a concentration gradient, the concentration of attractant will continue to increase, and inhibition of CheA by binding of attractant will be more powerful than acceleration of CheA by receptor protein methylation. Once, however, the bacterium reaches a region of fixed concentration, the rate of phosphorylation by CheA gradually increases, and the bacterium resumes its random walk.

Olfaction in insects

Chemoreception in animals works much as in bacteria but with important differences. There are receptor proteins which detect chemicals of interest, but instead of directly altering the rate of flagellar motion they produce an electrical signal by changing the

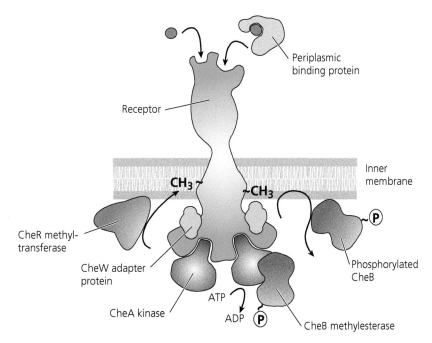

Figure 7.3 Proteins of adaptation for bacterial chemotaxis. Adaptation is mediated by methylation of receptor proteins. The CheR protein (methyltransferase) adds methyl groups to the receptor, and the CheB protein (methylesterase) takes them off. The rate of demethylation is accelerated by phosphorylation of the CheB protein by the CheA kinase. See text. Circle with P is phosphate group.

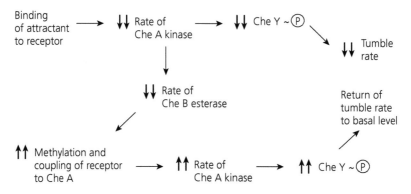

Figure 7.4 Mechanism of adaptation. Binding of attractant not only decreases the rate of production of phosphorylated CheY, but also decreases the rate of phosphorylation of CheB, leading to increased methylation of the receptor. Increased methylation facilitates the coupling of the receptor to CheA and causes the rate of the CheA kinase to increase, restoring the concentration of the phosphorylated CheY to its basal level.

conductance of the receptor-cell membrane. The molecules of transduction are completely different. The receptor proteins are either ion channels or G-protein-coupled receptors like those described in Chapters 3 and 4. There are no second-messenger proteins like CheY, though other substances such as Ca^{2+} and cyclic nucleotides can act as second messengers in their stead. Chemoreception in animals also adapts, but the receptor proteins are phosphorylated not methylated, and there seem to be additional mechanisms that modulate the ion channels and enzymes of the transduction cascade.

The most sophisticated and best understood form of chemoreception in animals is olfaction, the sense of smell. Remarkable progress has been made understanding olfaction in vertebrates, as described later in the chapter. I begin, however, with insects. Olfaction in insects is interesting in its own right, but it has the added attraction that it may hold the key to safe and effective pest control. Male moths can detect a remarkably small concentration of female sex pheromone, as we saw in Chapter 2, and this and similar phenomena in other species may be useful in controlling populations that cause disease or destroy crops.

In addition to pheromones, insects respond to a wide variety of more common chemicals. The sense of smell of *Drosophila* appears to be somewhat limited, perhaps most acute for compounds emitted from decaying fruit; but a honeybee is another thing altogether. The German behaviorist Karl von Frisch (1967) was able to train bees to respond to twenty-eight

different odors, mostly extracts from flowers or essential oils that humans can also sense as floral or spicy. These substances probably represent a small fraction of the number a worker bee can detect. The concentrations of odors required for detection were similar for bees and humans, sometimes greater for the bees but sometimes less. These odors provide important cues for the location and identification of sources of nectar.

The sense of smell of insects is localized to sensory bristles in the antenna and other appendages of the head (Lee and Strausfeld, 1990; Stocker, 1994; Shields and Hildebrand, 2001; Stocker, 2001). Although the antennae of male moths are impressively large (Figure 2.14), those of the fruit fly *Drosophila* are considerably more modest. In *Drosophila* the olfactory receptor cells are contained within sensilla primarily located on the third antennal segment, the funiculus (Figure 7.5A, *a*; see also Figure 6.2A), and on a separate structure called the maxillary palp (Figure 7.5A, *p*). The funiculus has three different kinds of sensilla—the basiconic, trichoid, and coeloconic (Figure 7.5B), with the basiconic coming in two sizes, large and small (Figure 7.5C). The maxillary palp, on the other hand, seems to have only the basiconic form of sensillum. All three types of sensory hairs are similar in structure (Figure 7.6). Each has a cuticular covering penetrated by many tiny pores of the order of 10 nm in diameter (see Figure 2.15B), leading into narrow channels called tubules.

Within the sensillum, there are one to four olfactory receptor neurons, as well as several accessory

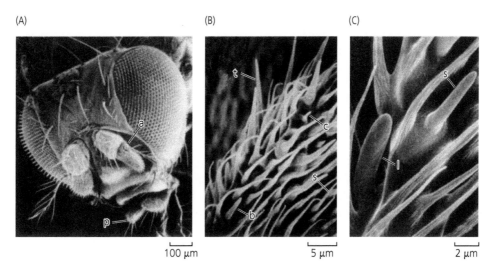

Figure 7.5 Insect olfaction: the olfactory bristles (or sensilla) of *Drosophila*. (A) Olfactory bristles in *Drosophila* are located primarily in the funiculus of the antenna (a) and maxillary palp (p). (B) Surface of funiculus, showing three classes of sensilla: basiconic (b), coeloconic (c), and trichoid (t). Also shown are hairs that have no nerve innervation and serve no sensory function, called spinules (s). (C) Two types of basiconic sensilla, the large (l) and the small (s). (A from de Bruyne et al., 2001; B and C from Riesgo-Escovar et al., 1997.)

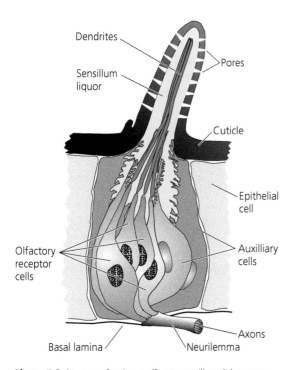

Figure 7.6 Structure of an insect olfactory sensillum. Odors enter the sensillum through pores in the wall of the bristle (Figure 2.15B), dissolve in the sensillum liquor, and bind to the dendrites of the olfactory receptor neurons.

cells not directly involved in chemoreception. The receptor neurons extend dendrites up into the bristle and are bathed in a fluid called the sensillum liquor. This fluid has a function similar to the mucous layer of the nose: odorant molecules must first dissolve in this fluid before they can be detected by the dendrites of the receptor neurons. The sensillum liquor contains small-molecular-weight, soluble proteins that bind odorants, appropriately called odorant-binding proteins. The first of these molecules to be discovered was a soluble protein binding with high affinity to the sex pheromone of the silk moth *Antheraea polyphemus* (Vogt and Riddiford, 1981). Many related proteins have since been described in other species of Lepidoptera, as well as in *Drosophila* (see Leal, 2013). Among the olfactory sensilla, there is a clear localization of different odorant-binding proteins to characteristic groups of sensory hairs, but more than one protein can be present in the same sensillum.

Odorant-binding proteins in the sensillum liquor can reach millimolar concentrations and seem likely to play some role in olfaction, though exactly what they do is far from clear. As we shall see, in fruit flies and most probably other insects as well, odorants must bind to membrane proteins called olfactory (or odorant) receptor proteins in order to produce a sensory signal. The soluble odorant-binding proteins in the sensillum liquor are not essential, because receptor proteins expressed in a heterologous cell system are capable of mediating transduction in the complete absence of binding proteins, as we show below. Nevertheless, mutant flies have been isolated lacking one of the odorant-binding proteins, called LUSH because the mutants are abnormally attracted to ethanol. These flies show defects in olfactory processing and behavior (Kim et al., 1998; Kim and Smith, 2001; Wang et al., 2001) and are completely incapable of detecting at least one kind of molecule (Xu et al., 2005). Odorant-binding proteins have been suggested to participate in the binding of the odorant to the receptor protein, or to act as carrier proteins, increasing the sensitivity of olfaction by facilitating the movement of the odorant through the sensillum liquor to the receptor cell dendrite.

Insect receptor proteins

Within the sensilla of the fruit fly antenna, there are three kinds of olfactory receptor proteins located in the membranes of the dendrites: OR (odorant receptors), IR (ionotropic receptors), and the Gr21a/Gr63a receptors responding to CO_2 and related to gustatory receptors mediating taste (see Chapter 8). The OR were first identified as seven-transmembrane proteins (Clyne et al., 1999; Vosshall et al., 1999, 2000) produced by a family of about sixty genes expressed primarily within the basiconic and trichoid sensilla. Each of the OR genes is expressed in only a limited group of sensilla, such that each receptor neuron has only one or a very few (Kaupp, 2010; Leal, 2013; Wilson, 2013). The patterns of expression within the antenna or maxillary palp for any particular gene are remarkably constant from animal to animal, suggesting that the organization of the periphery of the Drosophila olfactory system is genetically determined and fixed rather rigidly (see Vosshall, 2001). One gene, called Or83b or Orco, is much more widely expressed and functions as a co-receptor together with one or more of the OR genes, forming heteromeric complexes (Neuhaus et al., 2005) and facilitating the correct localization of the odorant receptors within the dendritic membrane (Benton et al., 2006).

Because the OR proteins have seven transmembrane sequences, they were initially thought to be G-protein receptors. They have, however, no amino acid homology to other G-protein receptors and are inserted into the membrane effectively upside-down, with the N-terminus exposed to the cytoplasm and the C-terminus extracellular. Instead of triggering a cascade leading to a change in a second messenger, the OR proteins function as novel ion channels (Sato et al., 2008; Wicher et al., 2008). Receptor responses are unaffected by knocking down or deleting all of the relevant G proteins (Yao and Carlson, 2010), and odorants can produce responses even in isolated membrane patches. In Figure 7.7, one of the OR (Or47a) together with the Orco co-receptor (Or83b) were expressed together in a *Xenopus* oocyte (as in Figure 1.9A). Outside-out patch recordings were then made of single channels before (A), during (B), and after (C) application of 300 μM pentyl acetate (PA), which is a known agonist for Or47a. Recall that in an outside-out patch, the external surface of the membrane faces the bathing solution (Figure 1.7). When pentyl acetate was perfused across the outside surface, channel openings (downward deflections) were more frequent. In part D of Figure 7.7, the current measurements were divided into bins, and the frequency of each current event was plotted on the ordinate versus the value of the current on the abscissa. The most frequent current event value was zero, when the channels were closed; but there were additional events having a current of about 1.2 pA in the presence of pentyl acetate. Further experimentation showed that the channels producing these events are non-selective cationic with a conductance of about 15–20 pS.

Figure 7.8 illustrates the recently determined cryo-EM structure of an assembled channel made entirely from the Orco receptor of a parasitic wasp. Orco by itself forms a homotetramer which, instead of six transmembrane α helices, has seven. Helix 7 (in red) is nearest the carboxyl terminus (Figure 7.8A). It is this part of the protein that forms the channel pore (Figure 7.8B) and is the most highly conserved among different OR proteins. Native channels are likely also to be tetramers, consisting of some combination of Orco and ORs. The close association of the four proteins of the channel may explain why Orco is necessary not only for channel

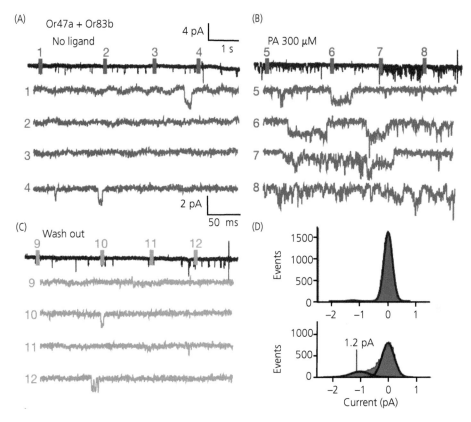

Figure 7.7 Insect OR proteins form heteromeric ligand-gated ion channels. Outside-out patch-clamp recording of Or47a/Or83b currents measured in oocyte membranes (A) before stimulation, (B) during stimulation with 300 mM pentyl acetate (PA), and (C) after removal of PA. The bottom traces of each panel indicate expansions of 300-ms current traces of single-channel recording at the positions indicated by the numbers. The data for (A–C) were all from the same cell. (D) Histograms of amplitudes of unitary current events before (blue) and during (red) application of PA from (A) and (B). Amplitude distributions were fitted with two Gaussian components (black lines). Channels gated by PA had a mean current of about 1.2 pA. (From Sato et al., 2008.)

activity but also for the assembling and trafficking of the ORs to the dendritic membrane.

A second family of insect chemoreceptors, called IR, also has sixty or so members. Only fifteen seem to participate in chemoreception, primarily in coeloconic sensilla dispersed across the antennal surface (Benton et al., 2009). Two seem more widely distributed and may function as co-receptors (Abuin et al., 2011). The membrane topology of IRs resembles that of ionotropic glutamate receptors of the CNS, with only three transmembrane sequences and a pore loop. There are large extracellular regions forming the odorant binding site, similar to the extracellular regions where glutamate binds to an NMDA or AMPA receptor (Rytz et al., 2013). The odors detected by IRs seem to be rather distinct from those detected by ORs (Silbering et al., 2011).

Most of the odorant receptors in *Drosophila* are therefore channels with odorant directly altering the rate of channel opening, much like an ionotropic synaptic receptor. There remains, however, considerable uncertainty whether this mechanism is entirely universal among insects. *Drosophila* is small and may not need to sense smell from long distances. It has wings and flies, and it may require an olfactory sense responding rapidly to changing odor concentration. Even in *Drosophila*, the antennal CO_2 receptors may not be ionotropic and may require a G protein (see Wicher et al., 2008; Yao and Carlson, 2010). In moths there is some evidence that detection of pheromones requires a metabotropic mechanism (Stengl, 2010). Metabotropic sensory transduction can have a much higher gain than ionotropic detection, because one receptor protein can trigger the opening

(A)

(B)

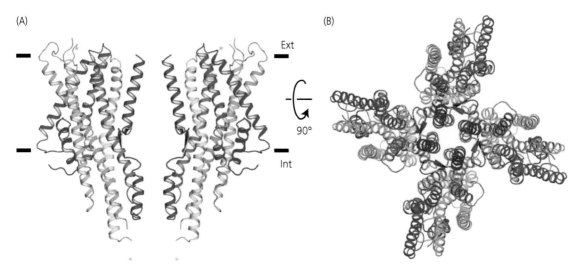

Figure 7.8 Cryo-EM structure of the insect olfactory receptor Orco. Ribbon representation from cryo-EM of Orco tetramer shown (A) from the side with only two of the subunits represented and (B) from the top. Each of the seven transmembrane α helices of the Orco subunits has been given a different color. Helix 7 (in red) forms the channel pore. Small asterisks indicate parts of the protein on extracellular and cytoplasmic sides that could not be modeled. Lines in (A) indicate membrane boundaries (Ext, external surface; Int, cytoplasmic surface). (From Butterwick et al., 2018.)

of many channels rather than just one. The detection of female sex pheromone is so sensitive (Figure 2.15C) that it would be surprising if it were mediated by direct gating of channels. Only time (and further experimentation) will tell.

Coding of olfaction in insects

Since the receptor neurons of insect sensilla are primary receptors producing action potentials which are relatively easy for physiologists to record with extracellular electrodes (see Figures 2.15C and 5.10), a lot of information is available about the response properties of these cells (see Wilson, 2013). Odorants often produce excitation and an increase in spike firing, but sometimes they generate inhibition and a decrease in action potential frequency (see Boeckh et al., 1965; Shields and Hildebrand, 2001). Some cells show excitation to one group of compounds and inhibition to another group of quite a different chemical structure.

In *Drosophila* it has been possible to make a rather complete inventory of the sensitivities of a large number of the sensilla, first in the maxillary palp (de Bruyne et al., 1999), then in the basiconic sensory hairs of the antenna (de Bruyne et al., 2001),

and later in coeloconic sensilla containing IRs (Yao et al., 2005; Abuin et al., 2011). These extraordinary studies have revealed some important principles of sensillum design. The population of sensilla of a given type (for example, the basiconic sensilla) can be separated into a defined number of subtypes, each containing olfactory receptor neurons of the same chemical specificity. Thus for the basiconic sensilla of the antenna (Figure 7.9A) there appear to be seven major types containing sixteen different kinds of receptor neurons, always in the same combinations. The *ab3* sensillum, for example, always contains one neuron giving large responses to pentyl acetate and ethyl butyrate, and a different neuron that gives a vigorous response to heptanone. These sensilla are not distributed randomly across the surface of the antenna but seem rather to occur in loosely organized clusters that are reasonably uniform from fly to fly (Figure 7.9B). The coeloconic sensilla containing IRs are similarly spread across the surface of the antenna, interspersed among the basiconic and trichoid sensilla (see Rytz et al., 2013).

Each odorant receptor protein is expressed in something like ten to a hundred olfactory receptor neurons within the sensilla of a *Drosophila* antenna. The axons of all the neurons expressing any one

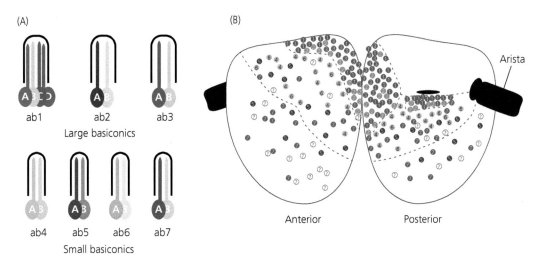

Figure 7.9 Mapping the classes of receptor cells in the olfactory sensilla of *Drosophila*. (A) Seven functional classes of basiconic sensilla of the *Drosophila* antenna as defined by the responses of their receptor neurons to odorants. (B) Map of distribution of functional types from (A). Each circle represents a recording from a single sensillum, and a question mark indicates recordings of uncertain classification. The front of the antenna is shown to the left and the back to the right; dorsal is at the top and medial is along the center of the diagram. (From de Bruyne et al., 2001.)

given receptor protein travel together in the antennal nerve to a part of the brain called the antennal lobe. There they make synapses in modules of synaptic processing called glomeruli. In addition to the axon terminals of the receptor neurons, the glomeruli contain the dendrites of interneurons and projection neurons connecting the glomeruli to higher centers in the brain. Since receptor neurons expressing the same receptor protein all converge onto the same glomeruli in the antennal lobe (Gao et al., 2000; Vosshall et al., 2000; Silbering et al., 2011), the glomeruli are specialized units of integration, designed for the processing of signals of chemically similar odorants.

Strong support for this olfactory map has come from imaging studies recording the responses of many glomeruli simultaneously. The activity of the glomeruli can be monitored by filling or genetically encoding cells in the antennal lobe with a Ca^{2+}-sensitive dye that fluoresces when the cells depolarize and fire action potentials (see Figures 6.3 and 6.6). Depolarization opens voltage-gated Ca^{2+} channels and Ca^{2+} binds to the dye, producing an optical signal that can be measured from many of the glomeruli simultaneously with a high-sensitivity, charge-coupled device (CCD) camera (see Silbering et al., 2012).

In the honey bee, whose antennal lobe contains nearly four times as many glomeruli as *Drosophila* (Figure 7.10A), odors produce activity in strongly responding glomeruli (Figures 7.10B and C) which tend to be adjacent or in close proximity to one another. The positions of the glomeruli responding to a given chemical are rather constant from individual to individual, suggesting that the synaptic connectivity of receptor neurons onto cells of the antennal lobe is also genetically determined. Because the honey bee can distinguish many more odors than it has receptor proteins or glomeruli, it uses the pattern of activity among the glomeruli to detect a particular odor. As we will see later in the chapter, the vertebrate olfactory system works in a similar way.

The olfactory code in insects is therefore generated by the specificity of the olfactory receptor proteins. Each receptor neuron expresses one or a very few of these receptor proteins, and the binding of an odorant produces excitation or inhibition, which is transmitted to specific glomeruli in the antennal lobe. The signals of the glomeruli are conveyed by projection neurons to higher centers in the brain, where they again produce specific patterns of activity among populations of neurons that seem to be different for

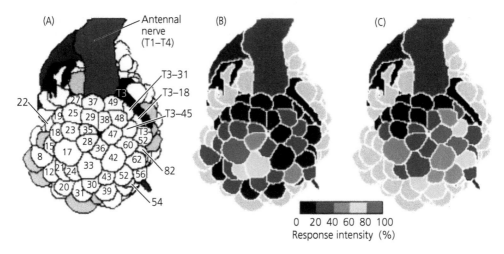

Figure 7.10 Olfactory coding in the antennal lobe of the honey bee. The surface of the brain of an anesthetized honey bee was superfused with a solution containing Calcium Green AM to fill the cells of the glomeruli with a calcium-sensitive indicator dye. The brain was then perfused with normal Ringer's solution and placed in a recording chamber on a microscope, and the antennae of the bee were stimulated with puffs of air containing odor. (A) Map of antennal lobe showing the most frequently recognized glomeruli visible on the brain surface. (B) Averaged pattern of glomerular activity to 1-octanol. (C) Averaged pattern of glomerular activity to clove oil. The scale at the bottom gives the color code for response intensity used in both parts (B) and (C). (From Galizia et al., 1999.)

different odors (Wang et al., 2001). The detection of a specific odor is produced by these patterns of activity, though how this occurs is still unknown and remains an area of active investigation.

Olfaction in vertebrates: the primary olfactory epithelium

In vertebrates, the sensory receptor cells are not dispersed in individual sensilla on an antenna but are rather placed next to one another in a continuous layer of epithelium within the interior of the nose (see DeMaria and Ngai, 2010; Kaupp, 2010). In rodents and many other mammals, there are several of these structures with somewhat different properties (Munger et al., 2009). The most important is the main or primary olfactory epithelium (Figures 1.2A and 7.11A), which is the one in our nose that we use to detect odor in the environment. It is the largest olfactory organ in mammals and has an area in humans of about 1–2 cm², containing approximately twelve million receptor cells. In a German shepherd, it is much larger with four billion cells. In addition to the receptor neurons, which are the most abundant cell type, there are supporting or sustentacular cells which, together with the glands of Bowman, secrete components of the mucus per-

haps assisting in the capturing and removal of odorants. Below the receptor cells and the sustentacular cells there are stem cells called basal cells. There are two types of basal cells, the horizontal and globose, both capable of cell division and proliferation. These cells slowly replace the sustentacular and receptor cells damaged by exposure to the environment. Finally, there is a small population of microvillar cells, which are not illustrated in Figure 7.11A.

The receptor cells in the main olfactory epithelium are bipolar neurons, having at their apex a dendritic knob, from which numerous cilia project into the mucus (Figure 7.11B). These cilia provide the membrane that contains the olfactory receptor proteins and most if not all of the other molecules directly responsible for the transduction of smell. The collective membrane surface provided by the cilia can be immense—in humans about 20 cm², but in dogs as much as 5–10 m² (50–100 ft²). Below the dendritic knob is a long process called the dendrite, followed by a swelling of the cell containing the soma and nucleus. Vertebrate olfactory receptor cells are primary sense cells with axons emerging below their nuclei. The axons of the receptor neurons join one another in bundles called fascicles to form the first cranial nerve. This nerve travels a short distance

(A)

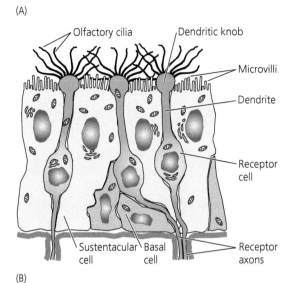

(B)

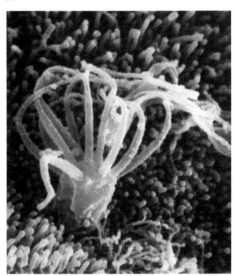

Figure 7.11 Olfactory epithelium. (A) Schematic cross-section of vertebrate primary olfactory epithelium. (B) Scanning micrograph of a dendritic knob and dendrites of a human olfactory receptor neuron. Magnification 18,500×. (B from Morrison and Costanzo, 1990.)

upward to the brain and terminates in the two olfactory bulbs, which are described later in the chapter.

Olfactory receptor proteins

The sensation of odor begins with the dissolving of air-borne chemicals into the mucus. The mucus contains several different kinds of odorant-binding proteins, quite different in structure from those in the sensillum liquor of insect sensilla (see Ronnett and Moon, 2002) but again probably assisting in the delivery of some odorants to the receptor proteins or in their subsequent removal. After diffusion through the mucus, the odorants bind to a receptor protein (see Mombaerts, 1999a; Buck, 2000; Touhara and Vosshall, 2009; Kaupp, 2010). Most of the receptor proteins expressed in the main olfactory epithelium are members of a large gene family first discovered by Buck and Axel (1991), who supposed that the proteins responsible for smell would be G-protein-coupled receptors specifically expressed in the olfactory epithelium. They used the polymerase chain reaction (PCR) to amplify RNA extracted from the nose of rats and identified a family of genes of G-protein-coupled receptors. They cloned and characterized eighteen members of this family and estimated that the entire olfactory epithelium may use hundreds of such genes. One receptor protein (called *I7*) was subsequently expressed within the rat olfactory epithelium in a large number of the receptor neurons by infection with an adenoviral vector engineered to contain the *I7* gene together with green fluorescent protein (GFP) (see also Krautwurst et al., 1998; Zhao et al., 1998). Responses were then recorded from cells that fluoresced with GFP, indicating that they had been successfully infected. In this way, a member of this gene family was shown to encode a functional olfactory receptor, which generates an electrical response to a restricted group of chemical compounds.

We now have a reasonably complete idea for many species of the total number of genes of the olfactory (or odorant) receptor family, which for simplicity we will refer to as vertebrate *ORs*. Mice have about 1300 *ORs*, of which about 300 appear to be pseudogenes, that is genes containing one or more disruptions in sequence such as insertions, deletions, and premature stop codons preventing expression of a functional protein. This leaves approximately 1000 functional genes, occurring in clusters on almost every chromosome. In the human, there are about 900 *OR* genes with a far greater percentage of pseudogenes, so that only about 350 sequences appear to be able to produce functional proteins. Human *ORs* are again found in clusters of variable size on nearly every chromosome. Fish have a sense of smell, as Aristotle first surmised (see Chapter 1), which has an important role in fish behavior. Fish smell water-borne odorants, which bind to a family of *ORs* much smaller than for mammals with probably no more than a hundred members (Ngai et al., 1993; Niimura and Nei, 2005). Receptors related to those in fish can also be found in the frog *Xenopus*, but this organism also has other *ORs* more closely related to those of mammals (Freitag et al., 1995; Niimura and Nei, 2005).

In addition to this large family of *ORs*, some cells of the main olfactory epithelium express a protein from a different family of G-protein receptors, called *trace amine-associated receptors* (TAARs) (Liberles and Buck, 2006). These proteins recognize small amines like those found in urine and are more closely related to serotonin and dopamine receptors in the brain than to ORs. There are fifteen TAARs in mouse and six in humans. Mice are generally repelled by the smell of amines in urine, particularly the urine of predators, and this aversion is entirely removed when the *TAAR* genes are deleted (Dewan et al., 2013).

A considerable body of evidence indicates that each olfactory receptor neuron probably expresses only a single *OR* or *TAAR* gene. A cell that expresses a *TAAR* does not also co-express an *OR* (Liberles and Buck, 2006). Once the receptor cell decides which protein to express, the rest of the receptor genes are silenced (see Dalton and Lomvardas, 2015). In rodents, expression of *ORs* is mostly confined to one of four broad zones of epithelium and is then random within this zone (see Mombaerts, 1999a). The reason for this arrangement is not known but may be related to the gradual increase in both the size of the nasal epithelium and the diversity of receptor proteins during the evolution of terrestrial vertebrates. Within the receptor cell itself, functional receptor protein is located mostly on the membrane of the cilia, uniformly distributed from the dendritic knob out to the ciliary apex (see for example Lowe and Gold, 1991).

The mechanism of transduction

Both the ORs and the TAARs are G-protein-coupled receptors and transduce odor in the same way (Figure 7.12). The activation of the receptor protein is coupled to a specific G protein with an α subunit called $G\alpha_{olf}$ (Jones and Reed, 1989). Transgenic mice lacking $G\alpha_{olf}$ are largely unable to respond to odor and in most cases die at a young age, since they are unable to nurse (Belluscio et al., 1998). Receptor cells are present in $G\alpha_{olf}$-deletion animals with receptor proteins and many of the other transduction enzymes. The receptor cell axons even make seemingly normal connections to the olfactory bulbs in the CNS, but neither the receptor neurons nor the cells in the bulb show much of a response to an odor.

Activation of odorant receptor causes an exchange of GTP for GDP on the guanosine nucleotide binding site of $G\alpha_{olf}$ (Figure 7.12), and $G\alpha_{olf} \bullet GTP$ then stimulates a specialized adenylyl cyclase called adenylyl cyclase 3 (AC3), which is highly expressed in olfactory epithelium (Bakalyar and Reed, 1990). This protein again seems to be absolutely necessary for most olfactory transduction, since transgenic animals lacking AC3 give no electrical response to odorant stimuli and are largely unable to discriminate common odors in behavioral tests that normal mice are able to perform with little difficulty (Wong et al., 2000).

Even before the role of AC3 was discovered, odors were known to produce an increase in cAMP from biochemical measurements made with isolated olfactory epithelium (Pace et al., 1985; Sklar et al., 1986). The increase in cAMP within the cytoplasm of the olfactory receptor cilium causes an increase in the conductance of the plasma membrane by binding to cyclic-nucleotide-gated channels (see Chapter 4). This result was first obtained by Nakamura and Gold (1987b), who placed a patch pipette on the cilium of an isolated olfactory receptor cell and pulled off an inside-out patch, much as Fesenko et al. (1985) had done for rod outer segments (Figure 4.10A). Perfusion of the cytosolic side of the patch with either cAMP or cGMP produced an increase in a non-selective cationic conductance. Subsequently, Firestein et al. (1991) showed that single-channel responses to odors could be detected in on-cell recordings from intact olfactory receptor cells (Figure 7.13), and these channel openings were identical to those produced when the cyclic-nucleotide concentration in the cell was increased.

In rodents, three different cyclic-nucleotide-gated channel subunits are expressed in the main olfactory epithelium, called CNGA2, CNGA4, and CNGB1b. All three assemble to produce a functional channel,

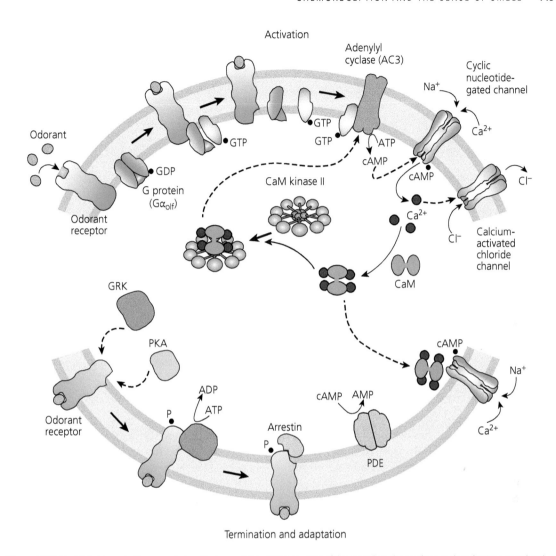

Figure 7.12 Postulated transduction cascade for olfactory receptor cells. Activation of the cascade is shown above and mechanisms postulated for termination and adaptation are shown below. See text.

which is tetrameric (Figure 4.11). The CNGA2 subunit is absolutely required for channel function, because transgenic animals lacking this protein behave much like animals lacking adenylyl cyclase: responses to odor are almost entirely absent from the main olfactory epithelium (Brunet et al., 1996). The other two channel subunits substantially alter the properties of the fully assembled channel, for example its cAMP affinity and single-channel conductance (see for example Bonigk et al., 1999; Shapiro and Zagotta, 2000).

When cAMP binds to the cyclic-nucleotide-gated channels, the channels open. Channel opening then produces an increase in Na^+ conductance and depolarization of the membrane potential, leading to an increase in the frequency of action potentials in the olfactory receptor-cell axon. In addition to Na^+ and other monovalent cations, the cyclic-nucleotide-gated channels have a surprisingly large permeability to Ca^{2+}. This has the consequence that odorants produce a large influx of Ca^{2+} into the cilium, which can be detected as an increase in

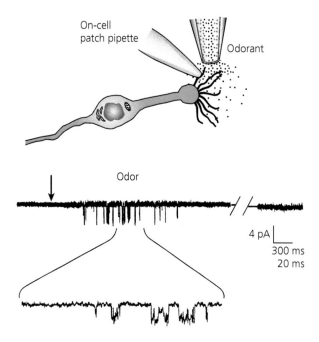

Figure 7.13 Odor gates single-channel opening in a membrane patch of an olfactory receptor neuron. A pulse of odor 150 ms in duration (beginning at arrow) was puffed onto a single olfactory receptor neuron isolated from the nasal epithelium of a salamander. Single channels were recorded with on-cell patch clamp. The record below shows activity at higher time resolution. (From Firestein et al., 1991.)

fluorescence if the cell has been filled with a Ca²⁺-indicator dye (Figure 7.14A). The Ca²⁺ concentration rises with about the same time course as the inward current produced by the opening of the channels (Figure 7.14B), and its amplitude increases almost proportionately; however, the decline of Ca²⁺ is noticeably slower than the current—compare the amplitude of the two signals in Figure 7.14 at 4.5 s, the time indicated by the dashed lines. This difference probably occurs because Ca²⁺ is only gradually removed from the cilium, largely by Na⁺-dependent transport (Reisert and Matthews, 1998). This transport is known to be produced in mouse primarily by the membrane protein NCKX4 (Stephan et al., 2012), which exchanges 4 Na⁺ ions from outside the cell for one K⁺ and one Ca²⁺ inside.

This increase in intracellular Ca²⁺ is significant, because Ca²⁺ can also act as a second messenger in transduction in addition to cAMP to provide additional amplification of the olfactory signal. This amplification is important because, as we will see later in the chapter, our sense of smell is primarily combinatorial. Like insects, we can detect many more odors than we have olfactory receptors, and we accomplish this feat by combining the signals of several olfactory receptors with different affinities much like the honey bee for octanol or clove oil (Figure 7.10). For this mechanism to work, most olfactory receptors cannot be rigidly selective for single odorants but must have a rather low affinity for compounds at their binding sites. As a result, odorants are unlikely to bind much longer than a millisecond and cannot in general produce a very large or long-duration change in cAMP (Bhandawat et al., 2005). To boost the amplitude of the response, olfactory cilia contain a chloride channel gated by Ca²⁺ called Ano2 (Anoctamin 2, also TMEM16B; see Stephan et al., 2009). The opening of the cyclic-nucleotide-gated channels produces an influx of Ca²⁺, which increases the intracellular free-Ca²⁺ concentration and gates the opening of Ca²⁺-activated Cl⁻ channels. The olfactory neurons have a high internal concentration of Cl⁻ primarily from Na⁺- and K⁺-dependent Cl⁻ transport (Reisert et al., 2005), and the reversal potential for the Cl⁻ conductance is therefore near zero mV. As a consequence, the opening of

(A) Calcium

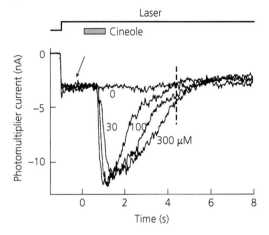

(B) Current

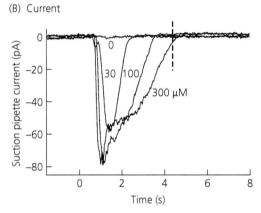

Figure 7.14 Odors produce an increase in intracellular Ca²⁺ concentration. Simultaneous recording of calcium dye fluorescence (A) and current (B) from a salamander olfactory receptor neuron to the odorant cineole at the indicated concentrations. Calcium was recorded by loading the neuron with the calcium-sensitive fluorescent dye fluo-3 and then illuminating the cilia with light from a laser. Fluorescence in absence of odor (arrow in A) was mostly produced by the resting Ca²⁺ concentration of the cell. Current in B was recorded from the cell body with a suction electrode. Dashed lines indicate values of Ca²⁺ and current at a time of 4.5 s after the beginning of stimulation with cineole. (From Reisert and Matthews, 2001.)

the Cl⁻ channels and increase in permeability to Cl⁻ further depolarizes the membrane potential.

The Cl⁻ conductance can make an important contribution to the response of an odorant, as can be seen in the recordings of Figure 7.15. In these experiments (Li et al., 2016), the cell body of an iso-

lated frog olfactory receptor neuron was drawn into a suction pipette with the cilia exposed to the external solution. The cell was stimulated by perfusion with the odorant (+)-limonene, which is the major component of the oil of citrus-fruit peel. The total current is shown for each of the odorant concentrations as the black traces. The contribution of the cyclic-nucleotide-gated channels to the current (shown in red) was assessed by blocking the Cl⁻ current with the compound niflumic acid. The remainder, which is due to Cl⁻ efflux, is shown in green. At all but the highest odorant concentrations, the current produced by the Cl⁻ conductance is responsible for most of the odorant response. Similar experiments have been done for mouse olfactory receptors (Li et al., 2018).

The opening of the Cl⁻ channels by Ca²⁺ has a number of useful consequences (see Reisert and Reingruber, 2018). First, as Figure 7.15 shows, it boosts the total current produced by an odorant, increasing the response particularly for small signals and therefore increasing the sensitivity of the receptor neuron. Second, the Cl⁻ channels improve the stability of the odorant response and make it possible for the cell to respond when the extracellular Na⁺ concentration changes or becomes very low, for example in fresh-water fish or in mammals following dilution of the mucus with water after swimming. Provided some Ca²⁺ enters the cell, the Cl⁻ channels can open and Cl⁻ can flow outward even when very little Na⁺ is available to flow inward. Finally, the Cl⁻ current increases the duration of the response. This effect occurs because the Cl⁻ channels remain open as long as Ca²⁺ is available to gate them. Once the cyclic-nucleotide-gated channels close, Ca²⁺ ceases to flow into the cell, but the Cl⁻ channels continue to remain open as Ca²⁺ is slowly removed from the cilium (Figure 7.14). The importance of the Cl⁻ current has sometimes been called into question, because the knocking out of the gene for the Cl⁻ channel Ano2 seems to have little effect on olfactory behavior (Billig et al., 2011). This result may, however, be more an indication of the difficulty of doing behavioral measurements on mouse. Recent results like those shown in Figure 7.15 (see also Li et al., 2018) demonstrate the indisputable importance of the Cl⁻ current for amplification of the odorant response.

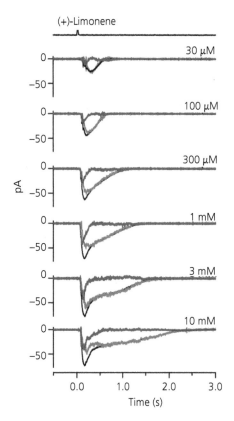

Figure 7.15 Contribution of Ca²⁺-activated Cl⁻ channels to response to odor. Currents were recorded from isolated frog olfactory neurons, as in Figure 7.14B, with a suction electrode. Traces show total current (black) at different (+)-limonene concentrations. The component of current produced by the cyclic-nucleotide-gated channels (red) was determined by blocking the Ca²⁺-activated Cl⁻ channels with niflumic acid. The red traces were then subtracted from the black traces to give the component of current produced by the Ca²⁺-activated Cl⁻ channels (green). The total current is dominated by Cl⁻ current at all but the highest odorant concentrations. (From Li et al., 2016.)

Desensitization and adaptation

The response even to the most potent odor fades with time. The smell of Chanel No. 5 is greatest at the beginning, but with continued exposure we—like bacteria swimming in a pool of attractant—will adapt and no longer detect the perfume. A new wave of sensation can, however, still occur if we draw closer to the source of the odor or its concentration is suddenly increased.

The lower half of Figure 7.12 summarizes schematically the numerous mechanisms that have been

proposed to regulate response termination and adaptation (Kaupp, 2010; Reisert and Zhao, 2011). Part of the decrease in sensitivity during adaptation may be produced by desensitization of the receptor proteins themselves. Olfactory receptor cells contain both a β-adrenergic receptor kinase (βARK) and a β-arrestin (Dawson et al., 1993), and both are capable of inactivating the receptor proteins. The kinase first phosphorylates the receptor, and arrestin then binds and prevents any further activation of G proteins (Figure 4.1). We will see in Chapter 9 that receptor phosphorylation provides an important mechanism of receptor desensitization and inactivation for the visual protein rhodopsin. This may not be true of olfactory receptor proteins, however, because odorant binding affinity is in general rather low. Binding of odorant to an olfactory receptor is probably too brief in most cases to promote phosphorylation and arrestin binding (Bhandawat et al., 2005). Receptor desensitization may therefore be utilized only when odorants are presented at high concentration for a long duration.

There are several other mechanisms that have been suggested to produce adaptation, each employing a different molecular scheme. They all share one common feature: all seem to require Ca²⁺. Part of the evidence of a role for Ca²⁺ comes from experiments like the one shown in Figure 7.16. A long exposure to an odorant (indicated by the upper trace filled with blue) was given to an olfactory receptor neuron whose response was recorded with whole-cell patch clamp. The response labeled "Control" shows that the inward current of the receptor neuron normally adapts during the maintained presence of the odorant. If, however, the cell contains the Ca²⁺ buffer BAPTA to slow the change in Ca²⁺ concentration, the recovery of the response is greatly prolonged and adaptation is mostly prevented. A similar effect is produced by knocking out the NCKX4 exchanger to retard Ca²⁺ extrusion from the cell (Stephan et al., 2012).

Both cyclase and phosphodiesterase activity can be modulated by Ca²⁺. Ca²⁺ binds to calmodulin and activates CaM kinase II, which is highly expressed in olfactory receptor neurons and can phosphorylate AC3, reducing its activity (Wei et al., 1998). Inhibition of CaM kinase II can prolong recovery (Leinders-Zufall et al., 1999), but adaptation to brief

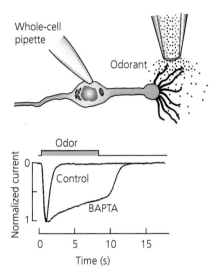

Figure 7.16 Buffering Ca²⁺ slows the time course of adaptation. Isolated olfactory neurons from a salamander were voltage clamped, and the current was recorded to prolonged exposure to the odorant n-amyl acetate (the trace labeled "Control"). The trace labeled "BAPTA" was from a cell also exposed to odorant whose cytoplasm was preloaded with the Ca²⁺ buffer BAPTA. (From Leinders-Zufall et al., 1999.)

stimuli is unaffected, and the role of cyclase regulation is still unclear (Reisert and Zhao, 2011). Olfactory receptor neurons have two phosphodiesterases: PDE1C in the cilia regulated by Ca²⁺-calmodulin (Yan et al., 1995), and PDE4A which is insensitive to Ca²⁺-calmodulin and is expressed in the cell body including the dendritic knob but not in the cilia (Cherry and Davis, 1995; Juilfs et al., 1997). Knocking out the gene for the ciliary Ca²⁺-sensitive PDE has almost no effect on the waveform of the response to odors, indicating that Ca²⁺ regulation of PDE1C is unlikely to play a significant role in termination or adaptation. When, however, both PDE genes are knocked out, response recovery is significantly prolonged, though there is little effect on adaptation to successive stimuli (Cygnar and Zhao, 2009). Other evidence suggests that phosphodiesterase activity, though necessary for hydrolysis of cAMP and termination of the response, is not an important component of the mechanism of adaptation (Boccaccio et al., 2006).

This brings us to the cyclic-nucleotide-gated channels. In an important series of experiments, Kurahashi and Menini (1997) showed that a brief increase

in cAMP concentration could adapt the neuron to a second increase in cAMP concentration in a manner that was indistinguishable from adaptation produced by stimulation with an odor. If the receptor neuron can adapt to an increase in cAMP just as it does to an odor, then adaptation is most likely produced at the place where the cAMP is acting, namely the cyclic-nucleotide-gated channels. These experiments seemed initially compelling because Chen and Yau (1994) showed that the effective affinity of the olfactory cyclic-nucleotide-gated channel for cAMP is reduced by Ca²⁺/calmodulin. A Ca²⁺-dependent decrease in the gating of the channel by cAMP would produce a decrease in responsiveness to repeated stimulation and perfectly explain the Kurahashi and Menini experiment.

This mechanism was then tested by genetically deleting the binding site for Ca²⁺/calmodulin on the olfactory channel. Response decay was modestly prolonged, but there was no effect on adaptation to successive stimuli (Song et al., 2008). Either modulation of the channel is not responsible for adaptation or the channels are regulated by some unknown pathway that does not involve Ca²⁺/calmodulin. Adaptation of olfactory receptor neurons is seemingly under the control of a change in free-Ca²⁺ concentration, and several Ca²⁺-dependent mechanisms have been described; but we still do not know which of these mechanisms (if any) is actually responsible for adjusting sensitivity in the presence of a maintained odor.

Coding in the principal olfactory epithelium

One of the first successful attempts to understand how olfactory receptor cells code for individual odors was made by Firestein et al. (1993), who recorded responses of salamander olfactory neurons. In the recordings shown in Figure 7.17, three cells were each perfused with the same rather high concentration of three odorants. These different cells, presumably containing different olfactory receptor proteins, showed a range of responses. Some cells responded to only one or two of the compounds and some to all three, each with different sensitivities. The amino acid composition of the receptor proteins is most variable in the membrane-spanning regions, which probably form the odorant binding sites (see Mombaerts, 2004). Each of the different receptor proteins with different binding sites would be expected to have a somewhat different selectivity, in some cases highly specific for chemical compounds of a certain structure, but in others cases binding a larger range of molecular conformations.

More information about the coding of odorant sensitivity has come from three kinds of experiments (see Mombaerts, 2004; Malnic, 2007; Malnic et al., 2010). In the first, one odorant-receptor protein along with GFP was expressed in a large number of receptor cells by adenoviral infection (Zhao et al., 1998), and the responses of the neurons expressing that protein were then studied in detail (Araneda

et al., 2000). Alternatively, a transgenic mouse was made with a gene for one of the receptor proteins linked to expression of GFP (Bozza et al., 2002; Grosmaitre et al., 2006). Experiments of this kind for the receptor protein I7 show that it binds aliphatic aldehydes with carbon backbones of seven to ten carbons, with the largest response to octanal. Replacement of the aldehyde with a carboxylic acid or an alcohol group significantly diminished the response.

A second way to investigate the selectivity of receptor proteins is to record the response of neurons to a spectrum of odors *in situ*, either with whole-cell recording or with a fluorescent dye to measure the increase in free-Ca^{2+} concentration produced by excitation. Single-cell PCR can then be performed to determine the sequence of the receptor

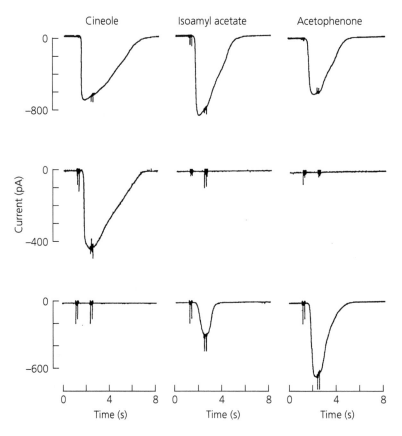

Figure 7.17 Odorant coding in single olfactory receptor neurons. Whole-cell patch recordings were made from dissociated salamander olfactory receptor neurons at a holding potential of −55 mV. Responses shown are to the odorants cineole, isoamyl acetate, and acetophenone, perfused onto the receptors at concentrations of 0.5 mM. The duration of odorant exposure was 1.2 s, indicated by the downward-going electrical artifacts produced by the perfusion system at the beginning and end of the stimuli. Each row shows a response from the same cell for each of the three odorants. (From Firestein et al., 1993.)

protein, which can be correlated with the odor sensitivity of the cell (Malnic et al., 1999). Alternatively, the receptor gene can be cloned from a dissociated cell responding to odorant, and the odor sensitivity of the receptor protein can be studied in detail after expressing the gene with a viral vector back into the olfactory epithelium (Touhara et al., 1999). Much of interest about odorant receptor protein specificity has also been learned from large-scale screening of Ca^{2+} responses in dissociated olfactory neurons without sequencing of the receptor proteins (Nara et al., 2011).

A third method avoids electrophysiology and expresses olfactory receptor proteins in a cultured cell line such as HEK293 cells. The receptor proteins are difficult to express, because they are often retained in the endoplasmic reticulum and degraded. One way to circumvent this difficulty has been to fuse most of the receptor protein to the amino-terminal sequence of rhodopsin, a protein that can be targeted more successfully to the plasma membrane (Krautwurst et al., 1998; Kajiya et al., 2001). Even better results can be obtained if the receptor proteins are expressed along with *receptor-transporting proteins* (RTP1 and RTP2) and/or *receptor expression enhancing protein* (REEP1), which greatly facilitate the incorporation of ORs into the plasma membrane. Responses can then be recorded by measuring changes in cAMP (Saito et al., 2004, 2009; Keller et al., 2007).

Some general features of the olfactory code have been learned from these studies. The majority of receptor proteins are like I7, narrowly tuned to a group of chemicals of similar shape and with similar functional groups. A small proportion seem, however, to be much more broadly tuned. Some of the narrowly tuned receptor proteins respond highly specifically to stimuli important to the survival of the animal, such as cadavarine, which is the smell of rotting flesh (Nara et al., 2011). Most odors bind, however, to many receptor proteins, though with varying sensitivity. As a result, different odors bind to different combinations of receptor proteins, and it is probably this combinatorial response that provides the code we use to identify a distinct smell.

The olfactory bulb

The axons of the olfactory receptor neurons terminate in the CNS in bilaterally symmetric olfactory bulbs.

Like the antennal lobes of insects, olfactory bulbs contain centers of neural processing called glomeruli, which are globular clusters about 100 μm in diameter where the axons of the receptor cells synapse onto neurons of the CNS (see Mori et al., 2006; Johnson and Leon, 2007; Murthy, 2011). All of the receptor neurons expressing any given OR exit the olfactory epithelium and converge in most cases onto only two glomeruli in each bulb. A single glomerulus may receive input from many thousands of such neurons (see Mombaerts, 1999b). The convergence of receptor axons onto the glomeruli has been visualized either with *in situ* hybridization (Ressler et al., 1994; Vassar et al., 1994) or by linking the expression of a receptor protein to an intracellular marker such as β-galactosidase (Mombaerts et al., 1996), GFP (Potter et al., 2001), or a trans-synaptic marker that can be detected by binding to an antibody (Zou et al., 2001).

A typical result is given in Figure 7.18. Part A shows the olfactory epithelium of a mouse stained with the blue X-gal compound produced by β-galactosidase, which in this case had been linked to an OR gene called *P2* (from Mombaerts et al., 1996). Axons originated from many neurons expressing this gene, seemingly randomly distributed across one of the four zones of olfactory epithelium. These axons converged onto two of the 1800 or so glomeruli in each of the two olfactory bulbs (Figure 7.18B), one on the medial side and the other on the lateral side. Figure 7.18C (from Potter et al., 2001) shows axons in a different experiment from neurons expressing the *M72* gene converging onto one lateral and one medial glomerulus in the left and right olfactory bulbs, as shown in the diagram of Figure 7.18D. The stereo pair in Figure 7.18E shows *M72*-labeled axons and a single glomerulus at higher power.

The picture that emerges is that axons from neurons expressing one of the 1000 or so genes for receptor proteins in mouse converge usually onto two (but occasionally onto just one) of the 1800 glomeruli in each of the two bulbs. The positions of the glomeruli for any one receptor protein are remarkably constant in the left and right bulbs for a single mouse, and they are just as reliably positioned in different mice, with an error on average of about the width of a single glomerulus (Soucy et al., 2009). This remarkably precise placement of the glomeruli

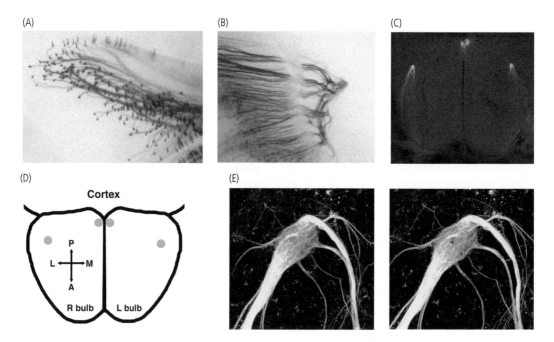

Figure 7.18 Specificity of connections of olfactory receptor neurons with glomeruli in the olfactory bulb. (A, B) Olfactory receptors expressing the gene *P2* and labeled with X-gal. Individual neurons randomly distributed in the epithelium converge onto just a few glomeruli in the bulb. (C, D) Olfactory receptors expressing the gene *M72* were engineered also to express GFP. Two glomeruli are stained in each olfactory bulb: one lateral and one medial. Orientation indicated in (D): P, posterior; L, lateral; M, medial; A, anterior. R bulb and L bulb are right and left olfactory bulbs. (E) Stereo pair of glomerulus for *M72* at a higher power. Magnification 200×. (A and B from Mombaerts et al., 1996; C–E from Potter et al., 2001.)

seems to be determined in part by gradients of signaling proteins during development, by the position of the relevant olfactory receptor neurons in the four zones in the epithelium, and by the chromosomal locus of the gene for the OR (see Murthy, 2011). The OR itself seems also to have a role (Feinstein and Mombaerts, 2004), particularly when individual olfactory receptor neurons are replaced by dividing basal cells as the epithelium is renewed during the life of the organism (Gogos et al., 2000).

Does the map of the glomeruli tell us anything about the processing of olfactory signals? One clue may have come from a systematic investigation of the sensitivity of different glomeruli to a large number of different odorants, either from microscopic methods marking active neurons or by imaging techniques on the surface of the olfactory bulb, much as for the antennal lobe of the honey bee (see Figure 7.10). The upper surface of the bulb can be exposed, and the activity of either the receptor axon terminals or the cells in the glomeruli can be monitored with a

Ca^{2+} indicator dye or with intrinsic optical signals produced by cell activity. These studies have revealed systematic changes in the sensitivity of glomeruli according to the chemical nature of the odor.

A demonstration of this approach is given in Figure 7.19. Part A illustrates the method. The picture to the left shows a television image of the dorsal surface of one of the olfactory bulbs of a rat, containing about 10 percent of the glomeruli and with prominent blood vessels running along the surface. When the cells within the image were excited, there were subtle changes in this video image, produced by increased oxygen consumption and changes in hemoglobin absorption, as well as by small changes in cell volume which altered the scattering of the light illuminating the surface of the bulb. These changes could be exploited in the following way. The rat was stimulated by an odor, and the optical image without the odor was subtracted *pixel by pixel* from the image in the presence of the odor. The difference in the optical image can then show which cells responded. The image to the right in Figure 7.19A gives the result of this subtraction when a rat was stimulated with the smell of peanut butter. Two well-defined glomeruli, now appearing as dark areas, were excited among those present in this part of the olfactory bulb. One was in the upper left, and one was in the lower right.

This method can be used to make maps of glomeruli responding to chemicals having related structures. Glomeruli stimulated by aliphatic acids (Figure 7.19B), for example, are all located in the anteromedial part of the bulb but change position as the length of the carbon chain of the molecule is increased. Aliphatic alcohols (Figure 7.19C), on the other hand, are clustered more laterally, with again a similar dispersion as the carbon-chain length is altered. These patterns of stimulation are to some extent a function of the concentration of the odor, since, for most odors, the stronger the concentration, the larger the number of different glomeruli that are activated.

Many investigations of this kind have revealed clusters of glomeruli responding to acids, alcohols, aldehydes, ethers, ketones, and hydrocarbons, in some cases systematically arrayed as in Figure 7.19 according to the number of carbon atoms and length

of the aliphatic chain (Mori et al., 2006; Johnson and Leon, 2007). It is possible that glomeruli responding to similar odors are placed in contiguity to facilitate the integration of odor signals. This arrangement may, for example, allow inhibitory interneurons synapsing on adjacent glomeruli to enhance differences in responses to similar chemicals. The difficulty with this notion is that maps like those in Figure 7.19 are actually rather coarse. Glomeruli responding to similar compounds are often found interspersed among other glomeruli responding to quite different chemicals (Soucy et al., 2009). It is difficult to understand how such a seemingly random arrangement would facilitate odor processing.

Perhaps the placement of glomeruli is simply a result of the evolution of the olfactory system. As OR genes were duplicated, often producing responses to similar compounds, their respective olfactory receptor neurons may have projected to similar places in the bulb. In some cases, single amino acid mutations may have produced rather larger changes in chemical selectivity, which could account for the interspersed glomeruli (Murthy, 2011). That is not to say that the arrangement of glomeruli within the bulb has no influence on odor detection, just that, at present, we have no idea what this influence might be.

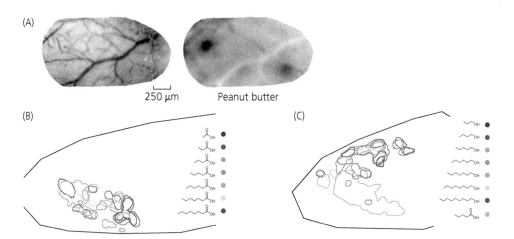

Figure 7.19 Intrinsic optical imagining of the olfactory bulb. (A) The surface of the olfactory bulb from an anesthetized rat (left). Note the pattern of blood vessels. Subtracted image showing response to the smell of peanut butter (right). (B, C) Maps of the surface of the rat olfactory bulb constructed with the same method as in (A). Maps show areas stimulated by aliphatic acids (B) and alcohols (C) of different chain length, whose chemical formulas are given to the right. (A from Rubin and Katz, 1999; B and C from Uchida et al., 2000.)

The accessory olfactory system and vomeronasal organ

Our sense of smell is largely produced by the system just described, consisting of sensory neurons expressing single ORs and synapsing together onto specific glomeruli in the main olfactory bulbs. The code is combinatorial. Some ORs are quite selective and others rather less so (Nara et al., 2011), such that single chemicals generally stimulate more than one OR and several glomeruli. The combination of signals from the different glomeruli seems to provide our perceptions of particular odors. A code designed in this way can detect many more chemicals than ORs and can respond to odors we have never before experienced, such as those of a novel organic synthesis or a newly marketed perfume.

In addition to this system, there are others that seem to be mostly responsible for the high-sensitivity detection of specific chemicals that have some role in affecting behavior, such as sexual determination, kin recognition, mating, suckling, and instinctive fear or aggression (see Liberles, 2014). These chemicals are called pheromones. I have already mentioned the TAARs, which are expressed within the main olfactory epithelium, and which sense amines in urine and mediate interspecific aversion. The main olfactory epithelium also contains a small number of receptor neurons that do not contain $G\alpha_{olf}$, AC3, or the cyclic-nucleotide-gated channel subunits CNGA2 or CNGB1b. These cells instead express guanylyl cyclase D, which is a member of the family of membrane-bound guanylyl cyclases (see Figure 4.2). These proteins have an extracellular binding domain that can act as a receptor, a single transmembrane domain, and a C-terminal domain that catalyzes the synthesis of cGMP. Cells expressing guanylyl cyclase D also express the cyclic-nucleotide-gated channel subunit CNGA3, which binds cGMP with higher affinity than cAMP. These receptor neurons project to a small group of glomeruli in the caudal region of both of the main olfactory bulbs, called the necklace glomeruli. The necklace glomeruli may contribute to the sensing of peptide components of the urine (Leinders-Zufall et al., 2007) and/or the detection of CO_2 (Hu et al., 2007).

In addition to the main olfactory epithelium, the nasal cavity of many mammals and other vertebrates contains other clusters of olfactory neurons, of which the largest and most important is called Jacobson's organ, or more usually the vomeronasal organ (Munger et al., 2009). In many mammals (though apparently not in us), this organ is located below the main olfactory epithelium just underneath the vomer bone of the nose and forms a self-enclosed pouch normally isolated from the air entering the main nasal cavity. Arousal can produce coordinated vasodilation and constriction of the surrounding tissue, effectively pumping air through a duct into the organ. When a horse snorts in the presence of mares in heat by extending its neck and curling its lip to expose its teeth, it is sending air preferentially into the vomeronasal organ. In snakes and lizards, pheromones are delivered to the vomeronasal organ by the tongue. Chemicals adhering to the tongue are deposited near the openings of the vomeronasal organ after tongue retraction, and the forking of the tongue may provide two-point sampling for the detection of gradients of pheromone concentration (Schwenk, 1994).

The vomeronasal organ contains an epithelium with bipolar neurons similar in morphology to the olfactory receptor cells of the main epithelium, but the sensory membrane is elaborated by microvilli rather than by cilia, and it is bathed in a fluid analogous to the mucus but secreted by the vomeronasal glands (see Touhara and Vosshall, 2009; Stowers and Kuo, 2015; Silva and Antunes, 2017). The receptor cells of the epithelium are nearly equally divided into two kinds (Figure 7.20A): those whose nuclei are located apically (closer to the sensory surface formed by the microvilli), containing a large family of G-protein receptors called the V1Rs, and expressing the G-protein subunit $G\alpha_{i2}$; and those whose nuclei are located more proximally, containing the G-protein-coupled but otherwise unrelated family of V2R receptor proteins and expressing the G-protein subunit $G\alpha_o$. Neither group of cells expresses the G-protein subunit $G\alpha_{olf}$ of the main olfactory epithelium.

The *V1R* family of receptor genes was discovered by creating cDNA libraries for individual, isolated vomeronasal neurons and looking for genes that were highly expressed in only a small fraction of the cells (Dulac and Axel, 1995). The *V1R* gene family has about 300 members, of which nearly 200 can

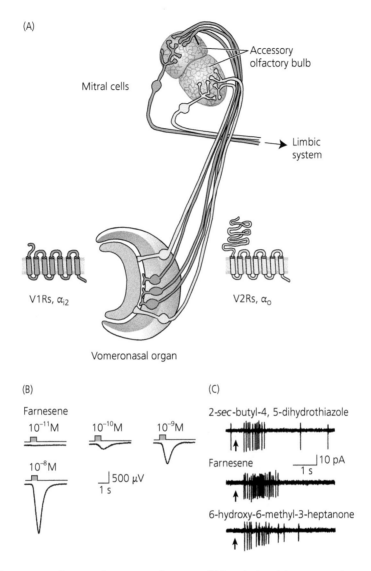

Figure 7.20 Mammalian vomeronasal organ and responses to pheromones. (A) Organization of the vomeronasal organ. See text. (B) Extracellular voltage change recorded from the surface of a mouse vomeronasal organ produced by application of a mixture of the pheromones E,E-α-farnesene and E-β-farnesene at four different concentrations. (C) Action potentials recorded from single neurons to three putative pheromones (all at a concentration of 10^{-7} M) in a slice of vomeronasal organ. (A after Pantages and Dulac, 2000; B and C from Leinders-Zufall et al., 2000.)

form functional proteins (Rodriguez et al., 2002). Although highly divergent in amino acid sequence, the proteins of this superfamily all share a similar structure, with a short amino-terminal sequence and a binding pocket probably located within the transmembrane sequences, much as for ORs of the main epithelium.

The more than a hundred functional V2R receptors are an entirely different group of proteins (Herrada and Dulac, 1997; Matsunami and Buck, 1997; Ryba and Tirindelli, 1997). They all have a large amino-terminal sequence that is extracellular and probably contains the odorant binding site, much like metabotropic GABA and glutamate receptors in the brain. This difference in structure would seem to suggest that the chemicals these receptors detect are quite different from those that bind to the V1R proteins. It is often suggested that the V1Rs bind small, volatile compounds (for example Boschat et al., 2002) and the V2R bind water-soluble phero-mones, including for example proteins in urine (Chamero et al., 2007). Although some difference in receptor affinity would be expected from the differ-ence in receptor structure, the agonists for only a very few of the V1Rs or V2Rs have been identified, and we still know little in detail about the binding specificities of the two receptor families.

In a few cases where binding partners have been identified, vomeronasal cells seem to be much more sharply tuned than receptors in the main olfactory epithelium, responding only to specific chemical compounds and not to other molecules even con-taining similar functional groups. Sensitivity is quite high, such that responses can be detected for con-centrations of pheromones as low as 10^{-11} to 10^{-10} M (Figure 7.20B), which then generate trains of action potentials communicated to the CNS (Figure 7.20C). Responses of this kind might be expected from cells responding to pheromones, which can trigger stereotypic and often innate behavioral responses even at very low concentration.

Most of the receptor cells of the vomeronasal organ express either a V1R protein or a V2R protein. A few express one of the ORs also found in the main olfactory epithelium (Levai et al., 2006), or a formyl peptide receptor (Liberles et al., 2009; Riviere et al., 2009) which binds formylated peptides secreted by bacteria or mitochondria during infection or tissue damage. By analogy to the main olfactory epithe-lium, a single cell might be expected to express only a single receptor gene. This supposition seems for the most part to be true, but a few V2R receptors are more widely expressed in the basal vomeronasal organ and are co-expressed with other V2Rs (Mar-tini et al., 2001). They may form dimers similar to the dimers of bitter receptors in the mammalian tongue (see Chapter 8).

Regardless of the kind of receptor protein they express, neurons of the rodent vomeronasal organ all seem to lack $G\alpha_{olf}$, AC3, or CNGA2 (Berghard et al., 1996) and use some alternative mechanism of transduction. Considerable evidence indicates that they nearly all express the TRP channel protein TRPC2, which is localized to the microvilli responsible for producing the electrical signal (Liman et al., 1999). Despite considerable effort (see for example Lucas et al., 2003; Chamero et al., 2017), we still can-not say whether TRPC2 is directly responsible for producing the response of the vomeronasal neurons, or how TRPC2 channels are gated. What we can say is that deletion of mouse TRPC2 produces large changes in mating behavior and aggression, indi-cating that this protein is likely to have some fairly consequential role in the production of the vomero-nasal olfactory response (for example Leypold et al., 2002; Stowers et al., 2002; Kimchi et al., 2007). Vome-ronasal organs also express Ca^{2+}-activated Cl^- channels within the sensory microvilli, where they probably help to amplify the receptor current much as in the main olfactory epithelium (see Yang and Delay, 2010; Kim et al., 2011; Dibattista et al., 2012; Amjad et al., 2015).

The axons of the neurons expressing V1R and V2R proteins coalesce and course between the two main olfactory bulbs, terminating in bilaterally sym-metric accessory olfactory bulbs. Each accessory olfac-tory bulb is in turn divided in two (Figure 7.20A), with the anterior part receiving input from recep-tors expressing V1R receptors and $G\alpha_{i2}$, and the posterior part from those expressing V2R receptors and $G\alpha_o$ (Jia and Halpern, 1996). When the neurons expressing a particular receptor protein are labeled, their axons can be seen to converge onto six to thirty glomeruli in each accessory olfactory bulb, more than for a single receptor in the main olfactory bulbs (Belluscio et al., 1999; Rodriguez et al., 1999; Del Punta et al., 2002). The connections of the neurons in the accessory olfactory bulb may, however, be just as precise as those for the main olfactory bulb (Del Punta et al., 2002). The cells of the accessory olfactory bulb do not project to the olfactory cortex but rather to the amygdala and hypothalamus of the limbic system, where they may contribute to unconscious responses to chemical cues.

Summary

The ability of an organism to detect chemicals in its environment, called chemoreception, arose very early in evolution. Chemicals bind to receptors, which may respond to a broad range of odorants to distinguish a variety of chemical structures, or to a narrow range to mediate signaling essential for the reproduction or defense of the organism. In its simplest form, chemoreception is found even in bacteria, where attractants and repellents produce a change in movement called chemotaxis. Chemotaxis is mediated by a small family of receptor molecules, which bind substances like sugars and amino acids and regulate the phosphorylation of a second-messenger protein called CheY. CheY then directly modulates the flagellar motor of the bacterium. When CheY binds to the motor, it produces an event called a tumble that causes the bacterium to stop and then change its direction of motion. In the presence of an increasing concentration of attractant, less CheY is phosphorylated, there are fewer tumbles, and the bacterium swims in a more continuous motion toward the nutrient. This response adapts: a maintained concentration of attractant causes the tumble frequency gradually to return nearly to its basal level. As a result, the bacterium, having arrived at a food source, does not then swim away but remains in the vicinity.

In animals, the most sensitive and sophisticated form of chemoreception is olfaction, our sense of smell. Olfaction has been studied in most detail in insects, particularly in *Drosophila*, and in mammals, particularly in mouse. The great majority of receptor proteins in *Drosophila* belong to two families of ion channels, the ORs and IRs, expressed in separate groups of sensilla on the antenna and maxillary palp. The binding of an odorant gates the opening of a cationic channel, directly generating the depolarization and action potentials that are conveyed to the CNS. The organization of the sensilla and map of expressed receptor proteins is stereotypic and genetically determined, with little variation from organism to organism within the same species. All of the neurons expressing one of the receptor proteins converge in the antennal lobe onto glomeruli, which provide the units of olfactory processing in the CNS. The code is largely combinatorial: any one chemical binds to a distinct group of receptor proteins and stimulates a certain subset of glomeruli, and this combination of responses signals detection to the CNS.

In mammals, the main olfactory epithelium contains receptor neurons that express olfactory receptors primarily from a large family of about 1000 functional proteins in mouse, though probably only 350 in man. Each protein seems to have a somewhat different specificity for odorant molecules, in some cases quite specific for certain chemicals and in other cases of broader selectivity. A few receptor neurons express G-protein receptors from the TAAR family, which respond to amines in urine. Olfactory neurons expressing either OR or TAAR proteins contain a specialized G-protein α subunit called $G\alpha_{olf}$ that activates adenylyl cyclase 3, increasing the synthesis of the second messenger cAMP. The cAMP then gates the opening of cyclic-nucleotide-channels, producing an increase in membrane permeability to Na^+, depolarization of the membrane potential, and an increase in the firing of action potentials.

The cyclic-nucleotide-gated channels also allow the entry of Ca^{2+}, which acts as an additional second messenger gating the opening of Ca^{2+}-activated Cl^- channels, considerably increasing the sensitivity and amplitude of the electrical signal. The Ca^{2+} also produces adaptation of the olfactory response, though the particular Ca^{2+}-dependent mechanisms that are responsible for adaptation have not been elucidated. In addition to this scheme using adenylyl cyclase and cAMP, there is increasing evidence that at least some olfactory neurons employ an alternative scheme requiring membrane-bound guanylyl cyclase, the second messenger cGMP, and different ion-channel subunits.

In mammals as in insects, each receptor neuron appears to express only a single kind of receptor protein. All of the neurons expressing this protein converge onto one or two centers of processing called glomeruli in each of the two olfactory bulbs. The glomeruli seem to be arranged so that signals produced by chemicals of related structure are processed by neighboring glomeruli. This arrangement may facilitate inhibitory interactions among cells processing chemicals of similar structure, but it may also have been an accident of the evolution of receptor proteins by gene duplication, with proteins of similar amino acid sequence having similar trajectories into the CNS. In either case, the olfactory code seems again to be combinatorial, with the signals of several glomeruli indicating the detection of particular odorants. The neurons of the olfactory bulb send the results of their calculations to the olfactory cortex, where our conscious sensations of smell probably arise.

Insects have two kinds of olfactory receptors: odor generalists which mediate odor discrimination of the kind honey bees use to detect flowers, and odor specialists responsible for the ultra-sensitive detection of pheromones such as sex attractants. Vertebrates also have odor generalists and odor specialists, and both may be present within the main olfactory epithelium. Many species, however, also have a separate olfactory organ primarily for phero-mone detection called the vomeronasal organ. The receptor cells in the vomeronasal organ resemble those of the main olfactory epithelium, but the sensory membrane is formed from microvilli instead of cilia. The receptor proteins come from two different families of G-protein-coupled receptors each con-taining 100–200 functioning members (in mouse), unrelated in amino acid sequence to one another or to the receptor proteins of the principal olfactory epithelium. The transduction cascade seems not to employ $G\alpha_{olf}$ or adenylyl cylase but uses instead TRPC2, though the details of channel gating are still uncertain. The responses of the receptor neurons can be extremely sensitive and highly specific to particu-lar chemical structures, as would be appropriate for pheromone detection. Axons of the receptor cells terminate in glomeruli of the accessory olfactory bulbs, which together with higher centers in the amygdala and hypothalamus seem to have an important role in producing unconscious responses to chemical messages.

CHAPTER 8

Taste

Taste, like olfaction, is the detection and identification of chemical compounds in the environment, but these two senses are in other respects quite different. The detection of odor is used for a variety of purposes, including not only the sensing of food sources but also the identification of a mate or prey, the recognition of territories, and the detection of friend or foe. Taste, on the other hand, serves only one role: the acceptance or rejection of food (see Liman et al., 2014; Freeman and Dahanukar, 2015; Roper and Chaudhari, 2017; Scott, 2018). Although we can sense and distinguish a very large number of different odors, we appear to have only five major categories of tastes: the familiar sweet, sour, bitter, and salty, as well as *umami* (from the Japanese word meaning delicious), the taste of amino acids in many proteinaceous foods like meat broth and cheese. It is possible that we have a limited ability to distinguish different kinds of sweet and bitter compounds, but most of the apparent intricacy of taste (and most of the pleasure we derive from eating) comes from the aroma of the food, that is from the nose, not the tongue.

Though the tongue may seem unsubtle and prosaic, transduction in gustation is almost bewildering in its complexity. This should probably come as no surprise. We shouldn't really expect to detect Na^+ in the same way as sucrose or monosodium glutamate, and indeed the mechanisms of detection are completely different. In vertebrates, there is now excellent evidence that sweet, umami, and bitter are mediated by G-protein-coupled receptors. Saltiness and sour, on the other hand, seem to be transduced by ionotropic mechanisms, by channels that produce depolarization of the receptor cell. The species of receptor or channel the animal uses may very well be influenced by its diet; we might not expect an animal with a low salt diet such as a herbivore to use the same Na^+ channels as an animal with high Na^+ intake.

Another surprising feature of taste is its insensitivity. If we are astonished by the ability of the eye to detect single photons of light (Figure 2.13), and of olfactory receptor sensilla in moths to sense a single molecule of sex pheromone (Figure 2.15), we may be just a little disappointed by the high concentrations of sucrose and NaCl required to jolt our tongue into recognition. Part of the reason taste is insensitive is that the gustatory organs are usually located near or inside the mouth, so that detection at a distance is not necessary. There is, however, another important reason: the insensitivity of detection is actually adaptive. Food is sensed as particularly sweet only when the sucrose concentration is relatively high, because foods high in sucrose are a particularly good energy source. Salt sensation (in man) is similarly insensitive, in part because we are adapted to the taste of our own saliva, which has a NaCl concentration rarely less than 10 mM and much higher during salivation. Food only seems salty to us when the concentration of NaCl is even greater. Bitter compounds are an exception. They can be detected at low concentration, probably because bitter foods are mostly foods we reject. Small amounts of alkaloids and other bitter substances can actually be poisonous.

It is sometimes said that smell is the detection of chemicals that are air-borne, and taste, those that

Sensory Transduction. Second Edition. Gordon L. Fain, Oxford University Press (2020). © Gordon L. Fain 2020.
DOI: 10.1093/oso/9780198835028.001.0001

are water-borne. Though generally true for terrestrial insects and vertebrates, this distinction is unlikely to be helpful for a very simple organism such as a bacterium (see Chapter 7), or for a round worm like *C. elegans*, living within the soil and sensing substances in the environment with a small group of neurons using a large family of G-protein-coupled receptors (Troemel, 1999; Zhang et al., 2014). This organism can detect at least a hundred different chemicals, some volatile and some water soluble, analogous to the odors and tastes of vertebrates. Though some chemosensory neurons seem to sense only one sort or the other, some have receptors for both, and there are no organs comparable to the nose or tongue. The water solubility of the compound seems less important to the organization of chemoreception in this organism than whether the compound signals attraction or repulsion.

The notion that tastes are water-borne and odors air-borne also clearly fails for aquatic animals such as fish, which have olfactory receptors homologous to our own located within the snout, but gustatory receptors in taste buds inside the oral cavity. Although all the chemical stimulants recognized by fish are water-borne, smell and taste remain fundamentally different. The olfactory system is important in feeding and reproduction, in the detection of routes for migration and spawning, in the recognition of kin and predators (Laberge and Hara, 2001; Hamdani el and Doving, 2007). Taste, on the other hand, serves only one role: is it yummy?

Gustation in insects

Taste is an important sense for insects but is likely to be quite variable from species to species, because insects occupy so many ecological niches and utilize a variety of food sources, from the nectar of flowers to blood and carrion. Taste has been most thoroughly investigated in flies, whose gustatory sensilla are found not only in the principal mouth parts of the animal but also on the legs, especially on the forelegs (see Figure 2.8A).

Flies find food by walking around on a substrate (Figure 8.1; Dethier, 1976). Stimulation of leg receptors with sucrose, for example, causes the fly to extend its proboscis. If the bristles on the labellum confirm the presence of an attractive substance, the

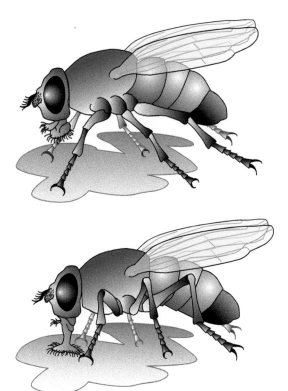

Figure 8.1 Taste and feeding in the house fly. Flies have gustatory sensilla on their legs and sense nutrients by walking over the substrate. They then extend their proboscis to initiate feeding. (After Dethier, 1976.)

muscular pharynx begins to pump the food up through the proboscis (Figure 8.2). Sensory receptors in the pharynx subject the food to further testing, which if favorable directs the pumping of the food into the esophagus, where it is then carried by peristaltic action throughout the rest of the digestive system. In addition to the legs and mouth parts, there are taste sensilla on the margins of the wings of unknown function, as well as on the ovipositors of females, perhaps used to select sites for the deposition of eggs.

A single taste sensillum in a fly contains a variable number of neurons, usually two to four chemoreceptor cells, and a single mechanoreceptor responding to deflection of the bristle (see Figure 2.8B). The chemoreceptors generate robust action potentials that are easily recorded with an extracellular electrode, placed directly on top of the bristle and containing a salt or sugar solution (Hodgson et al., 1955). Another

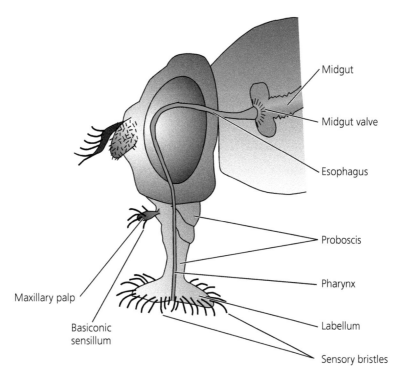

Figure 8.2 Principal parts of the gustatory system in the mouth of a fly. Tastes are detected by sensory bristles on the labellum, causing the pharynx to pump food into the proboscis and midgut. (After Dethier, 1976; Stocker, 1994.)

method is to make a small crack in the side of the bristle and place the recording electrode there, leaving the top of the sensillum free for the application of the stimulating fluid (Morita, 1959).

The labellum of *Drosophila* has over thirty sensilla, which can be placed into three major classes according to their length: long (L), intermediate (I), and short (S). L and S sensilla have four chemoreceptors (I sensilla have only two), and extensive experimentation (summarized in Dethier, 1976; Pollack and Balakrishnan, 1997; Liman et al., 2014; Freeman and Dahanukar, 2015) has shown that each neuron has a different chemical selectivity. In L sensilla, one neuron responds to sugars, another to salt, a third to water or low osmolarity, and a fourth—sometimes called the second salt receptor—to high salt. Bitter compounds are not processed in L sensilla but only by the cells of I and S sensilla. In an exhaustive study of the selectivity of the labellum, Weiss and collaborators (2011) showed that none of the L sensilla give any response to bitter

compounds, and no response could be detected for two of the S sensilla. All of the other S sensilla respond to bitter with broad selectivity. The two subclasses of I sensilla also respond to bitter but with a more narrow selectivity; each is activated by different groups of bitter compounds (Figure 8.3).

The responses of the taste sensilla are mediated by several groups of receptors, of which the largest is the GR family with sixty-eight genes in *Drosophila* but over a hundred in some mosquitoes. The GRs are distantly related to insect ORs, and they were also initially thought to be G-protein receptors. They have, however, the same upside-down topology as ORs (Zhang et al., 2011; Xu et al., 2012) and are now known to form ion channels. Evidence for an ionotropic mechanism was first obtained for the sugar response of the fly labellum by Murakami and Kijima (2000). In Figure 8.4, a patch pipette was sealed onto the membrane of the sensory dendrite, the part of the cell responsible for sensory transduction. The cell was stimulated with sucrose (blue

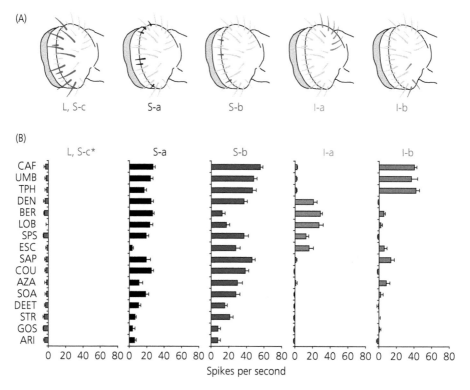

Figure 8.3 Detection of bitter by *Drosophila* labellum sensilla. (A) Distribution of five classes of sensilla on the labellum. (B) Mean responses of all sensilla of a given functional class ± s.e.m. L and S-c* sensilla did not exhibit any observable physiological responses to any tested bitter compounds. S-a and S-b classes responded broadly to a wide range of compounds; I-a and I-b classes responded more narrowly. Tested compounds (top to bottom): CAF, caffeine; UMB, umbelliferone; TPH, theophylline; DEN, denatonium benzoate; BER, berberine chloride; LOB, (–)-lobeline hydrochloride; SPS, sparteine sulfate salt; ESC, escin; SAP saponin from quillaja bark; COU, coumarin; AZA, azadirachtin; SOA, D-(+)-sucrose octaacetate; DEET, N-diethyl-*m*-toluamide; STR, strychnine nitrate salt; GOS, gossypol from cotton seeds; and ARI, aristolochic acid. (From Weiss et al., 2011.)

bar), which in the whole-cell recording to the left produced a steady inward current and an increase in the rate of rapid downward deflections, which are poorly clamped action potentials generated by the axon of the cell. The pipette was then pulled off, forming an outside-out patch, and the sensory membrane was stimulated again. Action potentials were no longer present because the sensory membrane was now separated from the rest of the cell, but there was a robust increase in inward current. Additional experiments showed that this current is produced by the opening of non-selective cationic channels having a conductance of about 30 pS.

In an outside-out recording, the extracellular side of the membrane faces the bathing solution. The sucrose perfused across the extracellular surface of

the outside-out patch of Figure 8.4 must therefore have been binding directly to a receptor on the membrane to produce a response in the absence of the rest of the cell. A response could even be recorded in the presence of GDPβS, which inhibits G-protein activation by binding to the guanosine-nucleoside binding site of Gα to prevent GTP binding. Similar results have been obtained by Sato and co-workers on expressed GR proteins (Sato et al., 2011). G proteins may have some role in insect sugar detection (see for example Ueno et al., 2006), but the basic mechanism is ionotropic like insect olfaction.

Drosophila ORs are generally expressed singly in any given sensillum together with the Orco receptor, but GRs are commonly expressed in larger numbers in the same cell (Weiss et al., 2011; Fujii et al.,

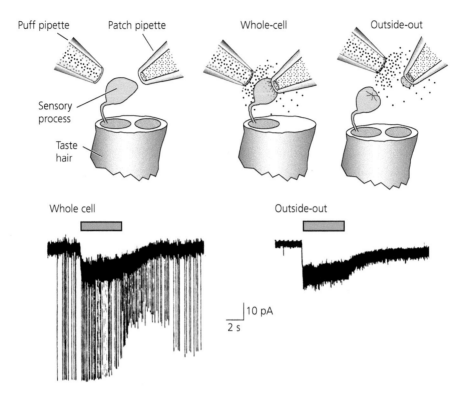

Figure 8.4 Response of fly labellum dendrite to sucrose. A taste sensillum was cut in its middle and briefly exposed to low Ca²⁺ solution, causing the sensory process of a dendrite to swell and round up. A patch pipette was then sealed onto the sensory process and used to make a whole-cell or outside-out recording. Sucrose was applied from a second, puff pipette. The duration of sucrose application is indicated by the blue bars above the whole-cell and outside-out recordings. (From Murakami and Kijima, 2000.)

2015). Responses to sugar have been shown to be mediated by at least eight of the *GR* genes, of which several are expressed in different combinations in different labellum sensilla. The larger group of bitter GRs are also expressed in combinations of as many as twenty-nine proteins in a single neuron, though the more narrowly responding bitter neurons tend to express a smaller number. The reason so many receptors are expressed together may be that insects do not need to discriminate between different kinds of sugars or different bitter compounds detected by different receptor proteins. They may need to know only whether the substance detected by the labellum is edible or potentially harmful.

In addition to the GRs, receptors from other protein families are expressed on the legs and mouth parts of insects. In Chapter 7 we saw that some insect olfactory receptors are members of an IR family resembling in structure ionotropic NMDA and AMPA glutamate receptors. Some members of this IR family are also expressed in gustatory sensilla. The *IR76b* gene, in particular, is required for the detection of attractive low concentrations of salt (Zhang et al., 2013). This gene is expressed in all of the L sensilla. When the *IR76b* gene is deleted, a low-salt response can no longer be recorded.

The detection of water requires an ion channel from the DEG/ENaC (degenerin/epithelial) family (Cameron et al., 2010; Chen et al., 2010c), which we previously encountered in our discussion of touch sensitivity in Chapter 5. DEG/ENaC family members have two membrane-spanning domains with intracellular amino and carboxyl termini, and they form channels as trimers. In *Drosophila* these channels are encoded by pickpocket (*ppk*) genes. Of the more than thirty *ppk* genes in the fruit fly, one called *ppk28* is expressed in water-sensing neurons in the labellum and not in neurons sensitive to sugars or bitter compounds. When the *ppk28* gene is deleted, sensitivity to water or low osmolarity is entirely abolished.

Mammals: taste buds and the tongue

Although present information indicates that all of taste in insects is produced by ionotropic receptors, in vertebrates including mammals there are different mechanisms for different taste modalities—some ionotropic but some clearly metabotropic (see Liman et al., 2014; Roper and Chaudhari, 2017). All of these modalities are mediated by taste receptor neurons, which are found in specialized clusters called taste buds present on the tongue but also elsewhere within the oral cavity, for example on the palate and epiglottis. Most of the taste buds of the tongue are found in small eruptions of the lingual surface called papillae, of which there are three kinds (Figure 8.5A): the fungiform near the front of the tongue, the foliate located on the lateral border, and the circumvallate, which are the largest of the papillae and occur in a V-shaped row at the very back of the tongue. Although in some species such as rodents there is evidence for spatial separation of the different taste modalities across the surface of the tongue, it is unclear whether a localization of this kind is present in humans. Taste buds in the three different kinds of papillae appear to be able to mediate most or all of the five principal taste modalities.

Taste buds (Figure 8.5B) are found along the exposed dorsal surface of the papilla. A fungiform often contains only a single taste bud, but a circumvallate has many thousands. The taste bud is a tightly organized structure having of the order of fifty to a hundred cells, resembling an onion with a small pore at its apex (called the taste pore or taste pit). The entire taste bud in a mammal is typically only about 50 µm in diameter. Taste receptor cells like olfactory receptor neurons can generate action potentials (see for example Chen et al., 1996), but whereas olfactory receptor cells have axons that enter the CNS, taste receptor neurons communicate their responses at the base of the taste bud onto distinct branches of cranial nerves, primarily the chorda tympani and the glossopharyngeal nerves. These nerves then carry the gustatory signals into the CNS.

Most of the cells within the taste bud are thin and elongate, extending the greater part of the length of the bud from base to tip. In mammals, these cells can be differentiated into three classes usually referred to as type I, type II, and type III, which can be distinguished according to their membrane specializations, shape of nucleus, electron density of cytoplasm, and presence of electron-dense granules or dense-core vesicles (Murray et al., 1969; see Roper and Chaudhari, 2017). At the base of the taste bud there are a few cells more nearly spherical in shape, called basal cells or sometimes type IV. These cells are progenitors, which continuously divide and completely renew the other cells every 1–2 weeks.

The different cell types appear to have different functions. The most numerous are type I, which may be sensitive to salt (Vandenbeuch et al., 2008) or may be supporting cells that do not participate directly in sensation. The next most common are the type II, which have receptors for sweet, umami, and bitter. The type II cells also express other proteins thought to be involved in transduction such as PLCβ2, IP$_3$R3 receptors, and the TRP channel TRPM5 (Zhang et al., 2003b; Clapp et al., 2004). Curiously, these cells do not make synapses onto cranial nerves and lack calcium channels and presynaptic proteins such as SNAP25 or synapsin (Clapp et al., 2006; DeFazio et al., 2006), which would ordinarily be required for synaptic vesicle release. There is, however, evidence that type II cells may release ATP onto purinergic receptors (Finger et al., 2005).

The only cells known to make typical chemical synapses are the type III, which are the least numerous and may comprise only a few percent of the total cell number (Murray et al., 1969). These cells can be recognized by the presence of dense-core vesicles containing 5-hydroxytryptamine (5-HT or serotonin), and there is evidence that type III cells transduce responses to sour (Huang et al., 2006). Salt detection seems to have its own group of cells (Chandrashekar et al., 2010), but it is still unclear whether these are a subgroup of type III cells (Lewandowski et al., 2016), type I cells (Vandenbeuch et al., 2008), or some still uncategorized cell type.

Taste transduction: metabotropic receptors

Of the five principal modalities of taste reception in the mammalian tongue, there are three whose receptors are metabotropic: bitter, sweet, and umami. Not every

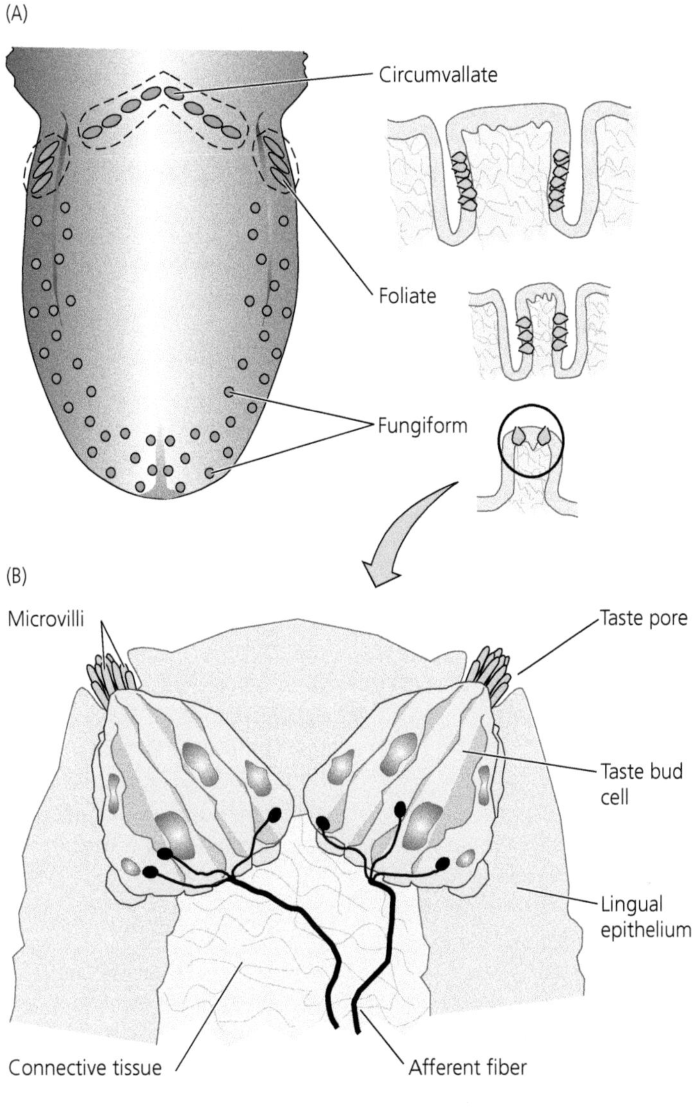

Figure 8.5 Structure and location of taste buds in the mammalian tongue. (A) Location and shape of three classes of papillae on the tongue. (B) Structure of taste buds from fungiform papillae. Those for other papillae are similar.

animal has all three. Cats lack any sensation of sweet, and pandas with their purely vegetarian diet lack taste receptors for amino acids. All three are, however, present in humans and in mice, which are the two mammals most commonly used in gustatory research.

Bitter

The first gene for a bitter receptor was discovered in humans by examination of a section of chromosome implicated in the detection of the bitter-tasting compound 6-n-propyl-2-thiouracil (PROP) (Reed et al., 1999). The gene for a single G-protein-coupled receptor from this locus was then used to search the rest of the human DNA data base, revealing a large family of *T2R* (or *Tas2R*) genes in both humans and rodents, whose members are expressed in taste buds (Adler et al., 2000; Matsunami et al., 2000). These genes code for G-protein receptors resembling rhodopsin and the β-adrenergic receptor (see

Figure 4.3A) with seven membrane-spanning, α-helical domains, and a small N-terminal extracellular sequence. The binding site for bitter compounds is thought to be contained within the membrane-spanning helices. There are more than forty *T2R* genes in rodents and twenty-five in humans, with some receptors having narrow and some broader specificity, much as in *Drosophila*. A single cell typically expresses several different receptors (Adler et al., 2000; Mueller et al., 2005).

Taste-bud neurons expressing T2R proteins also express a G-protein α subunit called gustducin or $G\alpha_{gus}$ (Figure 8.6A–C). $G\alpha_{gus}$ was first cloned from taste tissue (McLaughlin et al., 1992) and is a member of the α_i/α_o family that also includes transducin, the G protein of vertebrate phototransduction described in Chapter 9. Bitter compounds have been shown to activate $G\alpha_{gus}$ (Ming et al., 1998; Chandrashekar et al., 2000), and transgenic mice in which the $G\alpha_{gus}$ gene has been eliminated show less vigorous cranial nerve-fiber responses to bitter compounds and a reduced capacity for the detection of bitter in tests of taste preference (Wong et al., 1996).

Sweet

The detection of sweet compounds in mammals is also mediated by G-protein-coupled receptors. The first genes were cloned from a rat circumvallate papilla cDNA library (Hoon et al., 1999) and were called *T1R1* and *T1R2*. These genes code for proteins that belong to the same family of G-protein receptors as the V2R receptors of the vomeronasal organ (or GABA and glutamate metabotropic receptors in the brain), with large extracellular amino-terminal sequences containing the receptor binding site. The sequence of a third member of the *T1R* family was then obtained by genetic linkage to the locus *Sac*, which was known in mice to regulate preference for sugar and saccharin. Four groups nearly simultaneously published the sequence of a protein called T1R3, which also has a large extracellular amino-terminal sequence (Kitagawa et al., 2001; Max et al., 2001; Montmayeur et al., 2001; Sainz et al., 2001). Expression studies on cultured cells then showed that a response to sugar depends upon joint expression of T1R2 and T1R3 and does not require T1R1 (Nelson et al., 2001). The T1R2 and T1R3 proteins are obligatorily co-expressed in taste-bud cells

and function as a heterodimer (Nie et al., 2005). The knocking out of either gene greatly reduces the response to sweet (but see also Damak et al., 2003; Zhao et al., 2003). Sweet detection is also reduced in animals lacking $G\alpha_{gus}$ (Ruiz-Avila et al., 2001). Cells expressing T1R receptors do not also express the bitter T2R receptors (Figure 8.6D), indicating that bitter and sweet are transduced by separate populations of type II cells often found side-by-side in the same taste bud.

Umami

Although T1R1 has no role in sugar detection, it is now thought to function in umami. A response to glutamate can be produced in a cell line after simultaneous expression of *T1R1* and *T1R3*. The protein products of these mouse genes combine to form a dimeric receptor with broad sensitivity to amino acids (Nelson et al., 2002). The deletion of either of the *T1R1* or *T1R3* genes substantially reduces but does not completely eliminate responses to amino acids (Zhao et al., 2003; Kusuhara et al., 2013), perhaps indicating the presence of an additional receptor (Chaudhari et al., 2000). Umami detection seems also dependent on the presence of $G\alpha_{gus}$, maybe in combination with transducin Gα (He et al., 2004).

Transduction cascade

All three metabotropic modalities appear to signal through a similar transduction cascade, utilizing predominantly $G\alpha_{gus}$ together with the G-protein subunits β_1 or β_3 and γ_{13} (Huang et al., 1999). Activation of G protein stimulates PLCβ2 (Zhang et al., 2003b) to produce the second messenger IP_3, and IP_3 then binds to IP_3R3 receptors (Clapp et al., 2004) to release Ca^{2+} from internal stores (see Figures 4.2, 4.7, and 4.8). An increase in Ca^{2+} was first shown to be produced by bitter-tasting compounds many years ago (Akabas et al., 1988), and similar results have been obtained by many investigators for sugars and amino acids (for example Caicedo et al., 2002). The knocking out of the genes for either PLCβ2 (Zhang et al., 2003b) or the IP_3R3 receptor (Hisatsune et al., 2007) reduces taste responses to substances stimulating any of these three metabotropic modalities.

The increase in Ca^{2+} then gates a TRP channel. TRPM5 channels are co-expressed with both T1R and

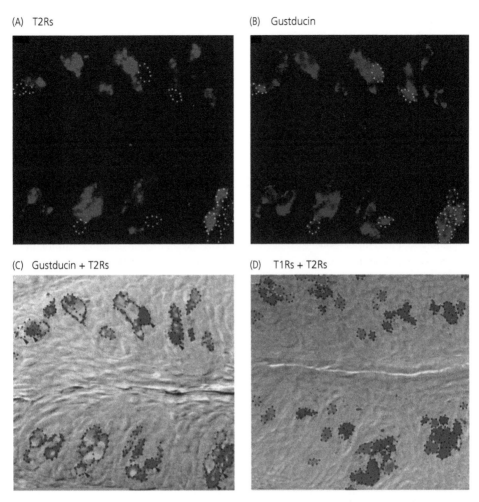

(A) T2Rs

(B) Gustducin

(C) Gustducin + T2Rs

(D) T1Rs + T2Rs

Figure 8.6 Localization of expression of genes in taste-bud cells. Fluorescent probes were used to localize the expression of genes for taste receptors and the G-protein α subunit gustducin with *in situ* hybridization. Dashed lines indicate approximate areas of taste receptor cells. (A) A mixture of ten probes for different genes was used to label cells expressing T2R proteins. (B) Probe to gustducin ($G\alpha_{gus}$). (C) Overlay of images in (A) and (B), with yellow indicating cells expressing both T2Rs and gustducin. Arrows indicate cells that express gustducin but not T2R proteins. (D) Double labeling with probes for T2Rs (green) and T1Rs (red). (From Adler et al., 2000.)

T2R receptors; when the gene for TRPM5 is deleted, responses to bitter compounds, sugars, and amino acids are all greatly reduced (Zhang et al., 2003b). Moreover, increases in Ca^{2+} can gate the opening of TRPM5 channels. This phenomenon has been explored in most detail by Zhang and co-workers (2007), who genetically labeled taste cells expressing TRPM5 with the fluorescent marker GFP. They then showed that increases in Ca^{2+} or IP_3 in the taste-cell cytoplasm produced an increase in conductance—but only if TRPM5 was present. Ca^{2+} but not IP_3 had the same effect on excised inside-out patches of mem-

brane (Figure 8.7A and B). Recall that for an inside-out patch, the cytoplasmic surface of the membrane faces the bathing solution. Because the Ca^{2+} in Figure 8.7 was applied directly to the cytoplasmic surface in the absence of the rest of the cell, the increase in current was most likely gated by Ca^{2+} binding to the interior of the taste-cell membrane. IP_3 didn't have a similar effect, presumably because the excised patches had no endoplasmic reticulum containing IP_3 receptors. The recording in Figure 8.7C shows that the TRPM5-mediated response is produced by single channels with a unitary conductance of about 20 pS, which Zhang and

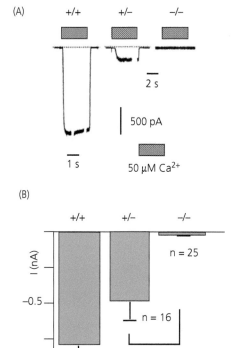

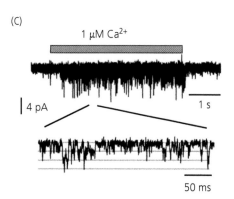

Figure 8.7 TRPM5 channels in taste-bud cells are activated by Ca²⁺. Inside-out patches of isolated mouse circumvallate taste-bud cells were perfused on the intracellular surface. (A) Patch currents from *TRPM5⁺/⁺*, *TRPM5⁺/⁻*, and *TRPM5⁻/⁻* mice in response to 50 μM Ca²⁺. Membrane potential was held at –80 mV. Blue rectangles above traces indicate duration of application of Ca²⁺. (B) Mean ± s.e.m from experiments in (A). ** $p < 0.01$. (C) Single TRPM5 channels activated by 100 μM Ca²⁺ in an excised patch from a *TRPM5⁺/⁻* mouse (V_m –80 mV). Lower record shows channel openings on a faster time base. (From Zhang et al., 2007.)

colleagues showed to be non-selective cationic. The activation of these channels would increase the Na⁺ permeability of the taste cells, producing a depolarization somehow transmitted to the cranial nerves.

Figure 8.8 summarizes our present knowledge of the most likely mechanisms for transduction in the bitter, sweet, and umami modalities. Activation of G protein stimulates PLCβ2 to make IP₃, which releases Ca²⁺ into the cytoplasm and gates the opening of TRPM5 channels. There are, however, many remaining uncertainties. Although Gα_gus is likely to participate in much of bitter, sweet, and umami transduction, other Gα proteins may have a role (see for example Tizzano et al., 2008). The mechanism of activation of PLCβ2 by G-protein requires further investigation. Other schemes involving cyclic nucleotides cannot be entirely excluded (see Margolskee, 2002), but results from PLCβ2 and TRPM5 gene-deletion experiments (Zhang et al., 2003b) give strong support to the principal elements of the cascade in Figure 8.8.

Taste transduction: ionotropic detection

Present evidence indicates that both salty and sour modalities are likely to use ionotropic mechanisms.

Salty

The mechanism of transduction for salt is in principle rather simple to describe. Consider a taste-bud cell with a large concentration of Na⁺ channels on the membrane facing the inside of the mouth (Figure 8.9). These Na⁺ channels should not resemble those responsible for action potentials, which require a depolarization to open and which rapidly inactivate. What is required instead is a channel selective for Na⁺ but open at the resting membrane potential and remaining open as the membrane potential changes during the response of the cell.

If a channel of this sort were present in a taste-bud neuron, an increase in Na⁺ concentration in the extracellular medium would produce a depolarization. For a cell permeable at rest to both Na⁺ and K⁺, the membrane potential is given by Eq. (3.5),

$$V_m = \frac{RT}{F} \ln \frac{\alpha [Na]_o + [K]_o}{\alpha [Na]_i + [K]_i}$$

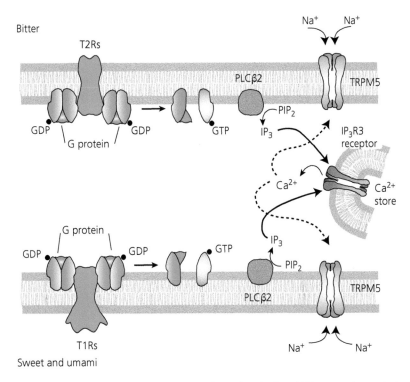

Figure 8.8 Components of cascades for transduction of bitter, sweet, and umami. Both T2R and T1R receptors in taste-bud cells are thought to activate PLCβ2 effector enzymes, which release IP$_3$ into the cytosol. The IP$_3$ is then thought to bind to IP$_3$R3 receptors to release Ca^{2+}, which gates the opening of TRPM5 channels.

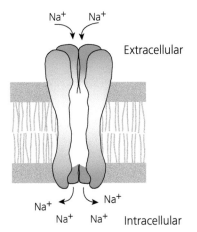

Figure 8.9 Na$^+$ channels in taste-bud cells detect salt. Taste-bud cells that mediate sensitivity to salt may have open channels selectively permeable to Na$^+$. An increase in extracellular Na$^+$ would then depolarize the cell directly.

where α is the ratio of the Na$^+$ and K$^+$ permeabilities. Whatever the value of α, an increase in the extracellular Na$^+$ concentration [Na]$_o$ will produce a more positive value of V_m provided the extracellular Na$^+$ concentration is greater than the intracellular Na$^+$ concentration. The greater the number of Na$^+$ channels open at the resting membrane potential and located in the membrane facing the exterior, the greater will be α, and the bigger will be the change in membrane potential produced by an increase in extracellular Na$^+$.

Na$^+$ channels with properties of this sort are present in the membrane of at least some taste-bud cells. Figure 8.10A is a whole-cell, voltage-clamp recording for a cell from a taste bud isolated from the tongue of a hamster (Gilbertson et al., 1993). At the beginning of the recording, the bud was bathed in saline solution containing 140 mM NaCl. The

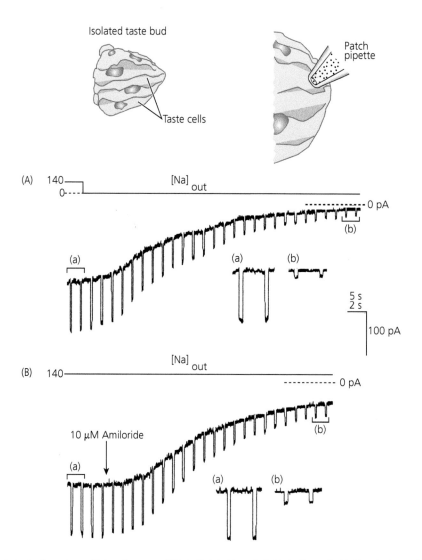

Figure 8.10 Taste-bud cells can have a high resting permeability to Na+ that is partially blocked by amiloride. Taste buds were isolated from hamster fungiform papillae. Patch electrodes were sealed onto single cells, generally at the periphery of the bud, and currents were recorded with whole-cell voltage clamp. Insets show current responses to brief voltage pulses, used to measure the time course of conductance change. (A) Replacement of 140 mM NaCl in the extracellular solution with the impermeant cation N-methyl-D-glucamine. A steady inward current initially present in 140 mM NaCl is gradually decreased as the NaCl washes out, and the conductance of the cell is diminished. (B) Addition to the bathing solution of 10 μM amiloride, a blocker of ENaC channels. Current and conductance are again decreased, but only part of the current is blocked. (From Gilbertson et al., 1993.)

taste-bud cell had a large resting inward current, which is almost entirely abolished when the Na+ in the solution is replaced with the large impermeant cation N-methyl-D-glucamine. The brief downward deflections show current responses to 15-mV pulses and are a measure of the conductance of the cell. The conductance decreased when Na+ was removed.

What kind of Na+ channels are these? Some seem to be members of the mammalian DEG/ENaC family of proteins found in many epithelial cells and used for the transport of Na+ ions. We have previously encountered channels from this family in the touch-sensitive microtubule cells of C. elegans (Figure 5.3) and osmolarity-sensitive, water-detecting sensilla of the labellum of *Drosophila*. ENaC proteins have just the properties a taste cell would require. They are almost entirely insensitive to membrane voltage and come in several varieties, all permeable to alkali metal cations but some highly selective for Na^+ over K^+ and some not quite so selective. Many of the channels in the DEG/ENaC family can be blocked by micromolar concentrations of the diuretic drug amiloride (see Figure 5.4C). If therefore ENaC proteins have a role in taste, amiloride might reduce the response of taste-bud cells to salt. In the experiment shown in Figure 8.10B, Gilbertson and co-workers perfused amiloride onto the taste-bud cell. Amiloride had much the same effect as removing Na^+, decreasing both the resting inward current and the cell conductance.

Even stronger evidence of a role for DEG/ENaC channels in salt detection has come from experiments on mice in which the gene for the channel has been deleted. Epithelial ENaC channels in mammals have three subunits, called α, β, and γ, which are all required for channel function. In the experiment shown in Figure 8.11 (from Chandrashekar et al., 2010), recordings were made of the massed response of axons coming from the tongue in wild-type mice and mice in which expression of the α subunit of the ENaC channel was prevented. In wild-type mice, only part of the response could be inhibited by amiloride, predominantly at low concentrations of salt. This amiloride-sensitive component could also be eliminated by deleting ENaCα.

This part of the response is indicated by the orange shading in Figure 8.11.

At higher concentrations of salt, the response of the nerve was not affected by either deletion of ENaCα or application of amiloride. This part of the response is indicated in the figure by blue shading. At such high concentrations, Na^+ becomes no longer attractive but aversive, presumably to prevent the consumption of too much salt. The mechanism of this aversive response is not known but may be conveyed at least in part by cells mediating aversive responses to bitter and sour (Oka et al., 2013; but see Wu et al., 2015).

The data in Figure 8.11 suggest that the mouse tongue has an amiloride-sensitive salt response produced by ENaC channels, responding predominantly at low salt concentrations and signaling acceptance. There is also an amiloride-insensitive component due to some other kind of channel, responding at higher salt concentrations and signaling rejection. This scheme is, however, most likely oversimplified. In both Figures 8.10 and 8.11, responses were recorded from the anterior part of the tongue subserved by the chorda tympani. In the circumvallate taste buds of the posterior tongue innervated by the glossopharyngeal nerve, amiloride seems to have little if any effect even for low salt concentrations (see for example Doolin and Gilbertson, 1996). Taste sensitivity for salty in humans seems to be completely insensitive to amiloride (see Halpern, 1998). There are likely to be other channels mediating an

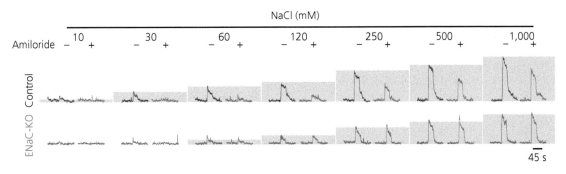

Figure 8.11 DEG/ENaC channel contribution to the mouse taste-bud response to salt. Massed nerve recordings from the chorda tympani nerve in the absence of amiloride from normal (control, black traces) and ENaC-KO mice (red traces), and in both groups of mice in the presence of 10 mM amiloride (blue traces). Shaded boxes indicate the amiloride-sensitive (pink) and amiloride-insensitive (blue) components. (From Chandrashekar et al., 2010.)

attractive salt taste in addition to amiloride-sensitive ENaC channels (see also Kretz et al., 1999), but the identity of these proteins is presently unknown.

Sour

We have known at least since the 1930s that taste buds can respond to acid (Pfaffmann, 1939). These early experiments also showed that the pH of the stimulating solution (the concentration of the H^+ or H_3O^+ ion) could only partly explain the sour response. Solutions that have a high concentration of organic acids, such as citric acid, taste more sour than a solution of HCl at the same pH. These organic acids can enter the cell in neutral form and dissociate, producing a change in cytoplasmic pH. Effects of this kind could be explained if the detection of sour were mediated by changes in pH inside the cell, which then altered the conductance of some other channel type, perhaps one permeable to Na^+ or K^+. Sour sensation may therefore involve at least two components, one an ionotropic response produced by some type of ion channel permeable to protons, and another a component sensitive to intracellular pH (Lyall et al., 2001; Ishimaru et al., 2006; Ye et al., 2016).

Many different kinds of channels have been proposed as the pH sensor (see Roper, 2007). A discrete population of taste-bud neurons have TRP channels different from the TRPM5 channels of Figure 8.8, which have been called polycystic-kidney-disease-like ion channels, or PDK1L3 and PDK2L1 (Huang et al., 2006; Ishimaru et al., 2006). These two proteins are co-expressed probably in type III cells (Huang et al., 2008), clearly distinct from the type II cells mediating sweet, umami, and bitter (see also Chang et al., 2010). If the cells in the tongue expressing these proteins are ablated, nerve responses to acid can no longer be recorded.

Although these results seemed initially to suggest that PDK1L3 and PDK2L1 form the channels responsible for sour sensation, two groups subsequently showed that the knocking out of the genes for these proteins had little or no effect on acid detection, measured with either nerve recordings or behavioral experiments (Nelson et al., 2010; Horio et al., 2011). Cells expressing PDK1L3 and PDK2L1 are sensitive to acid but do not use these proteins to mediate the sensitivity. The discovery of PDK1L3 and PDK2L1 expression nevertheless provided a powerful method of marking sour-sensitive cells by linking the expression of these proteins to a genetically encoded dye like green or yellow fluorescent protein. As a result, it became possible to isolate labeled cells for single-cell recording and show directly that neurons expressing PDK2L1 can respond to acid (Chang et al., 2010).

A major step forward was the discovery of the gene *Otop1*, which encodes a proton channel expressed in taste cells (Tu et al., 2018). *Otop1* was first identified as essential for the formation of the otoconia of the inner ear, though its function in this system is still not well understood. In taste cells, *Otop1* codes for a channel highly selective for protons, which can be specifically blocked by Zn^{2+}. Mutations in *Otop1* can reduce the proton current in taste cells.

Proton entry through *Otop1* channels cannot, however, be the only explanation for the sour response of the tongue, since sour-sensitive cells can also respond to weak acids even without a significant change in extracellular pH. In the experiment shown in Figure 8.12A (from Ye et al., 2016), an acid-sensitive cell marked with yellow fluorescent protein was perfused with a solution containing high extracellular K^+, to increase the current through K^+ channels. This solution also contained tetraetylammonium (TEA$^+$), to block voltage-gated K^+ channels like those responsible for the recovery of the action potential; and Zn^{2+}, to block the *Otop1* proton channels. When the cell was first exposed to this solution, it developed a large inward K^+ current as a result of the increase in extracellular K^+. When the external pH was then changed from 7.4 to 6.0, there was no response of the cell. When, however, the cell was perfused with 20 mM acetic acid without a change in external pH, the resting current was greatly diminished. This is apparently because the acetic acid could diffuse across the cell membrane in neutral form and then dissociate and acidify the cell interior.

Ye and co-workers (2016) showed that the K^+ current that is decreased in Figure 8.12A is produced by $K_{IR}2.1$, a member of a family of K^+ channels called inward rectifiers. $K_{IR}2.1$ is expressed in sour-sensitive neurons and can be blocked by acidification. In Figure 8.12B (also from Ye et al., 2016), $K_{IR}2.1$

(A)

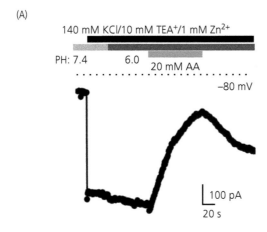

(B)

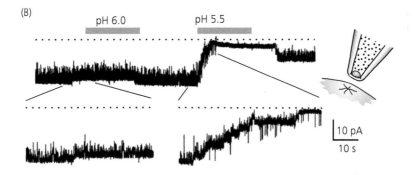

(C)

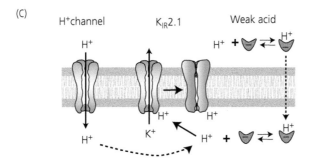

Figure 8.12 Transduction of sour in mouse taste-bud cells. (A) Whole-cell patch recordings were made from cells expressing PKD2L1 together with yellow fluorescent protein. External solution contained TEA and Zn^{2+} to block voltage-gated K^+ channels and *Otop1* proton channels. Perfusion with high external K^+ at a holding potential of −80 mV produced a large inward current which was reversibly inhibited by 20 mM acetic acid at pH 6.0, but not by a solution without acetic acid also at pH 6.0. (B) Inside-out patch recording was made from a $K_{IR}2.1$-transfected HEK cell (a cell line). The bath solution contained 150 mM KCl, 1 mM EGTA, and 1 mM EDTA. The membrane potential was held at −80 mV. The application of pH 5.5 completely inhibited channel activity, which was partially restored by return to neutral pH. Lower traces show channel currents as indicated on an expanded time base. (C) Proposed mechanism of sour transduction. See text. (A and B from Ye et al., 2016.)

channel activity was recorded in an excised, inside-out patch. Perfusion of low pH completely inhibited channel opening. Blocking the K^+ current would excite the sour-sensitive neuron, because in a resting cell in the absence of acid, the K^+ current is large enough to keep the cell hyperpolarized and prevent the initiation of action potentials. Low pH would remove this hyperpolarization and produce excitation.

Recall that in the experiment of Figure 8.12A, Zn^{2+} was added to the solution to prevent protons from entering the cell directly through proton channels. This step ensured that acidification produced by acetic acid was the result of dissociation of the acid within the cell. If the extracellular pH was instead lowered without Zn^{2+}, the resting K^+ current could also be decreased even without a weak acid like acetic acid. In this case, the protons were entering the cell from the outside. The K^+ channels can therefore be blocked either by H^+ (or H_3O^+) entering the cell through proton channels, or by protons dissociating from weak acids. Our present concept of sour detection is summarized in Figure 8.12C.

The coding of taste

We can all agree that candy tastes distinctly different from lemon juice, and that neither tastes remotely like salt. The taste of monosodium glutamate is not so easily differentiated, perhaps explaining why umami was proposed as a taste modality only relatively recently, 2200 years after the first four. These distinctions are not absolute: sugar at a high enough concentration can taste bitter. We can nevertheless agree that the different modalities are mostly distinct, which suggests that the CNS has some way of recognizing and differentiating these basic tastes to guide our food choices.

We can also agree that receptors (and ion channels) for each of the five different modalities are mostly expressed in different populations of cells in taste buds. T1R sweet or umami receptors are not expressed in the same cells as bitter T2R receptors (Figure 8.6D). Sour-sensitive cells do not also respond to sweet, bitter, or umami (Huang et al., 2006), and cells responding to salt probably also constitute a separate population of taste-bud cells (Chandrashekar et al., 2010). These groupings are again not absolute: some umami cells seem also to have receptors for sweet (Kusuhara et al., 2013), and high aversive concen-

trations of NaCl have been proposed to signal through taste-bud cells sensitive to sour and bitter (Oka et al., 2013; but see Wu et al., 2015). As a general rule, however, it would seem that the five tastes are processed in mostly separate classes of cells.

We might therefore want to propose that the signals produced by different taste-bud cells are wired into the CNS as labeled lines (Yarmolinsky et al., 2009). This notion has been supported by experiments expressing a modified opioid receptor, which can be activated with high specificity by an entirely synthetic agonist called spiradoline. Wild-type mice are normally indifferent to spiradoline, but when the modified opioid receptor was expressed in cells normally sensitive to sugar, the mice consumed spiradoline (Zhao et al., 2003); and when the receptor was expressed in cells normally sensitive to bitter compounds, the mice avoided it (Mueller et al., 2005). These experiments suggest that signals specific to sweetness and bitterness are conveyed preferentially to centers in the brain controlling food preference.

We might therefore expect that axons emerging from the tongue would separately encode each of the different taste modalities. The first test of this hypothesis was made just before the beginning of the Second World War by Carl Pfaffmann in Adrian's laboratory at Cambridge (see Pfaff, 1985). Pfaffmann patiently dissected out single cranial nerve fibers and placed them on electrodes just as Adrian had done for touch-sensitive receptors from the skin (see Chapter 1 and Figure 1.3A). Pfaffmann then showed that nerve fibers from the tongue respond to taste stimuli with an increase in spike firing (Figure 8.13). Some fibers could respond specifically to acid, others to salt, and still others sugar. But Pfaffman also discovered that a significant proportion of the axons were responsive to more than one taste modality, that is to acetic acid and also (for example) to bitter-tasting compounds. This conclusion has been confirmed by many subsequent investigations (see for example Frank and Pfaffmann, 1969; Frank, 1973; Frank et al., 1988; Hanamori et al., 1988; Yoshida et al., 2006).

A more modern approach to this problem was taken by Barretto et al. (2015), who expressed the genetically encoded Ca^{2+} indicator GCaMP in the cells of the geniculate ganglion. The geniculate ganglion

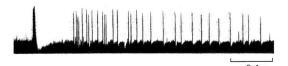

0.1 s

Figure 8.13 First recordings of extracellular action potentials of a single axon from the tongue. Response is to application of 0.05N acetic acid. Initial upward deflection is an artifact produced by solution application. (From Pfaffmann, 1939, with contrast reversed.)

contains the cells bodies from which the axons of the chorda tympani nerve arise. When the taste-bud cells in the tongue are stimulated, the axons of the chorda tympani nerve fire action potentials which travel down the nerve to the geniculate ganglion. The action potentials then invade the cell bodies, where they produce a depolarization and Ca^{2+} entry through voltage-gated Ca^{2+} channels. Barretto and co-workers recorded these increases in Ca^{2+} from single cells with GCaMP while stimulating the tongue with each of the five different modalities of taste stimuli. The results of their recordings are given in Figure 8.14A. Although some cells gave responses to more than one taste stimulus, about three-quarters of the cells were specific in their responses to only one of the five. Most of the cells showing sensitivity to more than one modality shared sensitivity to sour and bitter or umami and sweet.

Wu and co-workers (2015) then did the same experiment with nearly the same protocol (Figure 8.14B). At low concentrations of stimuli, they also reported that nearly three-quarters of the cells responded to only a single taste stimulus. At higher concentrations, however, over half the cells responded to more than one kind of taste. Nearly every possible combination was observed, though again the most common combinations were umami/sweet and sour/bitter. These results are reminiscent of responses in the olfactory bulb, where response specificity is also a function of stimulus concentration.

We might want to conclude from these experiments that the axons are indeed labeled lines, but that the labels are a little fuzzy and become less distinct at higher concentrations of stimulants. The responses of the axons may nevertheless be sufficiently specific to allow them to mediate our distinct sensations of different tastes. Perhaps the overlap is even adaptive in some way, increasing the attractiveness of food that is both meaty and sweet, or reinforcing the avoidance of substances that are both sour and bitter. Perhaps we shouldn't be troubled by the decrease in specificity at high concentrations. Taste is not, after all, the most subtle of our senses. The sloppiness may not matter.

We may instead be inclined to wonder how even this much specificity is achieved. The only cells to make classical chemical synaptic contacts onto the axons are the small number of type III cells, which other experiments show to be sensitive to sour. When, however, the type III cells are completely ablated, the responses of axons to sour are eliminated but those to sweet, umami, and bitter seem completely unaffected (Huang et al., 2006). The responses to sweet, umami, and bitter do disappear when purinergic receptors for ATP are genetically removed (Finger et al., 2005), suggesting that ATP may transmit signals of these modalities to the axons. But if ATP is the transmitter for both bitter and sweet, and if (as often) taste-bud cells mediating these two modalities are present in the same taste bud, how does the axon tell the difference? There is still a lot about the taste bud we do not understand.

Summary

The nose uses as many as a thousand different receptor molecules and can detect a rich repertoire of chemical signals which can have profound effects on behavior. Olfaction can be exquisitely sensitive, especially for the detection of pheromones. Taste by comparison is much less sensitive, restricted to only a few modalities both in insects and vertebrates, and governs only the acceptance of food.

In insects, the taste receptor neurons are located in sensilla primarily on the legs and mouth parts. Present evidence indicates that all of taste detection is ionotropic. Sensitivity to sweet and bitter is thought to be produced by receptors that are members of the GR family of receptor proteins, which we now know to form ion channels. Receptors from the IR and DEG/ENaC families are thought to be responsible for attraction to salt and water. Depolarization of sensillum neurons produces action potentials, which are conveyed directly into the CNS.

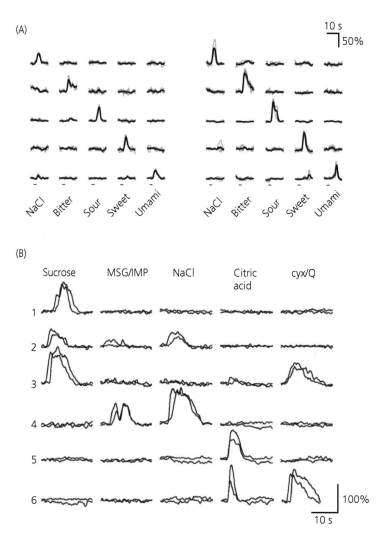

Figure 8.14 Coding of taste modalities in mouse geniculate ganglion neurons. Changes in fluorescence (ΔF) normalized to resting fluorescence (F) of genetically encoded Ca^{2+} indicator dye GCaMP3 from cell bodies of mouse neurons in the geniculate ganglion, whose axons form the chorda tympani nerve and innervate taste buds in the front of the tongue. (A) Data from one study show responses of ten ganglion neurons to five basic taste qualities. The ΔF/F values from four individual trials per stimulus (gray traces) were averaged and are shown as the black traces; horizontal bars below each column of responses mark the time and duration of the stimulus. Stimuli were NaCl (60 mM), bitter (quinine, 5 mM, and cycloheximide, 100–1000 μM), sour (citric acid, 50 mM), sweet (acesulfame potassium, 30 mM), and umami (monosodium glutamate, 50 mM, plus inosine monophosphate, 1 mM). (B) Data from second study. Representative examples of GCaMP3 signals as in (A), recorded from six geniculate ganglion neurons (from two mice) in response to taste stimuli. Neurons 1 and 5 responded only to one taste stimulus; neurons 2–4 and 6 responded to multiple taste stimuli. Stimuli were sucrose, 300 mM; monosodium glutamate, 100 mM, with 1 mM inosine monophosphate (MSG/IMP); NaCl, 250 mM; citric acid, 10 mM; and cycloheximide, 1 mM, plus quinine-HCl (cyx/Q), 0.3 mM. (A from Barretto et al., 2015; B from Wu et al., 2015.)

Taste in the mammalian tongue is mediated by cells of taste buds, located on the many papillae carpeting the oral surface. The taste buds contain receptor neurons, which are mostly specialized for detecting one of the five taste modalities. Detection of salt and acid is, as in insects, ionotropic. Taste-bud neurons responding to salt have channels selective for Na^+, which at least in some species can be amiloride-sensitive members of the DEG/ENaC family. An increase in extracellular Na^+ leads directly to membrane depolarization. In cells sensitive to sour, the depolarization seems to be produced by a combination of proton entry and block of a K^+ channel. Protons either enter the cell through membrane channels or are generated when weak acids diffuse into the cell in neutral form and dissociate within the cytoplasm.

Detection of sweet, umami, and bitter is metabotropic. The receptors are G-protein receptors from the T1R and T2R families. The T1Rs resemble metabotropic GABA and glutamate receptors with large extracellular binding sites and function as dimers: T1R2 and T1R3 together transduce signals for sweetness, whereas T1R1 and T1R3 bind amino acids and transduce signals for umami. The T2Rs resemble β-adrenergic receptors with binding sites within the interior of the plasma membrane; one taste-bud cell can express many bitter receptors.

The T1R and T2R receptors seem to transduce in a similar way. The taste stimulus binds to the receptor and activates a G protein, at least in part with the $G\alpha_{gus}$ subunit. Activated G protein stimulates PLCβ2 to produce IP_3, which binds to IP_3R3 receptors to release Ca^{2+} from internal stores. The Ca^{2+} then somehow gates TRPM5 channels to produce a depolarization. The knocking out of either the PLCβ2 or the TRPM5 channel gene greatly decreases the sensitivity of the tongue to sweet, umami, and bitter.

Although the transduction mechanisms for each of the five principal tastes are expressed in different cell types in the taste buds, the signals of the axons coming from the tongue sometimes reflect more than one modality. These signals can nevertheless produce more or less distinct sensations of the different tastes in a way we still do not understand. We are also unclear about the mechanisms of transmission of the taste-bud signals to the axons. Only one cell type in the bud makes conventional synapses onto the endings of the chorda tympani or glossopharyngeal nerves, but signals from the other cell types must also reach the endings, perhaps in part by non-vesicular release of ATP. Although considerable progress has been made elucidating the mechanisms of taste-bud function, there is still a lot to learn.

CHAPTER 9

Photoreception

Our bodies are exposed to a constant shower of electromagnetic radiation, in part man-made but in the great majority coming from the sun. The spectrum of this radiation encompasses high-energy gamma rays with wavelengths of the order of an atomic nucleus, to low energy radio waves with wavelengths of many kilometers. There is, however, a narrow band, from the near ultraviolet (300 nm) to the near infrared (1100 nm), which encompasses nearly three-quarters of the sun's energy. This band of wavelengths is further restricted by absorption by ozone and water vapor in the atmosphere, with the result that nearly half of the radiation reaching the surface of the Earth is within a range of approximately 400–700 nm (Wald, 1959). It is therefore hardly surprising that sensory proteins appeared very early in the evolution of life to absorb photons in this particular part of the electromagnetic spectrum, which we call visible light.

Proteins cannot themselves absorb visible light, because amino acids have their peak absorption in the ultraviolet. A protein molecule must be joined to a prosthetic group called a chromophore, which in combination with the protein is able to absorb illumination at longer wavelengths. Bacteria and other single-celled organisms use several different molecules to absorb light, but one family of compounds in particular was exploited very early in evolution by prokaryotes and green algae, and then subsequently by every animal from the coelenterates (Suga et al., 2008) to arthropods, mollusks, and all vertebrates including *Homo sapiens* (see Fain et al., 2010). These are the stereo-isomers of retinal. The archaeobacteria use all-*trans* retinal (Figure 9.1A).

Animals including vertebrates and man use the 11-*cis* isomer of this compound (Figure 9.1B), with a few minor variations: in some insects the 11-*cis* retinal contains an additional hydroxyl group on the third carbon of the ionone ring (and is called 3-hydroxy-11-*cis* retinal), and in fresh-water fish and some aquatic amphibians and reptiles there is an additional double bond in the ring, and the chromophore is called 3-dehydroretinal or sometimes 11-*cis* retinal$_2$ (see Fain, 2015).

Visible radiation has the fortunate property that the energy of the photon is too small to produce DNA mutation or tissue damage but is nevertheless in excess of 40 kcal mol^{-1} over the entire range of visible wavelengths, sufficiently large to exceed the activation energy of a chemical reaction. For both bacteria and animals, the absorption of a photon by retinal produces a *photoisomerization*, changing the chromophore from one isomer to another. In archaeobacteria, all-*trans* is converted to 13-*cis* (Figure 9.1A), whereas in animals, the 11-*cis* isomer is converted to all-*trans* (Figure 9.1B). These reactions change the shape of the chromophore and induce a change in the conformation of the protein to which the chromophore is attached, providing a signal that triggers light detection.

Proteins with chromophores are used as sensory pigments for a very good reason: the chemical change produced by photon absorption can be coupled to a metabotropic transduction cascade to produce an electrical response, whose sensitivity can be at the physical limit of a single quantum of light (see Figure 2.13). Pigments are not the only way of detecting electromagnetic radiation. Many organisms

Sensory Transduction. Second Edition. Gordon L. Fain, Oxford University Press (2020). © Gordon L. Fain 2020.
DOI: 10.1093/oso/9780198835028.001.0001

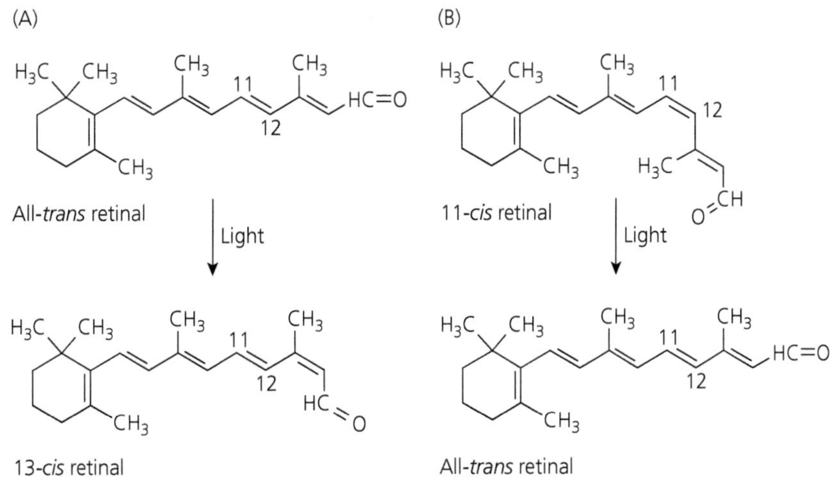

(A)

(B)

Figure 9.1 Retinal chromophores of visual pigments: photoisomerization. (A) In archaeobacteria, all-*trans* retinal is isomerized by light to 13-*cis* retinal. (B) In most animals, the chromophore is 11-*cis* retinal, which is isomerized by light to the all-*trans* form.

(including we ourselves) have receptors for infrared illumination, that is for heat, using quite a different mechanism. There are no chromophores or transduction cascades but rather proteins like TRPV1 (see Chapter 3 and Figure 3.6), which are ionotropic receptors with heat-sensitive channel gating. We defer a more complete description to Chapter 10.

Photopigment activation

Sensory proteins that use retinal chromophores are called *opsins*. As we saw in Chapter 4 (Figure 4.3A), the retinal in the X-ray crystallographic structure of vertebrate opsin lies in a binding pocket near the center of the opsin molecule. It is surrounded by seven α-helical, transmembrane domains of the protein, much as for epinephrine in the binding pocket of the β-adrenergic receptor. The retinal does not simply lie within the binding pocket but is *covalently attached* to the amino group of a lysine residue from the seventh of the α helices. Remarkably, in the sensory rhodopsins of archaeobacteria (see Spudich et al., 2000), the all-*trans* retinal chromophore also lies within a binding pocket in the center of the protein and is covalently bound, also to a lysine. Furthermore, the X-ray crystallographic structure of bacterial sensory rhodopsin has seven transmembrane α helices (Luecke et al., 2001) even though

bacterial sensory rhodopsins are not G-protein receptors (bacteria don't *have* G-protein receptors). The lysine that combines with the retinal is again on the seventh of the α helices.

The terminal nitrogen of the lysine reacts with the aldehyde of the retinal to form a double bond called a Schiff base. In most of the rhodopsins (see Rao and Oprian, 1996; Sakmar, 1998; Okada et al., 2001a; Sakmar et al., 2002), including those of archaeobacteria (see Hoff et al., 1997; Spudich et al., 2000), this Schiff base is protonated. Single charges are rarely if ever found buried in the middle of a protein, but pairs of charges called *salt bridges* are quite common. In most vertebrate rhodopsins, a salt bridge is formed by the protonated Schiff base lysine and a negative charge contributed by a glutamate residue from the third transmembrane helix (Figure 9.2).

The protonation of the Schiff base has a large effect on the wavelength of light absorbed by the visual pigment. The absorption peak of free 11-*cis* retinal depends somewhat upon the solvent into which the retinal is dissolved but is at about 360–380 nm in the near ultraviolet (Knowles and Dartnall, 1977). When the chromophore combines with opsin, the resulting pigment can have its absorption maximum shifted by as much as 200 nm toward longer wavelengths. A significant fraction of this shift in wavelength is produced by protonation of

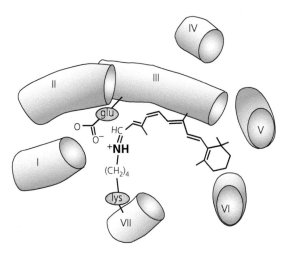

Figure 9.2 Attachment of chromophore to opsin. Retinal forms a covalent Schiff base attachment with a lysine from helix VII. This Schiff base in most visual pigments is protonated and exists as a salt bridge together with the negative charge of an acidic group from a glutamate in helix III.

the Schiff base (see Knowles and Dartnall, 1977). In humans and many other vertebrates, most of vision occurs between the wavelengths of 400 nm and 700 nm, from the blue to the deep red, as I have said. Vision in this part of the spectrum would hardly be possible if the absorption peak of our pigments were the same as that of the free chromophore.

The protonation of the Schiff base cannot be the only cause of the shift in wavelength; if it were, all pigments would have the same absorption peak. In human photoreceptors, there are four different pigments all structurally similar to one another with at least 40 percent identity of amino acid sequence (Nathans et al., 1986), all absorbing in somewhat different parts of the visible spectrum. The maximum absorption of our rod pigment is at about 500 nm in the blue-green, and our cone pigments absorb maximally at about 415–420 nm in the blue, 535–540 nm in the green, and 560–570 nm in the yellow (see Bowmaker et al., 1980; Nunn et al., 1984). Many amino acids in the vicinity of the chromophore can make subtle alterations in the electronic environment to tune the peak of absorption to different wavelengths. Furthermore, human middle and long wavelength cone pigments, and cone pigments in many other vertebrates, contain an extracellular binding site for Cl-, whose negative charge also

plays a significant role in moving the absorption maximum to longer wavelengths (see Ebrey and Koutalos, 2001; Stenkamp et al., 2002).

The protonation of the Schiff base has an important effect on the stability of the pigment. The sensitivity of sensory detection depends not only upon the sensitivity of transduction but also upon the noisiness of the receptor. Photoreceptors both in arthropods and in vertebrates are able to signal the absorption of a single quantum of light (Figure 2.13), but they would not be able to detect single photons reliably if rhodopsin even in darkness produced responses spontaneously. Because a single photoreceptor contains hundreds of millions of rhodopsins, even a very small rate of spontaneous conversion of the pigment to an active form would compromise the sensitivity of the cell. For this reason, rhodopsin in the dark is extraordinarily stable, with a half-life for spontaneous activation measured in the hundreds or even thousands of years (Yau et al., 1979; Baylor et al., 1980). Part of the reason for this stability is the formation of the salt bridge between the protonated lysine and the glutamate, because if this salt bridge is removed, for example by site-directed mutagenesis of the glutamate, the resulting visual pigment becomes spontaneously active (Robinson et al., 1992; Rao and Oprian, 1996; Spudich et al., 1997). The formation of the salt bridge cannot be the only cause of stability, because in some insects, birds, and even mammals (e.g. mouse), there are cones with photopigments sensitive to ultraviolet light, having a peak absorption close to that of free 11-*cis* retinal. In these pigments the Schiff bases appear *not* to be protonated (Shi et al., 2001; Fasick et al., 2002). There seem to be a variety of additional interactions probably including hydrogen bonds and hydrophobic forces within the protein structure which also contribute to stability (see Sakmar, 1998; Okada et al., 2001b; Sakmar et al., 2002; Filipek et al., 2003).

When rhodopsin absorbs a photon, the chromophore isomerizes and changes its molecular configuration. In vertebrate rhodopsin, the whole of the chromophore molecule rotates around the bond between carbons 11 and 12. This effect produces a rather large strain in the conformation of the protein component of the pigment. The strain is relieved by deprotonation of the Schiff base and movement of the α helices, in vertebrate rhodopsin predominantly helices V and VI (see Figure 4.3B), producing an intermediate called metarhodopsin or Rh*. It is this form of rhodopsin that triggers activation of the sensory cascade.

Phototransduction

What happens next depends very much upon the organism, because the mechanism of transduction is different for bacteria and animals, and for vertebrates and many invertebrates. In archaeobacteria, the cascade is quite similar to the one for chemotaxis (see Chapter 7 and Figure 7.2). Sensory rhodopsin is coupled to a transduction protein called Htr, which is linked to a CheA histidine kinase (see Hoff et al., 1997). Just as for chemotaxis, the CheA protein controls the phosphorylation of a CheY second messenger which binds to the flagellar motor and regulates the rate of flagella reversals, or tumbles. Adaptation is produced by methylation (see Figure 7.3), but it is the Htr protein that is methylated rather than the visual pigment.

In all multicellular animals so far investigated, rhodopsin is coupled to a heterotrimeric G protein. Transduction is produced primarily, if not exclusively, by binding of the $G\alpha$ subunit of the G protein to an effector enzyme. In insects, the $G\alpha$ subunit is a member of the $G\alpha_{q/11}$ family and is coupled to a phospholipase C (PLC), encoded in *Drosophila* by the *norpA* gene. In vertebrates, the $G\alpha$ subunit is a member of the $G\alpha_i/G\alpha_o$ family and is coupled to a phosphodiesterase (PDE6) (see Figure 4.5B). For *Drosophila*, the second messenger that gates the channels has not been identified. For vertebrates it is cGMP. In insects and other arthropods, light opens channels permeable to cations and produces a depolarization (Figure 1.5A). In vertebrate rods and cones, light *closes* channels that are also permeable to cations, producing a hyperpolarization (Figure 1.5B).

The variety of transduction mechanisms can perhaps best be illustrated by taking one particularly interesting example, the eye of the scallop *Pecten*. We described the peculiar anatomy of this eye in Chapter 2. It has two separate retinas, each containing its own population of receptor cells (Figure 2.11). Both have sensory membrane containing photopigment. In cells of the distal retina, the sensory membrane is elaborated from a modified cilium, whereas for cells in the proximal retina, the sensory membrane consists of microvilli (Barber et al., 1967). Hartline (1938) first showed that the physiology of the two retinas is also different. He recorded action potentials from the nerves coming from the retinas and discovered that cells in the proximal retina produce a burst of action potentials when the light goes on, but those in the distal retina are excited when light goes off.

The first intracellular recordings showed that light depolarizes the cells of the proximal retina much as in insects and other arthropods, whereas it hyperpolarizes the cells of the distal retina (Gorman and McReynolds, 1969; McReynolds and Gorman, 1970b). When the photoreceptors are voltage clamped (Figure 9.3A), light produces an inward (negative) current for the proximal cells but an outward (positive) current for the distal cells (Gomez and Nasi, 1997a). The inward current of the proximal cells has a reversal potential just positive of zero and is produced by an increase in permeability predominantly to Na^+ (McReynolds and Gorman, 1970b, 1974; Gomez and Nasi, 1996). The increase in Na^+ permeability causes the cell to depolarize—see Eq. (3.5). The outward current of the distal cells has a reversal potential near –70 mV, more negative than the dark resting membrane potential in these cells. It is produced predominantly by an increase in permeability to K^+, producing a hyperpolarization (McReynolds and Gorman, 1970b, 1974; Cornwall and Gorman, 1983a; Gomez and Nasi, 1994).

Although the sensory cascades for the two kinds of photoreceptors have not yet been elucidated in detail (see Nasi et al., 2000), they seem to have little in common. The photopigments have almost identical spectral absorption curves with maxima at about 500 nm (McReynolds and Gorman, 1970b; Cornwall and Gorman, 1983b). They are nevertheless distinct molecules with different amino acid sequences particularly in the interconnecting loop region between helices V and VI (Kojima et al., 1997). This is the part of the protein that is exposed to the cytoplasm and is thought to play an important role in activating the G protein. The proximal and distal retinas express different G proteins (Kojima et al., 1997; Gomez and Nasi, 2000): the proximal (depolarizing)

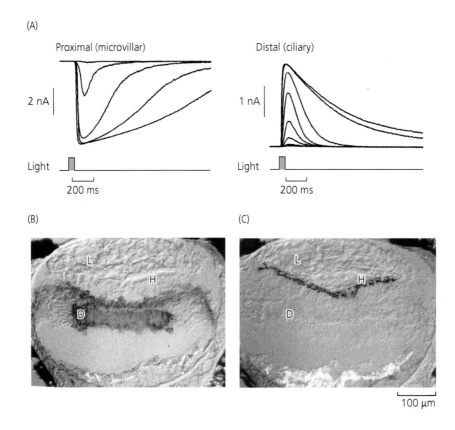

Figure 9.3 Phototransduction in the eye of the scallop. (A) Photocurrents of opposite polarity from voltage-clamped photoreceptors in the two retinal layers. The negative current recorded from the proximal microvillar photoreceptors ($V_m = -50$ mV) indicates an increase in conductance to Na$^+$ and other cations, whereas the positive current recorded from the distal ciliary photoreceptors ($V_m = -30$ mV) indicates an increase in conductance mostly to K$^+$. See Eq. (3.7). Anti-α_q (B) and anti-α_o (C) antisera were used for immunohistochemical localization of the G-protein α subunits. L, Lens; H, hyperpolarizing (distal) photoreceptor layer; D, depolarizing (proximal) photoreceptor layer. (A from Gomez and Nasi, 1997a; B and C from Kojima et al., 1997.)

photoreceptors express a Gα_q, like the depolarizing photoreceptors of *Drosophila* (Figure 9.3B), whereas the distal (hyperpolarizing) photoreceptors express a Gα_o (Figure 9.3C).

The channels are also different. The most commonly observed channel openings for the depolarizing photoreceptors have a conductance of about 48 pS, whereas those for the hyperpolarizing photoreceptors have a conductance of about 26 pS. The channels of the hyperpolarizing photoreceptors can be activated by cGMP and its analogs (del Pilar Gomez and Nasi, 1995) and blocked by a variety of compounds known to block cyclic-nucleotide-gated channels (Gomez and Nasi, 1997b). It would there-

fore appear that the distal photoreceptors in *Pecten* have some kind of cyclic-nucleotide-gated channel. These channels are, however, different from those of parietal photoreceptors (Chapter 4) or olfactory receptor cells (Chapter 7), because they have very little permeability to Na$^+$ and are instead selectively permeable to K$^+$. The channels of depolarizing photoreceptors seem *not* to be gated by cyclic nucleotides (Nasi and Gomez, 1991). Their mechanism of activation is still uncertain.

Because transduction can be so variable in the different photoreceptors of the animal kingdom, and because only a few species have been examined in detail, we are not yet in a position to draw any firm conclusions about the distribution of different G proteins, second messengers, and channels in the various phyla. The best we can do at present is to pick certain well-studied examples and explore them in detail. The choice is easy. The most thoroughly investigated photoreceptors are certainly those of two arthropods, the horseshoe crab *Limulus* and the fruit fly *Drosophila*; and the rods and cones of vertebrates.

The photoreceptors of arthropods

Like receptor cells in the proximal retina of scallop, all arthropod photoreceptors have sensory membrane produced by numerous microvilli which, together, form a structure called a *rhabdomere* (Figure 2.5). The organization of the rhabdomere varies considerably among different species. The ventral eye of the horseshoe crab *Limulus* and the compound eye of the fruit fly *Drosophila* provide useful contrasting examples.

Limulus is more closely related to spiders than to crabs. It has photoreceptors in many different places (Figure 9.4A). There are two large lateral eyes on either side of the body, homologous to the principal eyes of a lobster or dragonfly. There are in addition two median eyes on the top and in the center of the carapace, some of whose photoreceptors are sensitive to ultraviolet light (Nolte and Brown, 1969). There are even photoreceptors in the tail. The most useful cells for the study of transduction, however, have been the cells lying along the ventral nerve underneath the animal, located approximately in the center of the body. What these photoreceptors are doing in this unusual location is uncertain. Perhaps they help the animal detect changes in illumination when it is swimming—horseshoe crabs often swim upside down.

The ventral photoreceptors of *Limulus* are surrounded by glial cells and connective tissue, which can be peeled away to reveal the structure shown in Figure 9.4B (Stern et al., 1982). One part of the cell contains a dense skein of microvilli and is called the R (or rhabdomeric) lobe. It is somewhat analogous to the outer segment of a vertebrate rod or cone (Figures 2.7 and 9.13). The rest of the cell contains the nucleus and metabolic machinery of the photoreceptor and is called the A (or arhabdomeric)

lobe. The A lobe is analogous to the rod or cone inner segment.

The microvilli are a few microns in length and mostly restricted to the outermost surface of the R lobe. At the base of the microvilli, there is an extensive network of smooth endoplasmic reticulum (ER), which serves as a store for calcium (Payne et al., 1988). As we shall see, there is abundant evidence of a role of IP_3-gated calcium release in the mechanism of transduction, and the smooth ER at the base of the microvilli may be functioning much like the ER of other cells, sequestering calcium and releasing it when IP_3 binds to IP_3 receptors in the vesicle membrane (see Figures 4.7 and 4.8). The whole of the cell is of the order of 50 microns wide and several hundred microns long, with a volume fifty to a hundred times greater than even the largest vertebrate photoreceptor. The cells in the other eyes of *Limulus* are also large and ample targets for poking with pipettes. That is why *Limulus* has been so attractive to physiologists. The photoreceptors of *Limulus* were the first of any species whose responses were recorded with intracellular microelectrodes (Hartline et al., 1952), the first to provide evidence of single-photon sensitivity (Yeandle, 1958), the first to be studied with voltage clamp (Millecchia and Mauro, 1969a), and the first to implicate IP_3 and Ca^{2+} as important messenger substances in photoreceptor transduction and adaptation (Lisman and Brown, 1975; Fein et al., 1984).

Drosophila photoreceptors are much smaller but have the great advantage that they can be manipulated by the powerful tools of fruit-fly genetics. The photoreceptors are found in three small ocelli at the top of the head and in the two large compound eyes (Figure 9.5A), which each have 700–800 units called ommatidia with a clearly defined structure (see Hardie and Postma, 2008; Yau and Hardie, 2009; Fain et al., 2010). At the top of each ommatidium there is a cornea and a fluid-filled pseudocone that together act as a lens focusing light onto the rhabdomeres of the photoreceptors (Figure 9.5B). Below these structures are the Semper (supporting) cells and eight photoreceptors, six of which (called R1–R6) contain the same visual pigment with peak absorption at 480 nm in the bluish green. The other two cells, R7 and R8, have rhabdomeres more centrally located, with R7 lying above R8 so that

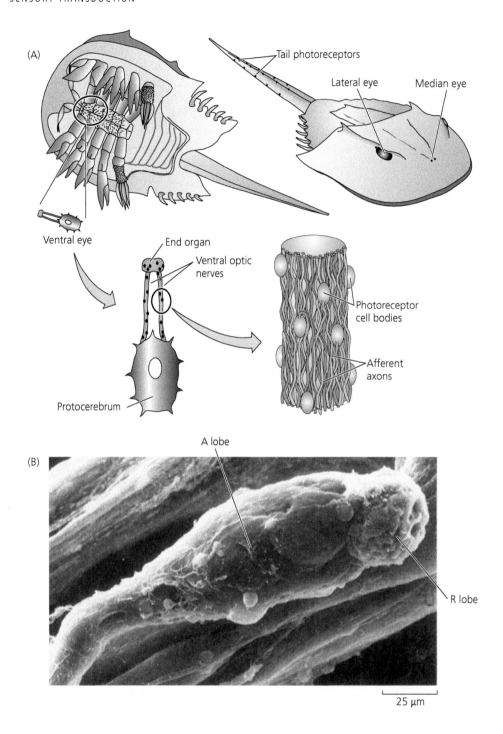

Figure 9.4 Photoreceptors of the horseshoe crab *Limulus*. (A) Photoreceptors are located in many places on the *Limulus* body. Those of the ventral eye are found dispersed along the ventral optic nerve underneath the animal. (B) Ventral photoreceptor denuded of glial cells and connective tissue. (A after Calman and Chamberlain, 1982; Dorlochter and Stieve, 1997; B from Stern et al., 1982.)

cross-sections of the ommatidium show the rhabdomere of one cell or the other but never both (Figure 9.5C). In most of the retina, R7 contains one of two ultraviolet-absorbing photopigments, with peak absorption either at 345 nm or 375 nm, whereas R8 contains one of two visible-absorbing pigments with peak absorption either at 437 nm or 508 nm. Different ommatidia express different pigments, but expression is coordinated so that when R7 expresses the 345-nm pigment, R8 expresses the 440-nm pigment, and similarly for the other two (see Montell, 1999; Minke and Hardie, 2000). The rhabdomeres of each photoreceptor contain of the order of 30,000 microvilli each 1–2 μm in length, which are continuous with the plasma membrane and make up over 90 percent of its area (Figure 2.5). The microvillar membrane contains most of the machinery for transducing light into an electrical signal.

In some insects, the rhabdomeres of individual photoreceptors in an ommatidium are fused to form a single structure called the rhabdome (see Chapter 5 of Cronin et al., 2014). In other species, including flies, the rhabdomeres of the different photoreceptors are physically separate from one another (Figure 9.5C). This separation has the consequence that each rhabdomere is directed toward a slightly different place in space. For R7 and R8, this difference poses no difficulty, because there is only one of each of these receptors per ommatidium. For the R1–R6 cells, however, the difference in orientation poses a problem. These photoreceptors need to combine their signals to increase sensitivity and lower noise, but how do they do this if each cell is pointed in a different direction? The animal solves this problem by combining signals from six photoreceptors—not in the same ommatidium but in different ommatidia, all pointed in a similar direction. In this way, signals are summed from cells detecting light from approximately the same spatial location.

The photoreceptors of *Drosophila* are much smaller than those of *Limulus* and more difficult to study with microelectrodes. *Drosophila* ommatidia can be dissociated by mechanical trituration (Hardie, 1991; Hardie et al., 1991; Ranganathan et al., 1991), and single cells can then be used—with difficulty—for patch-clamp recording. The advantages of this species for the study of phototransduction are never-

theless very great. The *Drosophila* genome is entirely sequenced, much is known about its genetics and development, animals are easy to grow and reproduce rapidly, mutant animals with disrupted phototransduction can be readily isolated, and the range of available techniques for studying genes and proteins—including altering or knocking out particular genomic sequences—is as powerful for *Drosophila* as for any other multicellular animal.

Transduction in arthropods

We begin with *Drosophila* because so much has been learned about arthropod vision from its genetics and molecular biology. Starting in the late 1960s and early 1970s, a large number of mutant *Drosophila* were isolated with abnormal visual behavior or defective photoreceptor light responses (see Pak, 1995; Minke and Hardie, 2000; Hardie and Postma, 2008). This work led to the identification of genes necessary for some aspect of phototransduction. The most important are given in Table 9.1, together with the proposed function of the gene product in the sensory cascade (Pak, 1995; Montell, 1999; Xu et al., 2000).

These studies have demonstrated that the G-protein α subunit responsible for activation in *Drosophila* is a member of the $G\alpha_{q/11}$ family. Mutations in the gene for this protein produce large decreases in light sensitivity (see Figure 2.13C and Scott et al., 1995; Scott and Zuker, 1998). Mutations in another gene called *norpA* produce *no receptor potential*. This gene was subsequently cloned and shown to encode a PLC abundantly expressed in the eye (Bloomquist et al., 1988). Activation of PLC produces a large and rapid decrease in PIP_2 (Hardie et al., 2001), generating the two second messengers IP_3 and diacylglycerol (DAG) (see Figure 4.6). In many cells, IP_3 binds to IP_3 receptors and triggers Ca^{2+} release (Figure 4.8). The genome of *Drosophila* seems to have only a single gene for an IP_3 receptor, and this gene is known to be expressed in the eye. Nevertheless, animals genetically engineered so that expression of this gene has been knocked out still have nearly normal photoreceptor responses (Acharya et al., 1997; Raghu et al., 2000b; but see Kohn et al., 2015; Bollepalli et al., 2017), throwing considerable doubt on the role of IP_3 in *Drosophila* phototransduction.

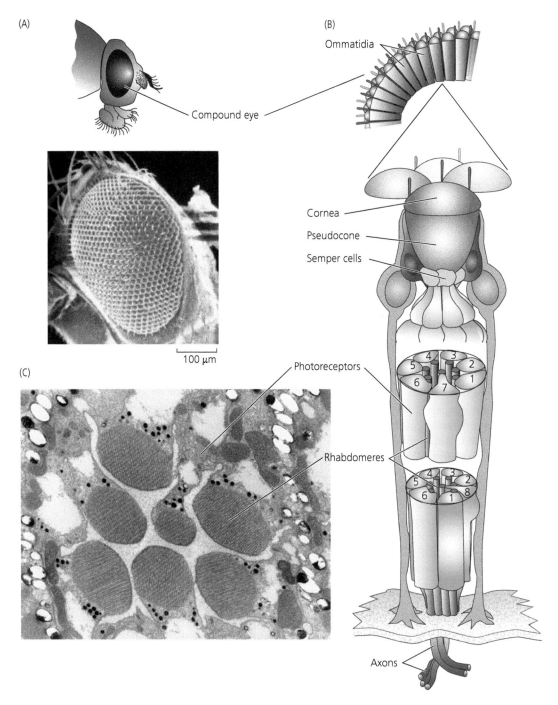

Figure 9.5 The compound eye of the fruit fly *Drosophila*. (A) Diagram and scanning electron micrograph of the compound eye. (B) The ommatidium. (C) Cross-section of the ommatidium. Magnification 11,000×. (A from Hodgkin and Bryant, 1978; B after Carlson et al., 1984; C courtesy of P. Raghu and R. Hardie.)

Table 9.1 Phototransduction proteins of *Drosophila*

Gene	Protein	Function
arrestin 1	arrestin 1	rhodopsin inactivation
arrestin 2	arrestin 2	rhodopsin inactivation
calx	Na$^+$/Ca^{2+} exchange protein	Ca^{2+} extrusion
cds	CDP-DAG synthase	PIP$_2$ metabolism
Ga$_q$	α subunit of G protein	activation
Gβ, Gγ	β and γ subunits of G protein	activation
inaC	protein kinase C	response turnoff
inaD	PDZ-containing protein	formation of transducisome/signalplex
ninaC	myosin III	response decay
ninaE	rhodopsin of R1–R6	visual pigment
norpA	PLC	activation
rdgA	DAG kinase	PIP$_2$/DAG metabolism
rdgB	phosphoinositol transfer-protein	PIP$_2$/DAG metabolism
rdgC	rhodopsin phosphatase	dephosphorylation of rhodopsin
trp	cation channel	light channel
trpl	cation channel	light channel

CDP-DAG, Cytidine diphosphate diacylglycerol.

The other possible second messenger, DAG, serves in many cells to activate protein kinase C (PKC). PKC seems to have some role in transduction, because mutations in a gene for PKC called *inaC* produce photoresponses with abnormal decay and adaptation. However, light responses in flies with eyes lacking PKC still show normal activation, so PKC cannot be essential for the production of the response. No genes have yet been isolated in screens of this sort for any aspect of cyclic-nucleotide metabolism.

These results show that activation in *Drosophila* photoreceptors is produced by binding of Gα$_q$•GTP to PLC, which hydrolyzes PIP$_2$ to generate IP$_3$ and DAG. What happens next is unclear. Although in *Drosophila* there is no evidence of a role of IP$_3$ receptors, the messenger of activation may be DAG itself (Raghu et al., 2000a; Estacion et al., 2001; Delgado et al., 2014) or some by-product of DAG metabolism, such as a polyunsaturated fatty acid like arachidonic or linolenic acid (Chyb et al., 1999). An important component of the transduction cascade may be DAG kinase (DGK), the enzyme that phosphorylates and inactivates DAG. This protein is encoded by the *rdgA* gene (Figure 9.6A), and mutations of

this gene cause the channels to activate even in darkness (Raghu et al., 2000a). In Figure 9.6B, responses were recorded first from a *norpA* hypomorph, which expresses only a very small amount of the *norpA* protein, which is PLC. Because the photoreceptor has so little of this essential effector enzyme, the light response of the hypomorph gave only a few small, sporadic quantum bumps. If this PLC hypomorph also has a mutation in the *rdgA* gene so that almost no DGK is expressed, the response is augmented by more than a hundred times (Hardie et al., 2002). The simplest interpretation of this observation is that the decrease in the concentration of DGK, the enzyme that inactivates DAG, leads to a buildup of DAG; this in turn augments channel opening. This result would seem to implicate DAG or one of its metabolites in channel gating.

Activation of PLC also results in depletion of PIP$_2$ and the release of protons, which have been proposed together to gate the opening of the light-dependent channels (Huang et al., 2010). An intriguing possibility is that depletion of PIP$_2$, which is a significant constituent of the microvillar membrane, reduces the surface area of the membrane and stimulates

(A)

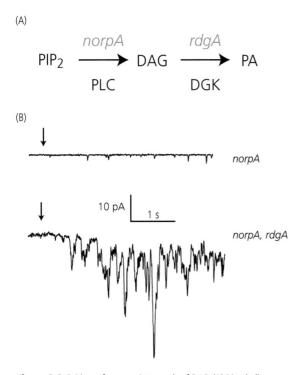

(B)

norpA

10 pA | 1 s

norpA, rdgA

Figure 9.6 Evidence for an excitatory role of DAG. (A) Metabolism of DAG in *Drosophila* photoreceptor. PIP$_2$, Phosphatidylinositol 4,5-bisphosphate; *norpA*, gene for *Drosophila* PLC (phospholipase C); *rdgA*, gene for DGK (diacylgylcerol kinase); PA, phospatidic acid. (B) Mutations in DGK greatly augment responses in PLC and Gα_q hypomorphs. Upper trace, a bright flash in a severe *norpA* hypomorph (*norpAP12*) expressing little PLC elicits no more than a few sporadic quantum bumps 1–2 pA in amplitude. Lower trace, the response to the same intensity is enhanced about one-hundred-fold in the double mutant *norpAp12/rdgA1* under-expressing both PLC and DGK. (From Hardie et al., 2002.)

the photoreceptor by mechanosensitive-channel opening (Hardie and Franze, 2012). In the experiment shown in Figure 9.7, an atomic force microscope was used to measure the change in the length of the phnotoreceptors. A 40-μm polystyrene bead was placed on top of the ommatidia in an intact retina, and the position of the bead was measured with a laser and photodiode. Light produced a graded decrease in the length of the rhabdomere (Figure 9.7A), presumably as more and more microvilli contracted. In response to a bright flash, the rhabdomeric length can shorten by as much as 0.5 μm. The decrease in length can precede the change in photoreceptor membrane potential (Figure 9.7B) and occurs over a similar range of light intensities. Moreover,

light responses can be produced in photoreceptors lacking native light-dependent channels, if the photoreceptors are supplied with extrinsic mechanosensitive channels. These channels are apparently activated directly by the membrane stretch produced when the microvilli contract.

In contrast to the results from *Drosophila*, there is considerable support in *Limulus* for a direct role of both IP$_3$ and Ca^{2+} in transduction. The ventral photoreceptors contain abundant receptors for IP$_3$ (Ukhanov et al., 1998), and if these receptors are blocked, for example with the compound heparin (Frank and Fein, 1991) or with 2-aminoethoxydiphenyl borate (2-APB) (see Wang et al., 2002), the light response is inhibited. Injection of IP$_3$ into the R lobe of the photoreceptor produces both a release of Ca^{2+} from internal stores and a depolarization like the one produced by light (Figure 9.8A). The increase in Ca^{2+} is rapid and seems to occur even before the opening of the light-dependent channels (Ukhanov and Payne, 1997; Payne and Demas, 2000). The open symbols in Figure 9.8B give the time course of the free-Ca^{2+} concentration measured with a fluorescent dye at four different light intensities, and the lines show the time course of the voltage response.

Other experiments seem, however, to support a role for cyclic nucleotides in producing the *Limulus* light response (Johnson et al., 1986; Bacigalupo et al., 1991; Chen et al., 1999b, 2001a; Garger et al., 2001). Injection of cGMP into the R lobe of a ventral photoreceptor produces a depolarization like that produced by light (Figure 9.9A), and the hydrolysis-resistant cGMP analog 8-Br-cGMP applied to inside-out patches from microvillar membrane gates the opening of channels, which are similar in their properties to channels recorded from on-cell patches stimulated with light (Figure 9.9B). The light-dependent increase in Ca^{2+} (Figure 9.8) has been proposed in *Limulus* to stimulate a guanylyl cyclase (Lisman et al., 2002), producing an increase in cGMP that opens the channels (see also Deckert et al., 1992).

In summary, everyone agrees that light produces the activation of a PLC in arthropod photoreceptors. The PLC then produces IP$_3$ and DAG. The IP$_3$ generates a Ca^{2+} increase in *Limulus*, but most of the available evidence indicates that the great majority of the increase in Ca^{2+} in *Drosophila* is produced by influx through channels in the plasma membrane and that IP$_3$-gated Ca^{2+} release has little if any role

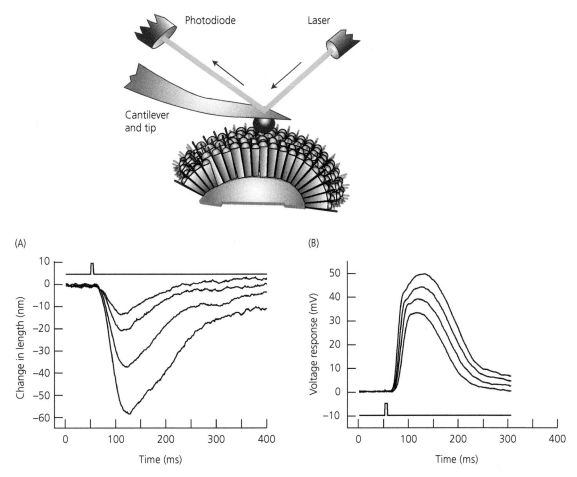

Figure 9.7 Illumination produces a change in length of the rhabdomere. Atomic-force microscopic measurements of the change in length of rhabdomeres in an intact retina. The polystyrene bead of the cantilever of an atomic force microscope was placed directly on top of the distal tips of the ommatidia in an excised *Drosophila* retina, and the change in position of the cantilever was measured with a laser and photodiode. (A) Contractions in a wild-type *Drosophila* retina in response to 5-ms flashes, with intensities from about 200 to 8000 effectively absorbed photons per photoreceptor. Timing of flashes is indicated by uppermost trace. (B) Voltage responses to the same stimuli from a whole-cell patch recording of a single dissociated photoreceptor. Absolute values of voltage are arbitrary. Timing of flashes is indicated by lowermost trace. (Courtesy of R. C. Hardie.)

in transduction. DAG, the other second messenger produced by PLC, typically activates a PKC. There is evidence of a role for PKC in both *Drosophila* (Hardie et al., 1993) and *Limulus* (Dabdoub and Payne, 1999) but no indication that PKC is required for the production of the light response. The channels may be activated by DAG itself or one of its metabolites. Light also depletes the microvillar membrane of PIP_2, which may trigger excitation in combination with proton production and membrane contraction. Some evidence suggests a role for cyclic nucleotides in *Limulus*, but there is no indication that cyclic

nucleotides are responsible for channel gating in *Drosophila*.

Photoreceptor channels in arthropods

The first voltage-clamp recordings from *Limulus* (Millecchia and Mauro, 1969a) showed that light produces an increase in conductance primarily to Na^+ with a reversal potential between 10 and 15 mV positive of zero. Light also produces an increase in conductance to Na^+ in *Drosophila*, and there are two molecular forms of light-activated channels with a

(A)

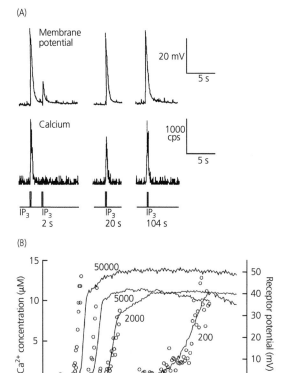

(B)

Figure 9.8 Calcium and IP₃ in *Limulus* ventral photoreceptors. (A) Effect of IP₃ injection on membrane potential and free-calcium concentration. Calcium was measured with the luminescent protein aequorin. cps, Counts per second measured by the photomultiplier tube from the luminescence of aequorin. Increase in counts indicates increase in the free-Ca^{2+} concentration. (B) Comparison of the time course of the increase in Ca^{2+} concentration measured with a fluorescent indicator dye (circles) with membrane potential (solid lines) at four different light intensities, given in units of effective photons next to each trace. (A from Payne et al., 1990; B from Payne and Demas, 2000.)

similar structure. The first to be identified and sequenced is encoded by the *trp* gene, which stands for *transient receptor potential*. This gene was the very first TRP-channel gene to be discovered and is the founding member of the large ion channel superfamily now known to be present throughout the animal kingdom, with for example twenty-seven distinct TRP channel genes in humans alone. The reason for the name TRP can be seen in Figure

9.10A (from Hardie et al., 2001). The response on the left is from a voltage-clamped wild-type *Drosophila* photoreceptor to a step of light 5 s in duration. The current is negative or inward as for the voltage-clamp currents of rhabdomeric photoreceptors in the proximal retina of *Pecten* (Figure 9.3A). The amplitude of the current is largest just at the beginning of the response, and the photoreceptor current then rapidly adapts and approaches a steady plateau level. In the *trp* mutant photoreceptors, on the other hand, the response decays back almost to the baseline even in the presence of maintained illumination (Figure 9.10A, right). There is little if any steady plateau response—compare the amplitude of the response near the end of the light stimulus in the two records.

The sequence of the *Drosophila trp* protein (Figure 9.11) shows many similarities to the thermoreceptor TRPV1 (Figure 3.6A) and mechanoreceptor NOMPC (Figure 5.10D). The amino-terminal end of the protein is cytosolic and contains ankyrin repeat units, but there are only four instead of the six in TRPV1 and twenty-nine found in NOMPC. The carboxyl end of the protein is also cytosolic and has several recognizable domains. Among the most interesting for our purposes are a calmodulin-binding domain suggesting possible modulation of the channel by Ca^{2+}-calmodulin, and a region near the carboxyl terminus for binding to a PDZ domain. We saw in Figure 2.5C that the INAD protein of *Drosophila* photoreceptors contains five protein-binding PDZ domains and forms a transducisome or signalplex, which is held in place within the microvillus (Li and Montell, 2000). The three-dimensional structure of the *Drosophila trp* protein has not yet been solved, but it is likely to be tetrameric like other channels of the TRP family (see Figures 3.6D and 5.12A, and Madej and Ziegler, 2018).

A second channel was subsequently discovered, whose gene was called *transient-receptor-potential-like* (*trpl*) (Phillips et al., 1992). The amino acid sequence of the *trpl* protein, though similar to that of *trp* and also containing ankyrin repeats and calmodulin-binding domains, lacks a region for binding to INAD and is not thought to associate with the transducisome. Both the *trp* and *trpl* proteins are localized to the rhabdomeres of the photoreceptors

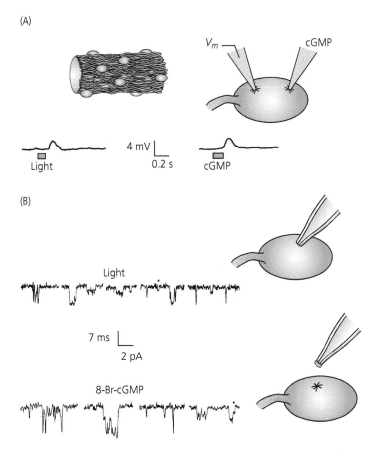

Figure 9.9 Evidence of cGMP-gated channels in ventral photoreceptors of *Limulus*. (A) Membrane potential was measured with an intracellular micropipette, and a second pipette was used to inject cGMP into the cell. Injection of cGMP produced a depolarization like that produced by light. (B) Comparison of light-activated single-channels recorded on-cell (above), and channel openings in an inside-out patch gated by the perfusion of 8-bromo-cGMP (below). (A from Johnson et al., 1986; B from Bacigalupo et al., 1991.)

(Huber et al., 1996; Niemeyer et al., 1996), but the *trp* protein is much more abundant (see Montell, 1999) and dominates the light response (Reuss et al., 1997).

The light responses of flies that are mutant for *trp* or for *trpl* behave differently, indicating that the *trp* and *trpl* channels have different properties (Niemeyer et al., 1996; Reuss et al., 1997). In particular, the light-dependent change in conductance recorded from these two mutant strains has a dramatically different permeability to Ca^{2+}. In the experiment shown in Figure 9.10B, voltage-clamp recordings were made with whole-cell patch pipettes containing a solution of 130 mM Cs^+ and no other permeable cat-

ions. This solution was dialyzed into the cell to replace the cations normally present. The external solution contained 10 mM Ca^{2+} and no other permeable cations. The reversal potential of the response was measured much as I described for hair cells in Chapter 3, and it was used to estimate the relative permeability of the channel to Ca^{2+} and Cs^+, that is P_{Ca}/P_{Cs}. As can be seen from the figure, the reversal potential was quite different for *trp* and *trpl* mutant photoreceptors. For wild-type flies, the value of the reversal potential was intermediate between the values for the two mutants. These experiments show that the TRPL channels left in the photoreceptors once the *trp* protein was deleted have a P_{Ca}/P_{Cs}

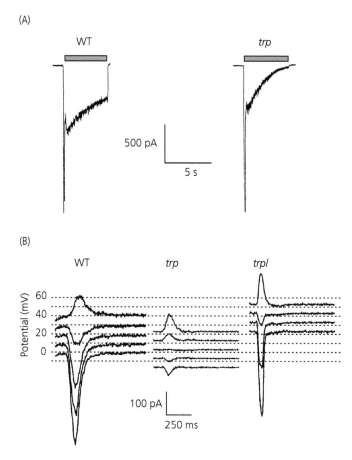

Figure 9.10 The TRP and TRPL channels of *Drosophila*. Currents were recorded from dissociated photoreceptors with whole-cell voltage clamp. (A) Photoreceptors from flies mutant for *trp* have transient light responses. (B) Measurement of the reversal potential of the light response from photoreceptors of three different lines of flies: WT (wild-type), *trp* (flies mutant for the *trp* gene and lacking functional TRP channels), and *trpl* (flies mutant for the *trpl* gene and lacking TRPL channels). Light responses to the same bright flashes were recorded for each photoreceptor at the different holding potentials indicated by the scale to the left. Notice that the reversal potential in *trp* photoreceptors is much less positive than for *trpl* or WT, showing that the TRPL channels left once TRP is removed have a much lower Ca^{2+} permeability than TRP or wild-type channels. (A from Hardie et al., 2001; B from Reuss et al., 1997.)

of only about seven, whereas TRP channels (in *trpl* mutant flies) have a surprisingly large P_{Ca}/P_{Cs} of about fifty-seven (Hardie, 2014). For both mutant strains and wild-type flies, Cs^+ and Na^+ were nearly equally permeant, that is P_{Cs}/P_{Na} was nearly one. TRP channels seem also to be present in the *Limulus* ventral eye (Bandyopadhyay and Payne, 2004), but their properties are likely to be somewhat different from those in *Drosophila*. The *Limulus* light-dependent conductance has considerably less Ca^{2+} permeability (see for example Brown and Mote, 1974).

The role of Ca^{2+} in the regulation of gain and turnoff

In both *Limulus* and *Drosophila*, light produces an increase in intracellular Ca^{2+} either by entry through the TRP channels from the extracellular medium or by IP_3-dependent release from intracellular stores. This light-induced increase in Ca^{2+} concentration may be as large as 100–200 μM and may act on transduction at some early stage. An experiment that supports such an interpretation is shown in

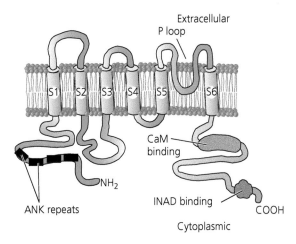

Extracellular
P loop

S1 S2 S3 S4 S5 S6

CaM
binding

NH₂

ANK repeats

INAD binding

COOH

Cytoplasmic

Figure 9.11 Predicted membrane topology of *the Drosophila trp* gene product.

Figure 9.12. This experiment utilized DM-nitrophen, often called caged Ca^{2+}. This molecule has the convenient property that it binds Ca^{2+} with high affinity, literally forming a cage around the calcium ion (Figure 9.12A). Bright ultraviolet light produces a photochemical reaction that causes the cage to fall apart, releasing the Ca^{2+}. In Figure 9.12B, a *Drosophila* photoreceptor was voltage-clamped with a whole-cell patch electrode (Hardie, 1995). Exposure to bright light (blue stimulus marker) produced an inward current with a latency of the order of 10 ms (note the rapid time base of the recording). Then, as the current was beginning to increase, a xenon flash lamp was triggered to give a brief but very intense exposure to ultraviolet light (arrows, Xe). In the first two records without DM-nitrophen, labeled "Controls," the flash lamp produced a brief deflection called the early receptor potential (ERP), caused by a movement of charge within the rhodopsin molecule itself; but there was no effect on the waveform of the light response produced by channel opening. In the third record, the patch pipette contained DM-nitrophen. When the effects of the xenon flash are compared in control photoreceptors and photoreceptors in which DM-nitrophen had been included in the patch pipette, the release of Ca^{2+} can be seen to produce an additional rapid increase in inward current, indicating a large and accelerated amplification of the photoreceptor response.

Ca^{2+} also has a role in response turnoff and light adaptation. In the experiment of Figure 9.12C, the delivery of the xenon flash-lamp exposure was given at a much later time after the stimulating light, and the xenon lamp then caused a rapid *decrease* in current. A Ca^{2+}-dependent desensitization is thought to be primarily responsible for light adaptation, both in *Limulus* (Lisman and Brown, 1975) and in *Drosophila* (Ranganathan et al., 1991; Minke and Hardie, 2000; Gu et al., 2005).

These results indicate that the rise in Ca^{2+} during the light response, produced either by IP_3-dependent release or by Ca^{2+} entry through the TRP channels, initially activates or facilitates the cascade by means of a non-linear boosting of the light response. In a rather short time, however, the increase in Ca^{2+} then accelerates the decay of the response and adapts the photoreceptor to maintained illumination. There are many possible sites of Ca^{2+} regulation of the cascade. When rhodopsin absorbs a photon, rhodopsin is converted to Rh*. The decay of Rh* occurs as for other G-protein receptors (Figure 4.1), by phosphorylation and binding of arrestin (Byk et al., 1993), though in *Drosophila* photoreceptors phosphorylation is not required for arrestin to bind and turn off the cascade (Vinos et al., 1997; Kiselev et al., 2000). Arrestin binding occurs much more slowly in the absence of external Ca^{2+}, and this Ca^{2+} dependence seems to be mediated by calmodulin and a protein called *neither inactivation nor afterpotential protein C* (NINAC), which is a form of myosin (see Table 9.1 and Liu et al., 2008). Moreover, arrestin can be phosphorylated by a kinase that is Ca^{2+} dependent (Matsumoto et al., 1994), and phosphorylation of arrestin may affect its association with rhodopsin (Alloway and Dolph, 1999). Dephosphorylation of rhodopsin is produced by the rdgC protein (Table 9.1), which is Ca^{2+} dependent. TRP channels have calmodulin binding sites and are inactivated by the binding of Ca^{2+}-calmodulin (Scott et al., 1997).

PKC, whose activity is boosted by an increase in Ca^{2+} concentration, also seems to have an important role in turnoff and light adaptation in *Drosophila* (Hardie et al., 1993). One of these roles appears to be the rapid down-regulation of PLC activity in response to Ca^{2+} influx through the TRP channels (Hardie et al., 2001). When there is little or no Ca^{2+}

(A)

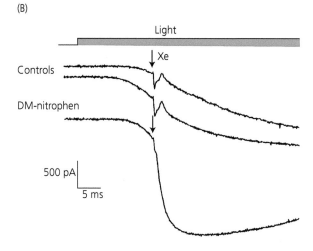

DM-nitrophen

(B)

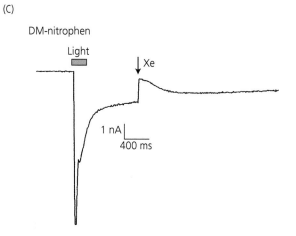

(C)

Figure 9.12 Effect of Ca²⁺ release on *Drosophila* photoreceptor light response. Whole-cell patch recording was used to load dissociated photoreceptors with DM-nitrophen (caged Ca²⁺), from which the Ca²⁺ was released with brief illumination from a xenon flash lamp (Xe). The patch pipette was also used to record light responses. (A) Mechanism of Ca²⁺ release from DM-nitrophen. (B) Release of Ca²⁺ just after the beginning of the light stimulation augments the photoreceptor response. Light stimulus is indicated by blue bar. Records labeled "Controls" lacked DM-nitrophen. Small response to the xenon flash was produced by a conformational change of rhodopsin in response to the very bright illumination of the flash lamp. Record below with DM-nitrophen produced a rapid amplification of the response. (C) Release at later times produced an abrupt decrease in the response. (B and C from Hardie, 1995.)

entry, as for example in *trp* mutants lacking Ca^{2+}-permeant TRP channels, PIP_2 hydrolysis by PLC continues unabated until there is almost no PIP_2 left in the microvillus. This collapse of PIP_2 concentration is probably the cause of the aberrant waveform of the *trp* mutant photoresponse (Figure 9.10A).

All of these effects may participate to some degree in the modulation of the cascade and in light adaptation, though there is still no consensus which of these mechanisms is most important or exactly what Ca^{2+} does. We return to this subject later for vertebrates, where Ca^{2+} also plays an important role as a second messenger in light adaptation.

Vertebrate rods and cones

There are two kinds of vertebrate photoreceptors, rods and the cones (Figure 9.13). Both have an outer segment with sensory membrane elaborated from a modified cilium and containing the visual pigment and all of the enzymes and channels required for transduction. The area of sensory membrane is greatly increased by numerous invaginations, which in rods detach from the plasma membrane as disks but in cones remain accessible to the extracellular solution as membrane lamellae. The repeat distance between disks or lamellae is about 30 nm and is rather uniform from species to species. The number of disks or lamellae is therefore mostly a function of the length of the outer segment: typical values are 1100 rod disks and 750 cone lamellae for the amphibian *Necturus* (Brown et al., 1963), and 1200 lamellae for a monkey cone (Dowling, 1965). The packing density of rhodopsin in the disk seems to be determined by the concentration necessary to maximize the probability of light capture. In a mammalian rod 25 μm long, something like two-thirds of the incident light will be absorbed at the wavelength of peak sensitivity of the photopigment.

The disks of rods, though independent of one another and separated from the plasma membrane by 10–20 nm, are nevertheless interconnected by fine filamentous material (Figure 9.14). The very edge of the disk forms a specialized structure called a rim, which is known to contain proteins not found in the rest of the outer segment. Molecules called peripherin and rom-1 are localized to the rim and seem to have an important role in the formation of the disks (see

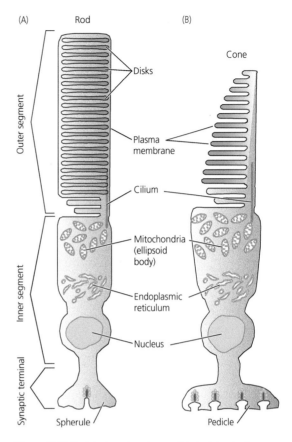

Figure 9.13 Vertebrate rods and cones. Principal structural features of vertebrate photoreceptors. (A) Rod. The outer segment is composed of disks detached from external plasma membrane. (B) Cone. The outer segment has membrane infoldings or lamellae instead of disks.

Molday, 1998; see also Burgoyne et al., 2015; Ding et al., 2015; Volland et al., 2015). This part of the disk also contains the ABCR/Rim protein, a transporter (or flippase) that moves retinal and lipid across the disk membrane from the inside surface of the membrane to the cytoplasmic surface (see Molday et al., 2009).

The metabolic part of the cell, called the inner segment, is also highly organized. In the region just adjacent to the outer segment there is a high concentration of mitochondria forming a condensed region called the *ellipsoid body*, visible in the light microscope. The plasma membrane just adjacent to the mitochondria contains a high concentration of Na^+/K^+ ATPase (Stirling and Lee, 1980). The ER and nucleus lie below the ellipsoid body, and at the proximal end of the cell there is a presynaptic terminal.

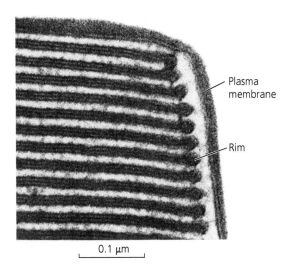

Figure 9.14 Low power electron micrograph of rod outer segment showing disk rim. Note fibrous protein between adjacent disks and between rim and plasma membrane. (Courtesy of Walter Schröder.)

Photoreceptors are secondary receptor cells lacking axons or voltage-gated Na^+ channels, and they do not in general produce Na^+-dependent action potentials (but see Kawai et al., 2001; Ohkuma et al., 2007). The change in membrane potential produced by light is communicated at specialized synapses onto second-order horizontal and bipolar cells. As for invertebrate photoreceptors (Figure 2.12A), the presynaptic terminals contain dense bodies that in rods and cones are called synaptic ribbons and resemble the presynaptic structures of electroreceptors (Figure 2.12B) and hair cells (Figure 6.5C).

In cross-section, the ribbons appear as dense rods surrounded by a halo of synaptic vesicles, which in some sections appear to be connected to the ribbons by fine filamentous material (Figure 9.15A). When the ribbons are followed through many serial sections, their shape can be reconstructed (Figure 9.15B), and they can be seen to occupy much of the synaptic ending and bind hundreds of synaptic vesicles. The major component of the ribbon is a protein

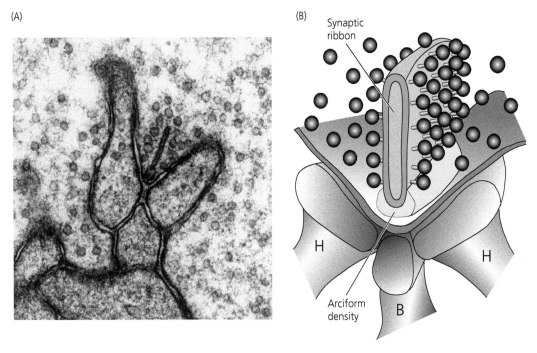

Figure 9.15 Photoreceptor synapse. (A) Electron micrograph of a synapse of a primate cone. Magnification 70,000×. (B) Schematic drawing of a photoreceptor synapse. H, Horizontal-cell process; B, bipolar-cell process. (A courtesy of S. J. Schein.)

called ribeye, which is associated with many other proteins including piccolo and bassoon (see Zanazzi and Matthews, 2009). These proteins together facilitate the movement of vesicles to their release sites just adjacent to another specialized structure, called the arciform density. Vesicle release is Ca^{2+} dependent and mediated by $Ca_V1.4$ voltage-gated Ca^{2+} channels. Because both rods and cones are depolarized in darkness and hyperpolarize to light, the release of synaptic transmitter is continuous in darkness and decreased by illumination (Dowling and Ripps, 1973; Cervetto and Piccolino, 1974). The synaptic transmitter is glutamate (see for example Ishida and Fain, 1981; Copenhagen and Jahr, 1989).

Transduction in vertebrate photoreceptors

The mechanism of activation in a vertebrate photoreceptor is now fairly clear (Figure 9.16A). The formation of Rh* produces a change in the conformation of the parts of the rhodopsin molecule exposed to the cytoplasm, primarily a shift of the sixth transmembrane domain outward toward the cytoplasmic surface of the lipid bilayer, and a smaller, similar movement of the fifth transmembrane domain (Figure 4.3B). These movements open up a binding site for a heterotrimeric G protein called transducin. Transducin binding triggers a change in the conformation of the guanosine nucleotide binding site on the transducin α subunit (Tα). GDP then falls off this binding site, and GTP binds in its place. Tα•GTP separates from the transducin β and γ subunits and comes off the disk membrane, diffusing within the cytoplasm of the photoreceptor between the disks of rods or membrane lamellae of cones (Kuhn et al., 1981).

Tα•GTP then binds to an effector enzyme, which for vertebrate photoreceptors is a cyclic nucleotide phosphodiesterase (PDE6). Rod PDE6 is a tetramer with α and β catalytic subunits and two inhibitory γ subunits (Figure 4.5B). Cone PDE6 is similar, but the two catalytic subunits (α') are identical and different in sequence from those of the rod. The inhibitory γ subunits are also distinct in cones. In the inactive rod or cone enzyme, the γ subunits prevent the catalytic subunits from hydrolyzing cGMP by blocking

access to the catalytic regions of the enzyme. The conformation of the γ subunits changes when Tα-GTP is bound, exposing the catalytic regions of PDE6 and greatly increasing the activity of the enzyme. Each catalytic subunit also contains a high-affinity, non-catalytic binding site for cGMP, and binding of cGMP to these sites can affect the nature of the interaction of the PDE6 inhibitory γ subunit with the α and β catalytic subunits (D'Amours and Cote, 1999). The affinity of cGMP for these non-catalytic sites is, however, fairly high, with the result that cGMP probably comes on and off too slowly to make much of a contribution to the photoreceptor light response (Calvert et al., 1998).

For every thousand rhodopsin molecules in the disk membrane of a rod, there are about a hundred transducins and something like ten PDE6 molecules (Figure 9.16B). As we saw in Chapter 2, rhodopsin can diffuse within the surface of the disk or cone lamella, and PDE6 and the inactive transducin heterotrimer are also attached to the membrane and membrane diffusible. Attachment to the membrane augments the chance of collision of these molecules—in effect, the disk or lamellar membrane acts as a catalyst. As a result, a single Rh* during its lifetime can collide randomly with many transducin molecules and produce many molecules of Tα-GTP, perhaps ten to fifteen in a mouse rod but several hundred in the much larger rods of amphibians (Leskov et al., 2000; Krispel et al., 2006; Reingruber et al., 2013; Yue et al., 2019).

Although the sensory cascade of vertebrate photoreceptors relies upon random collisions of membrane proteins instead of a highly organized transducisome/signalplex as in arthropod photoreceptors, transduction in rods and cones is surprisingly fast and efficient. Single-photon voltage responses in rods are of the order of 1 mV (Fain, 1975; Schneeweis and Schnapf, 1995)—smaller than for arthropod photoreceptors, but sufficiently large to produce reliable detection. The minimum latency of the photoreceptor to a bright light flash is only 7 ms for both rods and cones (Cobbs and Pugh, 1987; Hestrin and Korenbrot, 1990), not much different from that for *Drosophila* photoreceptors (see Figure 9.12B). These are the fastest G-protein, second-messenger cascades known to science.

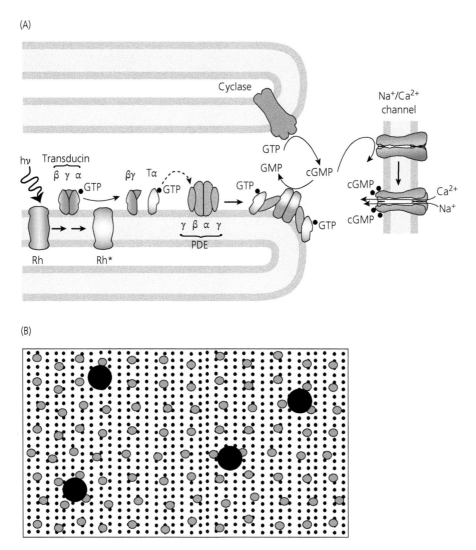

Figure 9.16 Vertebrate phototransduction. (A) Transduction cascade. (B) Relative proportions of rod transduction proteins. The small black dots represent rhodopsin molecules, the blue dots represent transducin, and the large black dots represent PDE6 molecules. Only about a third of the PDE6 molecules have been illustrated for clarity. The distribution of molecules would be much less regular in an actual disk membrane because all of these proteins can diffuse rather freely in the lipid of the disk. A disk membrane of this dimension would have adjacent to it an average of one free cGMP molecule. (A after Pugh and Lamb, 1993; B after Bownds and Arshavsky, 1995; Fain, 1999.)

Ion channels of rods and cones

Figure 9.17 gives the principal types of ion channels and transporters in a vertebrate photoreceptor. The cyclic-nucleotide-gated channels are mostly in the plasma membrane of the outer segment (see Kaupp and Seifert, 2002). In both rods and cones the channels are heterotetramers, consisting of α subunits called CNGA1 for rods and CNGA3 for cones, and β subunits called CNGB1 for rods and CNGB3 for cones. Both rod and cone channels have a stoichiometry of $\alpha_3\beta$ (Weitz et al., 2002; Zheng et al., 2002; Zhong et al., 2002). All of the channel subunits have different though related amino-acid sequences, with six transmembrane domains, a P region forming the channel pore, and a cytoplasmic carboxyl terminus with a

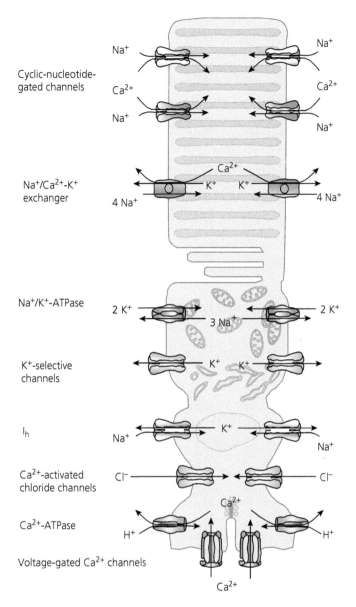

Figure 9.17 Channels and transporters in the membrane of a vertebrate rod. Major proteins responsible for ion conduction and transport are different in the inner and outer segment. See text.

binding site for cyclic nucleotide (see Figures 4.10B and 4.11). In rods but not cones, the β subunit has a large cytoplasmic amino-terminal region called the glutamic-acid-rich protein (GARP), which can also be secreted into the cytoplasm as a separate, soluble protein. GARP may play some role in protein–protein interactions in the outer segment (Korschen et al., 1999) or in channel gating (Michalakis et al., 2011).

Channels in both rods and cones are quite selective for cGMP over cAMP, unlike channels in olfactory receptors which bind both nucleotides nearly equally; but rod channels bind even cGMP with less affinity than olfactory channels bind either nucleotide.

The cGMP-gated channels in rods and cones, like those in the lizard parietal eye (see Chapter 4), are rather non-specifically permeable to monovalent cations and actually more permeable to Ca^{2+} than to Na^+. Since, however, the Na^+ concentration is so much higher than the Ca^{2+} concentration in the extracellular medium, only of the order of 10–15 percent of the current entering the cyclic-nucleotide-gated channels in a rod is carried by Ca^{2+}. Cones have channels different from those in rods, and nearly twice as much of the current or about 20–30 percent can be carried by Ca^{2+} (Perry and McNaughton, 1991; Ohyama et al., 2000). As we shall see, this Ca^{2+} influx has an important role in the physiology of the light response.

As in olfactory receptor cells, the entering Ca^{2+} is removed by a very active transport molecule, which uses the energy of both the Na^+ and K^+ gradients to move Ca^{2+} out of the cell (Cervetto et al., 1989; Lagnado and McNaughton, 1990). Four Na^+ ions move inward and one K^+ moves outward for every Ca^{2+} ion that is extruded. This stoichiometry has the result that four charges are moved inward and three charges outward for each cycle of the transporter, so that, like the Na^+/K^+ ATPase, Na^+/K^+-Ca^{2+} exchange is electrogenic. The inward current carried by the transporter can actually be recorded (see Figure 9.22), and this observation provided the first important clues for the role of Ca^{2+} in the physiology of the photoreceptor (Yau and Nakatani, 1984; Hodgkin et al., 1987). Remarkably, the exchange molecules are tightly bound to the channel at least in rods, with a fixed stoichiometry of two exchangers per channel (Schwarzer et al., 2000).

The inner segments of both rods and cones have a high concentration of Na^+/K^+ ATPase and an assortment of channels not directly activated by light (see Molday and Kaupp, 2000). There are K^+ channels that provide the principal K^+ permeability of the cell and make an important contribution to the resting membrane potential (Beech and Barnes, 1989), as well as voltage-gated channels called I_h (or HCN channels) permeable to both Na^+ and K^+ and activated by hyperpolarization (Fain et al., 1978; Hestrin, 1987; Wollmuth and Hille, 1992). There are also Ca^{2+}-activated Cl^- currents especially prominent in cones (Barnes and Hille, 1989; Jeon et al., 2013), which may help stabilize the membrane potential during synaptic transmission (Lalonde et al., 2008). Finally, at the synaptic terminal there are voltage-gated ($Ca_V1.4$) Ca^{2+} channels near the release sites, which regulate the exocytosis of the vesicles. The

Ca^{2+} entering the rod or cone at the synaptic terminal is removed primarily by a Ca^{2+} ATPase rather than by Na^+/K^+-Ca^{2+} exchange (Krizaj and Copenhagen, 1998; Morgans et al., 1998).

The photocurrent

In darkness, the cyclic-nucleotide-gated channels in both rods and cones are gated open by the resting concentration of cGMP in the outer segment. Because these channels are selectively permeable to cations, and in particular to Na^+ and Ca^{2+}, the light-dependent conductance has a reversal potential similar to that of hair cells (see Figure 3.14), somewhat positive of 0 mV. In darkness the resting membrane potential of a rod or cone is about –35 mV, with the consequence that there is a standing current through the channels in darkness which we can calculate from Eq. (3.7):

$$i_m = g(V_m - E_{rev})$$

The conductance g is positive (the channels are open), and $(V_m - E_{rev})$ is approximately (–35–0) or about –35 mV. The current i_m will therefore be negative, or inward, from the extracellular space into the outer segment. This standing inward current in darkness was first discovered by Penn and Hagins (1969) and was given the appropriate name *dark current*.

The large resting conductance of the cell to Na^+ produced by the cyclic-nucleotide-gated channels is responsible for the rather positive value of the resting membrane potential in darkness, which, as we have said, is of the order of –35 mV. This value is more depolarized than for most neurons and intermediate between the equilibrium potential for K^+ (E_K, –80 to –90 mV) and the reversal potential for the cyclic-nucleotide-gated channels, which is near zero. Light decreases the probability of opening of the cyclic-nucleotide-gated channels and decreases the Na^+ permeability. The decrease in Na^+ permeability causes the membrane potential of the rod to hyperpolarize—just the opposite of the effect of the increase in Na^+ permeability which occurs during the firing of an action potential.

Figure 9.18A shows the change in membrane potential produced by a series of brief light flashes of increasing intensity, recorded from a salamander rod with an intracellular microelectrode (from Baylor and Nunn, 1986). As the cyclic-nucleotide-gated channels close in the light, the membrane potential moves in a negative direction closer to E_K. The brighter the light, the larger the hyperpolarization. In very bright light there is an additional rapid relaxation in the voltage (arrow in Figure 9.18A). Hyperpolarization activates the I_h channels, and since these channels are also rather permeable to Na^+ (Wollmuth and Hille, 1992), the I_h current has a reversal potential near the dark resting membrane potential of the photoreceptor. As the I_h channels activate during the light response, they cause the membrane potential rapidly to depolarize back toward the resting potential, producing the "nose" at the beginning of the response. The I_h conductance is specifically blocked by low concentrations of extracellular Cs^+, and Cs^+ eliminates the rapid relaxation of the voltage response (Fain et al., 1978; Hestrin, 1987).

The current through the cyclic-nucleotide-gated channels can be measured by pulling the outer segment of the salamander rod into a suction electrode connected directly to a current-measuring amplifier. This method measures the current entering the outer segment, which is equal to the current passing through the cyclic-nucleotide-gated channels. These channels are mostly localized to the outer segment, and they seem to be the only functioning channels in this part of the cell. Suction-electrode recordings of light responses are shown in Figure 9.18B from the same cell for which voltage responses are given (Figure 9.18A). They show that a sustained dark current of about –35 pA flowing into the outer segment is decreased when the rod is illuminated with brief flashes. Bright light closes all of the cyclic-nucleotide-gated channels and reduces the current entering the outer segment to zero (Baylor et al., 1979b).

One disadvantage of suction-electrode recording is that it does not voltage clamp the cell. Since as stated in Eq. (3.8),

$$\Delta i_m = \Delta g(V_c - E_{rev}),$$

and since during voltage clamp $V_m = V_c$ and for the cyclic-nucleotide-gated channels E_{rev} is near zero, $\Delta i \sim \Delta g (V_m)$. As a result, changes in the value of membrane voltage during the light response like those shown in Figure 9.18A would be expected to influence the value of the current. This possibility was examined by Baylor and Nunn (1986), who recorded outer segment currents with a suction electrode from salamander rods that were simultaneously voltage clamped (Figure 9.18C). The upper traces show the total membrane current measured with voltage clamp. The current is initially zero, since in darkness the cell is at steady state, and the current entering the outer segment through the cyclic-nucleotide-gated channels is exactly balanced by current leaving the inner segment, mostly through K^+ channels. Because the inner-segment currents are unaffected by illumination, the time course of the voltage-clamp current in the upper traces reflects the time course of the decrease of the outer-segment conductance.

The lower traces show suction-electrode recordings made simultaneously from this same rod under voltage clamp. The amplitude of the suction-electrode currents is smaller than that of the voltage-clamp currents, in part because current is lost through the seal between the suction pipette and the cell, and in part because not all of the outer segment had been brought into the pipette. What is remarkable, however, is that the relative amplitude and time course of the suction-electrode currents in Figure 9.18C are quite similar to those of the currents recorded in Figure 9.18B, where no voltage clamp was used.

The reason for this correspondence is that the change in conductance Δg in Eq. (3.8) turns out to be voltage dependent. The voltage dependence isn't very large, but it is about the same magnitude and opposite in polarity to the change in the driving force, $(V_m - E_{rev})$. As a result, the photoreceptor current $\Delta g(V_m - E_{rev})$ shows very little dependence on voltage as V_m changes during the light response, because the change in V_m is nearly compensated by the voltage-dependent change in Δg. This has the happy consequence that suction-electrode recording, which is much easier than voltage clamping, gives a fairly faithful representation of the waveform of the light response of a salamander rod. Whether this circumstance is also true for cones or for photoreceptors in other species remains an open question.

Suction-electrode recordings have been made from the photoreceptors of many vertebrates, even from

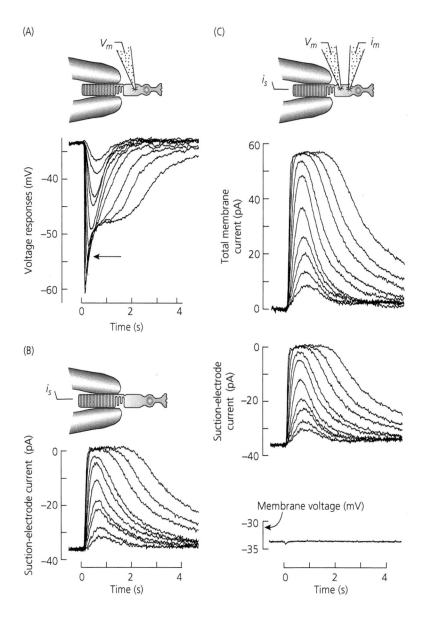

Figure 9.18 Electrical recording from a vertebrate rod. The responses of a salamander rod to a series of brief flashes given at t=0 of increasing intensity obtained by (A) intracellular recording, (B) suction-electrode recording, and (C) voltage clamp. Arrow in (A) indicates rapid repolarization of response produced by activation of voltage-gated I_h (HCN) channels. In (C), the upper records show total membrane current measured with the voltage-clamp circuit, and the lower records show suction-electrode currents measured from the same voltage-clamped rod from the outer segment. Lowermost trace shows that there was no change in the voltage of the rod during the voltage-clamp measurements. (After Fain, 1999 and 2014, with data published by permission from Baylor and Nunn, 1986.)

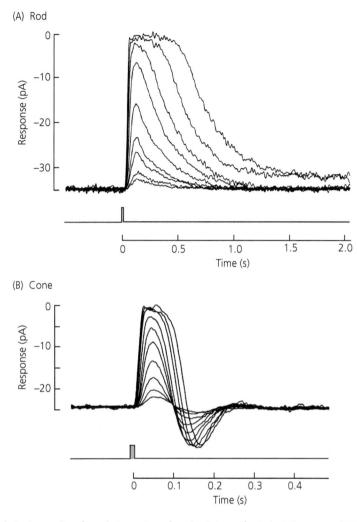

Figure 9.19 Suction-electrode recordings from photoreceptors of monkey (*Macaca fascicularis*). Responses of (A) rod and (B) long-wavelength-sensitive (L) cone to a series of flashes of increasing intensity. Note difference in time scale. (A from Baylor et al., 1984; B from Baylor et al., 1987.)

the very small rods and cones of mammals. Figure 9.19 illustrates the responses of monkey rods (Baylor et al., 1984) and cones (Baylor et al., 1987), each to a series of brief flashes of increasing intensity. Rods in mammals are typically about one-hundred times more sensitive to light than cones. Notice also the difference in time scale: responses of cones reach peak amplitude and decay much more rapidly than do those of rods.

Why are rods so much more sensitive? One possibility is the difference in anatomy—rods have detached, closed disks, whereas cones have open

membrane lamellae. This difference in anatomy seems, however, not to contribute to the difference in sensitivity. The evidence comes from lamprey, which (with hagfish) are cyclostomes and the last remaining representatives of vertebrates without jaws. Lamprey rods and cones have an identical anatomy; that is, both rods and cones have membrane lamellae without disks. They nevertheless respond very much like the rods and cones of other vertebrates. Lamprey rods are about seventy times more sensitive than lamprey cones, and lamprey rods have single-photon responses as large as those in mammals (Morshedian and Fain, 2015; Asteriti et al., 2015).

The difference in sensitivity between the two kinds of photoreceptors may rather result from molecular differences in the cascade. Rods and cones contain different isoforms of most of the proteins involved in sensory transduction (see Ingram et al., 2016), including the photopigment, transducin, PDE6, the cyclic-nucleotide-gated channel subunits, and the Na^+/K^+-Ca^{2+} transporter. Even when the proteins are the same, they can be expressed at very different levels (Cowan et al., 1998; Zhang et al., 2003a). Cone channels are more permeable to Ca^{2+} and can extrude Ca^{2+} more quickly by Na^+/K^+-Ca^{2+} exchange (Sampath et al., 1999).

Many attempts have been made to investigate the effects of each of these molecular differences, often by exploiting the tools of molecular biology to replace the rod protein with the cone protein and measure changes in photoreceptor sensitivity and response waveform (summarized in Ingram et al., 2016). No one change seems to be responsible for more than a factor of two or three of the difference in sensitivity. Because rods emerged after cones (Nickle and Robinson, 2007; Shichida and Matsuyama, 2009), it is likely that the rods evolved from cones by a series of small changes in many of the transduction proteins until the sensitivity of the rod was sufficient to produce single-photon responses enough above the noise of transduction to be detectable. All of these changes seem to have happened very early, before the lamprey (and the rest of the cyclostomes) separated from the other vertebrates in the late Cambrian. Darwin wondered how an organ as complicated as the human eye could possibly have evolved, and he speculated that eons of time must have been required. In truth, an eye like our own was present very early in vertebrate evolution. The physiology of vertebrate photoreceptors has remained nearly unaltered for 500 million years (Morshedian and Fain, 2017).

Shutting down the light response

Activation of the sensory cascade must be followed by response decay. The faster decay can occur, the more rapidly the cell can detect another stimulus, and the more accurately the visual system can

signal change and motion. All the steps in transduction must be returned to their initial condition: Rh* must be inactivated, $T\alpha$ must come off the inhibitory subunits of the PDE6 and recombine with $T\beta\gamma$, the cyclic-nucleotide concentration must be restored to its dark level, and the channels must re-open. The steps in decay are highly orchestrated and can be modulated, at least in part, by changes in Ca^{2+} concentration.

The inactivation of Rh* occurs as for other G-protein-coupled receptors (see Figure 4.1). Rhodopsin kinase phosphorylates serine and threonine residues on the carboxyl terminus of rhodopsin, and an arrestin protein then binds to phosphorylated rhodopsin, sterically inhibiting the binding of transducin. Phosphorylation (Bownds et al., 1972; Kuhn and Dreyer, 1972) and arrestin binding (Kuhn, 1978) of G-protein receptors were first discovered for vertebrate rods and have been extensively characterized. The carboxyl terminus of rhodopsin contains six or seven serine and threonine groups (depending on the species), which can all be phosphorylated (Wilden and Kuhn, 1982). Rh* inactivation seems to proceed by phosphorylation of at least three of these residues (Vishnivetskiy et al., 2007), with the phosphorylation of serine and threonine residues having somewhat different effects (Azevedo et al., 2015). If rhodopsin phosphorylation is prevented either by genetically altering rhodopsin to remove its carboxyl tail (Chen et al., 1995b; Mendez et al., 2000) or by disrupting the gene for rhodopsin kinase (Cideciyan et al., 1998; Chen et al., 1999a), photoresponses turn off abnormally and are greatly prolonged (Figure 9.20A, *Rh truncation*).

Rods have two molecular forms of arrestin, which are both splice variants of the same gene and have different affinities for rhodopsin (Palczewski et al., 1994; Burns et al., 2006). In mice for which the arrestin gene has been disrupted (–/–, Figure 9.20A), photoresponses again turn off abnormally and become prolonged (Xu et al., 1997), but the effect is not as great as that produced by removing the C-terminus of rhodopsin and preventing phosphorylation. The reason for this difference seems to be that phosphorylation even without arrestin binding can produce some inhibition of transducin binding (Wilden et al., 1986). The phosphorylation of rhodopsin and binding of arrestin happen rather quickly,

(A)

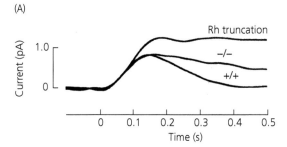

(B)

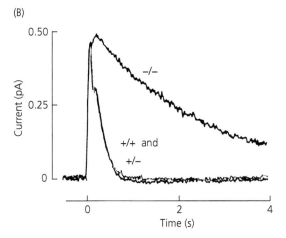

Figure 9.20 Components of rod response decay. (A) Average single-photon response of a rod recorded with a suction-electrode in wild-type mouse (+/+), in a mouse in which the gene for arrestin had been disrupted (−/−), and in a mouse in which the gene for rhodopsin had been altered so that the C-terminus of the molecule had been truncated, removing all of the sites of protein phosphorylation (Rh truncation). (B) Same as for part (A) but from mouse lacking photoreceptor RGS9 protein (−/−), a wild-type mouse (+/+), and a mouse heterozygous for the *rgs9* gene (+/−). (A from Xu et al., 1997; B from Chen et al., 2000.)

for a mouse rod probably with a time constant of no more than about 40 ms (Krispel et al., 2006; Chen et al., 2010a).

As for other heterotrimeric G proteins (see Chapter 4 and Figure 4.4), the inactivation of Tα-GTP and restoration of the Tαβγ complex requires the hydrolysis of GTP to GDP on the Tα guanosine nucleotide binding site. Although transducin by itself can hydrolyze GTP, the rate is rather slow. It is, however, greatly accelerated by GTPase-activating proteins (GAPs), and in particular by the protein RGS9 (He et al., 1998), which is abundant in rods and found at an even higher concentration in cones (Cowan et al., 1998; Zhang et al., 2003a). RGS9 is present in the outer

segment in a tight *GAP complex* together with a G-protein β subunit called Gβ5L (Makino et al., 1999; He et al., 2000), which is not the same as the β subunit of transducin, and the R9AP binding protein, which attaches the GAP complex to the disk or lamellar membrane (Hu and Wensel, 2002). The rate of turnoff is further accelerated by the PDE6γ subunit (Arshavsky and Bownds, 1992; Tsang et al., 1998), though PDE6γ seems to have little effect on its own and requires RGS9 (Chen et al., 2000). In mice in which the gene for the RGS9 protein has been disrupted (Chen et al., 2000), the recovery of the light response is again greatly retarded (Figure 9.20B).

To return the cyclic nucleotide concentration to its level in darkness, both rods and cones use membrane-bound guanylyl cyclases, of which two different forms are expressed in photoreceptors (Yang et al., 1995). Both have the structure of the membrane-bound guanylyl cyclases used as receptors for hormones (Figure 4.2). They are present in the outer segment as homodimers and are integrated into the disk or lamellar membrane (Yang and Garbers, 1997). Like other membrane-bound guanylyl cyclases, these proteins have an "extracellular" ligand-binding domain, which has no known ligand and is located in rods inside the disk (Figure 9.21A). This part of the protein is then connected by a single membrane-spanning domain to the cytosolic catalytic part of the protein, where cGMP is synthesized.

The activity of the guanylyl cyclase is tightly regulated by the cytoplasmic Ca^{2+} concentration (Koch and Stryer, 1988). In the dark, there is a large influx of Ca^{2+} into the outer segment through the cyclic-nucleotide-gated channels, which is balanced by efflux via the Na^+/K^+-Ca^{2+} exchanger. When the photoreceptor is illuminated, the channels close, which decreases the entry of Ca^{2+} into the outer segment. The exchanger seems not to be directly affected by light (Nakatani and Yau, 1988a; Koutalos et al., 1995) and continues to extrude Ca^{2+} until the decrease in intracellular Ca^{2+} concentration lowers the rate of the transporter enough for the cell to reach steady state. In effect, the closing of the channels by light causes a decrease in the Ca^{2+} concentration (Figure 9.22), from a dark level in a salamander rod of about 400–600 nM to as low as 5–10 nM when the channels are all closed (see Fain et al., 2001); and from about 250 nM to less than 25 nM in a mouse rod (Woodruff et al., 2002).

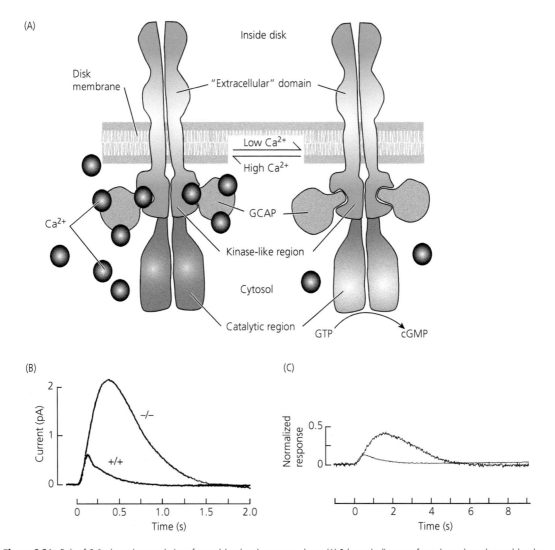

Figure 9.21 Role of Ca²⁺-dependent regulation of guanylyl cyclase in response decay. (A) Schematic diagram of membrane-bound guanylyl cyclase and GCAPs in rod disk membrane. (B) Average single-photon response of a rod recorded with a suction-electrode from a wild-type mouse rod (+/+) and from a rod for which both GCAP genes had been disrupted (–/–). Note much larger amplitude and slower decay after GCAP deletion. (C) Small-amplitude responses normalized to peak response amplitude for a salamander rod in Ringer solution (smaller response) and in 0-Ca²⁺/0-Na⁺ solution (larger response), which prevented Ca²⁺ entry and exit. (A after Polans et al., 1996; B after Mendez et al., 2001; C from Fain et al., 1989.)

The decrease in Ca²⁺ concentration alters the rate of the guanylyl cyclase via small-molecular-weight Ca²⁺ binding proteins called guanylyl-cyclase-activating proteins (GCAPs) (see Dizhoor et al., 2010). There are again two different molecular variants of GCAPs, but both appear to act in the same way. They associate with cytoplasmic binding sites on the guanylyl cyclase molecule near the disk membrane (Figure 9.21A). In the dark, when the Ca²⁺ concentration is high, the GCAPs inhibit the guanylyl cyclase. The decrease in Ca²⁺ concentration produced by the closing of the cyclic-nucleotide-gated channels causes the Ca²⁺ to come off the GCAPs, stimulating the cyclase to synthesize cGMP. This GCAP-dependent stimulation of the cyclase causes an accelerated return of cGMP concentration, which re-opens the channels. If the genes for both GCAPs are disrupted (Mendez et al., 2001), the initial

(A)

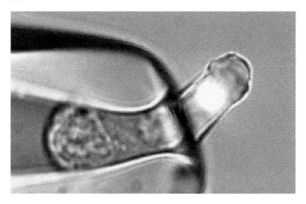

(B)

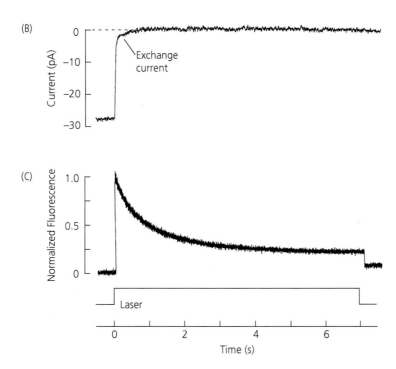

Figure 9.22 Light produces a decrease in photoreceptor intracellular Ca^{2+} concentration. (A) Isolated salamander rod loaded with the fluorescent Ca^{2+} indicator dye fluo-3 was held with its inner segment in a suction pipette so that its outer segment could be illuminated with visible light from an argon laser. (B) Turning on of the laser produced a rapid decline in suction-electrode current due to the closing of the channels. The more slowly declining component labeled "Exchange current" is the inward current produced by electrogenic Na^+/Ca^{2+}-K^+ exchange, which declined as Ca^{2+} decreased and reached a new steady-state concentration. (C) From the same rod, the time course of Ca^{2+} decrease measured from the fluorescence of the fluo-3 indicator dye. Fluorescence has been normalized to its peak value in darkness. (From Sampath et al., 1998.)

phase of the photocurrent is unaffected, but the channels continue to be closed for a longer time and re-open much more slowly (Figure 9.21B).

Virtually the same effect can be produced by preventing the light-dependent change in outer seg-

ment Ca^{2+} concentration. This can be done at least for a few seconds by rapidly perfusing the outer segments of rods or cones with a medium lacking both Ca^{2+} and Na^+ (Matthews et al., 1988). The removal of Ca^{2+} stops Ca^{2+} influx, whereas substituting Na^+ with another ion like Li⁺ or guanidinium⁺ stops the efflux of Ca^{2+} by the exchanger, since the exchanger requires extracellular Na⁺ to function. The photoreceptors still have responses to light in this solution, because both Li⁺ and guanidinium⁺ can permeate the cyclic-nucleotide-gated channels.

When both influx and efflux are blocked in this way, the Ca^{2+} concentration in the rod remains relatively constant (Fain et al., 1989; Matthews and Fain, 2001), and a light flash given to the rod in this solution (Figure 9.21C) produces a response that is prolonged in much the same way as disruption of the GCAP genes. Similar effects have been seen in cones (Nakatani and Yau, 1988b; Matthews et al., 1990). These experiments show that the change in outer segment Ca^{2+} concentration causes an acceleration of the return of outer segment current (see also Torre et al., 1986), and this effect of Ca^{2+} is almost entirely due to regulation of cyclase activity via the GCAP proteins (Burns et al., 2002).

Which of the steps of recovery is the slowest? That is, which step is the one that limits the rate of recovery of the light response? One possible approach might be to knock out each of the genes responsible for the various mechanisms of decay one by one. We could then see which of them alters the response waveform. The results in Figures 9.20 and 9.21 show, however, that this approach is unhelpful. When we reduce the rate of any of the steps that turn off the transduction cascade, we slow that step and make it the one that limits the recovery rate. The light response is prolonged whether we retard Rh* phosphorylation or the binding of arrestin (Figure 9.20A) or the rate of Tα•GTP hydrolysis (Figure 9.20B) or the activity of the guanylyl cyclase (Figure 9.21B). None of these results indicates which recovery step is rate limiting in a wild-type rod.

A better approach is to *increase* the rate of the steps of recovery. Suppose for example that in a normal rod the rate of Tα•GTP hydrolysis is very rapid, but phosphorylation of Rh* and binding of arrestin is much slower and limits the rate of return of the light response. If we increase the expression of the GAP molecules and make Tα•GTP hydrolysis even faster, we will not speed the rate of recovery. The response will still return only as Rh* is phosphorylated and arrestin binds. Increasing GAP expression and speeding Tα•GTP hydrolysis would

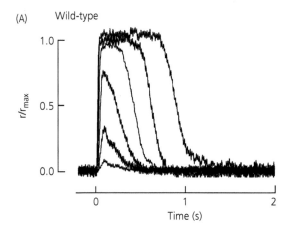

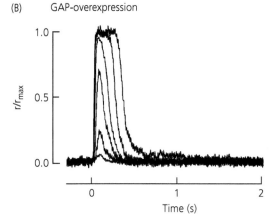

Figure 9.23 Effect of overexpression of GAP proteins on rod response decay. Superimposed responses from mouse rods to a series of flashes of increasing light intensity; the flashes were 10 ms in duration and were given at $t = 0$. Responses have been normalized to the peak amplitude of the response to the brightest intensity (r/r_{max}). (A) Wild-type rod. (B) Rod in mouse for which GAP-complex proteins had been overexpressed by four-fold. (From Krispel et al., 2006.)

only accelerate response recovery if Tα•GTP hydrolysis were the rate-limiting step. It should therefore be possible to determine the slowest step in recovery by systematically increasing the expression of the relevant genes.

Figure 9.23 illustrates an experiment of this kind (from Krispel et al., 2006). The expression of the R9AP binding protein was increased by a factor of four. Because the expression of the three GAP-complex genes is linked, overexpression of R9AP also increases expression of RGS9 and Gβ5L, and by approximately the same amount. Increasing GAP expression accelerates the decay of the response

(Figure 9.23B). In mice with variable expression of GAP genes, the greater the GAP expression, the greater the acceleration of response decay (Krispel et al., 2006; Chen et al., 2010a). The simplest explanation is that increasing GAP concentration in the outer segment accelerates the binding of the GAP complex to Tα•GTP and increases the rate at which the PDE is turned off. These experiments are representative of several studies that indicate that extinction of activated PDE6 by hydrolysis of Tα•GTP is rate limiting for recovery of the rod response (see also Sagoo and Lagnado, 1997; Tsang et al., 2006). The resynthesis of cGMP by the cyclase and binding of cyclic nucleotide to the channels are both so rapid that the electrical response effectively tracks the decline of PDE6 activity. In a mammalian rod, extinction of Rh* is also rapid and never rate-limiting under physiological conditions (Burns, 2010).

Light adaptation

Vertebrate photoreceptors adapt in the presence of steady light. Constant stimulation decreases sensitivity and resets the operating range of a photoreceptor, much as maintained hair bundle deflection does for a hair cell (Figure 6.11). This phenomenon can be seen in Figure 9.24, a suction-electrode recording from a salamander rod. In Figure 9.24A, the rod in darkness was stimulated with brief flashes at intensities that increased systematically by about a factor of four. The peak amplitude of these responses became larger as the light intensity increased (as in Figures 9.18 and 9.19) and trace out a response-intensity curve, giving peak amplitude of the response as a function of flash intensity. In Figure 9.24B and C, these same flash intensities were repeated for this same rod but in the presence of two different steady background lights. The backgrounds themselves produced a decrease in the outer segment current that slowly reached a steady plateau level. Flashes superimposed on this background produced a further closing of the channels but with decreased sensitivity, such that the whole operating range of the photoreceptor was shifted to higher light intensities.

The decrease in sensitivity is perhaps clearer in Figure 9.24D. Here, small-amplitude responses of a rod to brief flashes have been superimposed. Because the sensitivity of the rod changed in the presence of

the background light, the flashes in darkness and in the different backgrounds were not of the same brightness. The response amplitudes in each case have therefore been divided by the number of photons in the flashes and plotted in units of sensitivity. The largest response was recorded without a background light, and the others are for backgrounds of progressively increasing background intensity. As the background increased, sensitivity declined, and the waveform of the response was also altered. At each of the progressively brighter background intensities, the responses rose along approximately the same initial curve but began to decline at an increasingly earlier time. These recordings illustrate that one of the principal mechanisms for the sensitivity decrease in rods and cones is an acceleration in the time course of response decay (Baylor and Hodgkin, 1974).

Light adaptation in rods requires a diffusible second messenger, because the sensitivity of the whole of the outer segment can be changed when rhodopsin molecules are activated in only a small fraction of the disks (see Fain, 1986). A messenger also seems to regulate sensitivity in cones, and there is now considerable evidence that in both kinds of photoreceptors this messenger is Ca^{2+}. One way of demonstrating a role of Ca^{2+} is to use the approach of Figure 9.21C, perfusing the outer segment with a solution lacking both Ca^{2+} and Na^+, to eliminate or greatly reduce Ca^{2+} influx and efflux and minimize changes in outer segment Ca^{2+} concentration. Under these conditions, adaptation seems entirely eliminated (Matthews et al., 1988; Nakatani and Yau, 1988b). A large body of experimental work indicates that changes in Ca^{2+} are an important requirement for adaptation of the photoreceptor to light (see Fain et al., 2001).

What does the Ca^{2+} do? The answer to this question is still not clear, but one very important role of Ca^{2+} is the modulation of the activity of guanylyl cyclase (Figure 9.21). In the presence of a steady background light, there is a steady increase in the rate of hydrolysis of cGMP by the PDE6. This steady PDE6 activity decreases the cGMP, closes the channels, and decreases the intracellular free-Ca^{2+} concentration (Figure 9.22). As Ca^{2+} falls, the cyclase activity increases until the synthesis of cGMP equals its rate of hydrolysis. This increase in cyclase activity

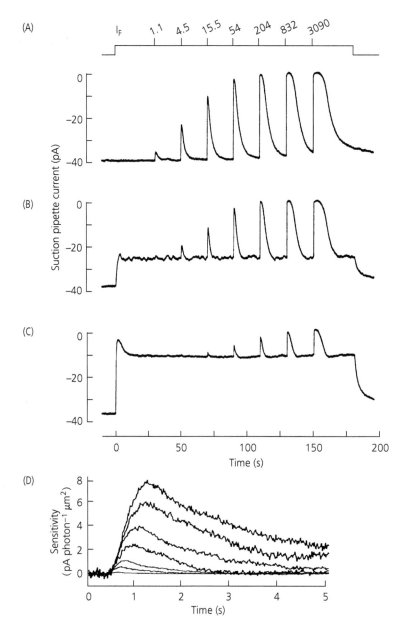

Figure 9.24 Light adaptation in a vertebrate rod. (A–C) Suction-electrode recording of responses of the same salamander rod to flashes given in darkness and in the presence of two steady background lights. I_F gives the number of photons in the flash in units of photons per micrometer squared (μm^{-2}). Background intensities were 1.7 (B) and 37.2 (C) photons μm^{-2} s^{-1}. (D) Suction-electrode recordings of responses of a different salamander rod in darkness (largest response) and in backgrounds of progressively increasing brightness. Responses have been plotted as sensitivities by dividing the current (in picoamperes) by the number of photons in the flash (in units of photons per micrometers squared). (A–C from Matthews, 1990; D from Fain, 1993.)

causes a fraction of the channels to re-open, preventing the light from saturating the rod and allowing the receptor to continue to respond even in the presence of maintained illumination.

When the Ca^{2+}-dependent regulation of the cyclase is eliminated by disrupting the genes for the GCAP proteins, much but not all of light adaptation is eliminated (Mendez et al., 2001). There are several other components of the transduction cascade that can be modulated by Ca^{2+}, but there is still considerable uncertainty about their contributions. The rate of rhodopsin phosphorylation by rhodopsin kinase has been shown to be altered by a small-molecular-weight Ca^{2+} binding protein (Kawamura, 1993; Chen et al., 1995a), usually referred to as recoverin. Although recoverin can modulate the rate of rhodopsin decay under physiological conditions (Chen et al., 2010a), the magnitude of this effect is rather small. The principal role of this protein seems rather to be the regulation of PDE6 (Fain, 2011), which adjusts the integration time and PDE6 spontaneous activity in background light (Morshedian et al., 2018). Ca^{2+} has also been shown to modulate the light-dependent channels (see Molday and Kaupp, 2000): in rods by binding to calmodulin as in olfactory receptors (Hsu and Molday, 1993), and in cones apparently by binding to some other Ca^{2+}-binding protein (Hackos and Korenbrot, 1997). The effect for the rod channel is small and seems to make little or no contribution to adaptation (Chen et al., 2010b), but the one for the cone channel may be of greater significance (Rebrik and Korenbrot, 1998; Rebrik et al., 2000).

Pigment renewal and the recovery of sensitivity after bright light

Light converts 11-*cis* retinal to all-*trans* retinal, and the chromophore must be re-isomerized to its 11-*cis* form before rhodopsin can again be re-activated by photon absorption. In microvillar photoreceptors, all-*trans* retinal remains covalently attached to the opsin, forming a thermally stable metarhodopsin which can also absorb light. In *Drosophila* the rhodopsin of the R1–6 photoreceptors containing 11-*cis* retinal absorbs maximally at 480 nm in the blue-green, and the corresponding metarhodopsin with all-*trans* retinal absorbs maximally at 570 nm in the

yellow (see Minke and Hardie, 2000). All of the other pigments in *Drosophila* including the ultraviolet-sensitive pigments also have metarhodopsins absorbing in the visible, at wavelengths between 460 and 520 nm. When arthropod metarhodopsin absorbs a photon, no light response is produced, but all-*trans* retinal is converted back to 11-*cis* (see for example Richard and Lisman, 1992; Liu et al., 2008). Most of pigment regeneration occurs by this energy-efficient mechanism, but some of the phosphorylated metarhodopsin is endocytosed and degraded (see for example Satoh and Ready, 2005). Arthropods accordingly also have a slower enzymatic pathway in retinal pigment cells to regenerate the 11-*cis* chromophore (Wang et al., 2010).

For the ciliary pigments of rods and cones, the bleaching of retinal from 11-*cis* to all-*trans* causes the chromophore to be released from the opsin, and most of the chromophore must then migrate to a different cell type to be regenerated. The mechanisms of pigment conversion are somewhat different for the two kinds of photoreceptors. For rods, most of the all-*trans* retinal is converted to 11-*cis* retinal by an enzymatic pathway (Figure 9.25 and see Fain et al., 2001; Lamb and Pugh, 2004). The all-*trans* retinal comes off the opsin protein and is reduced to all-*trans* retinol within the rod by a retinol dehydrogenase. All-*trans* retinol then leaves the photoreceptor and is carried through the extracellular space, perhaps in part by binding to interphotoreceptor retinol binding protein (IRBP). The all-*trans* retinol is then conveyed into an adjacent layer of cells called the retinal pigment epithelium (RPE). The RPE contains an enzyme called RPE65 which re-isomerizes the chromophore (Jin et al., 2005) and which is part of a complex of proteins converting all-*trans* retinol to 11-*cis* retinal. The 11-*cis* retinal is transported back to the photoreceptor, where it recombines with opsin and regenerates rhodopsin.

For cones, 11-*cis* retinal can also be produced enzymatically in the RPE, but much of the chromophore is re-isomerized by an alternative mechanism. We have long known that a second pathway must exist, because cone pigment can be regenerated in an isolated retina in the absence of the RPE (Goldstein, 1970; Hood and Hock, 1973). An important advance in our understanding was the demonstration that at

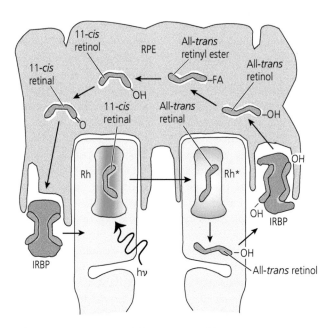

Figure 9.25 Principal enzymes and transport proteins responsible for regeneration of rhodopsin in a vertebrate rod and retinal pigment epithelial cell. RPE, Retinal pigment epithelium; FA, fatty acid; Rh, rhodopsin; IRBP, interphotoreceptor retinol binding protein; hν, light. (After Bok, 1993; Fain et al., 1996.)

least some of this regeneration can occur within the retinal Müller glial cells (Wang et al., 2009; Wang and Kefalov, 2009; see Wang and Kefalov, 2011). The Müller-cell pathway produces 11-*cis* retinol rather than 11-*cis* retinal (Mata et al., 2005), and this difference is important because cones can utilize 11-*cis* retinol but rods can't (Jones et al., 1989). As a consequence, the 11-*cis* chromophore made by the Müller cells can be preferentially utilized by cones to speed the rate of their recovery.

Recent evidence indicates that there is a protein in the Müller cells (and also in the RPE) which may be responsible for re-isomerizing at least part of the cone chromophore (Morshedian et al., 2019). This protein is called *retinal G-protein-coupled receptor* (RGR) *opsin* (Chen et al., 2001b). It is a member of the opsin family and is similar in structure to rhodopsin, again with a lysine in the seventh α helix; but it binds all-*trans* retinal instead of 11-*cis* retinal. Absorption of a photon then isomerizes all-*trans* retinal back to 11-*cis* retinal, which is subsequently converted to 11-*cis* retinol and conveyed to the cones. RGR opsin can act effectively as a photo-isomerase, much like metarhodopsin in arthropods.

Such a mechanism of photoconversion would be particularly appropriate for cones, which must continue to respond even in continuous bright light. Although it had long been thought that insects and vertebrates have quite different pathways for photopigment regeneration, both may use light together with enzymes in order to supply sufficient chromophore to their photoreceptors.

Cone pigment regeneration can be quite rapid, whereas enzymatic regeneration of rod pigment after bright light exposure is comparatively slow. For human rods, complete recovery can require as much as 30–35 minutes after the light is turned off. During this time, the sensitivity of vision is markedly depressed. This is the reason why, when we turn on the lights in the bathroom in the middle of the night to get an aspirin or a drink of water, we can barely find our way back to bed once the light is turned off. Part of the reason sensitivity is decreased is that there is less pigment to absorb photons, since a fraction of the pigment has been bleached and has lost its chromophore. There is then a somewhat smaller probability of absorption of a photon by the remaining unbleached rhodopsin. This decrease is, however, much too small to account for the loss of sensitivity, indicating that some other process must be occurring.

Stiles and Crawford (1932) first suggested that bleached pigment might desensitize the visual system by acting as an equivalent background light. In molecular terms, their suggestion would mean that some component of bleached pigment can stimulate the transduction cascade much like light, producing an activation of the PDE6, a decrease in Ca^{2+} concentration, and a modulation of the sensory cascade. It is in fact likely that virtually every bleaching intermediate can stimulate the cascade to some extent, including phosphorylated metarhodopsin and even opsin. The question is, which intermediate is most important?

The answer seems to depend upon the amount of visual pigment that has been bleached. For light that bleaches only a relatively small fraction of the pigment, the most important component of the equivalent background in a rod seems to be continued excitation of opsin by all-*trans* retinal (Hofmann et al., 1992; Jager et al., 1996). Sensitivity recovers as all-*trans* retinal is converted to inactive all-*trans* retinol and the retinol leaves the photoreceptor. For large bleaches, all-*trans* retinal is converted to all-*trans* retinol before most of the pigment is regenerated (Kennedy et al., 2001). The photoreceptors remain desensitized, and the equivalent background seems to be mostly produced by opsin itself, which stimulates the cascade, though with low probability (Cornwall and Fain, 1994). Opsin continues to activate the cascade and depress sensitivity until all of the photopigment has been regenerated.

Intrinsically photosensitive retinal ganglion cells

We have been accustomed to thinking that rods and cones are the only photoreceptors in the vertebrate retina capable of converting light into an electrical signal. We now know better: a small fraction of ganglion cells can do this trick too (Do and Yau, 2010; Schmidt et al., 2011; Hughes et al., 2012; Lucas, 2013). Ganglion cells receive signals from other retinal cells (see Figure 1.2B), and they send their axons into the optic nerve, which carries visual information from the retina into the CNS. A few of these cells can also respond to light directly, and this

population is especially important in non-image-forming visual tasks, such as pupillary light constriction and the setting of the circadian clock.

These light-sensitive cells are usually referred to as *intrinsically photosensitive retinal ganglion cells* (*ipRGCs*). Their light responses were first described by David Berson's laboratory (Berson et al., 2002), who labeled ganglion cells projecting to the superchiasmatic nucleus of the hypothalamus, a center in the brain known to house the master circadian pacemaker. When Berson and colleagues recorded from these cells (Figure 9.26A), they discovered responses to light even when all of the inputs from other retinal cells had been eliminated with synaptic blockers. At about the same time, Hattar and colleagues (2002) showed that these same cells express a visual pigment called melanopsin or Opn4, initially isolated from *Xenopus* melanophores (see Lucas, 2013). Melanopsin is similar in structure to other members of the opsin family but is more like the microvillar pigment of *Drosophila* and other invertebrates than the ciliary pigments of rods and cones. Melanopsin has a stable metarhodopsin intermediate, capable of regenerating the chromophore with light (Matsuyama et al., 2012). That is, absorption of one photon can isomerize 11-*cis* retinal to all-*trans* retinal and trigger the transduction cascade, ultimately depolarizing the membrane potential and generating action potentials (Figure 9.26A). Another photon can then re-isomerize all-*trans* to 11-*cis* and regenerate the visual pigment. Other mechanisms may also be capable of regenerating the chromophore by an alternative pathway that does not require illumination (Walker et al., 2008; Zhao et al., 2016).

The first kind of ipRGC to be identified is now called M1; it is the easiest to study because it has the highest melanopsin concentration, and it is also the most sensitive with the largest intrinsic light response. As a result, we know most about its mechanism of transduction, which resembles that of microvillar photoreceptors like those of *Drosophila* (Graham et al., 2008; Xue et al., 2011; Jiang et al., 2018). Melanopsin in an M1 ipRGC activates a G protein whose α subunit is a member of the $G\alpha_{q/11}$ family, and this G protein in turn stimulates a PLC like the *norpA* protein of *Drosophila*. The rest of the cascade is unclear but ultimately results in the opening of TRP channels, again much like in the fruit fly. The mechanism of channel gating seems not to involve Ca^{2+} release

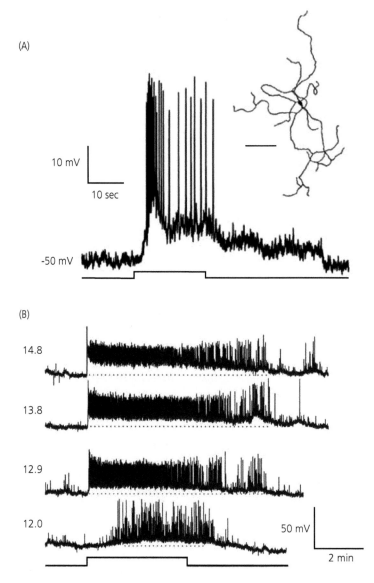

Figure 9.26 Responses of intrinsically photosensitive retinal ganglion cells. (A) Light response of an intrinsically photosensitive retinal ganglion cell (ipRGC) in the rat, recorded with patch clamp in an isolated retina in the presence of 2 mM $CoCl_2$, which blocks all synaptic input onto the ganglion cell. Timing of light stimulus is indicated by lowermost trace. Cell was identified by retrograde-labeling from the suprachiasmatic nucleus. Inset shows camera-lucida drawing of the recorded cell, scale bar 100 μm. (B) Voltage responses recorded in another cell to 4-minute exposures of increasing brightness showing maintained firing even to prolonged stimulation. The number to the left of each trace gives the light intensity in $\log_{10}$ photons per second per centimeter squared at a wavelength of 500 nm (blue-green). (From Berson et al., 2002.)

or IP_3. There are an additional five classes of melan-opsin-expressing cells called M2–M6 (see for example Ecker et al., 2010; Schmidt et al., 2014; Quattrochi et al., 2018), each with rather different properties. There is increasing evidence that these cells do not transduce like M1s, but can use a variety of mechanisms, apparently including cyclic nucleo-tides (Jiang et al., 2018; see also Sonoda et al., 2018).

Even the best-responding of the ipRGCs are orders of magnitude less sensitive than rods and cones. Although the single-photon response to an excited melanopsin can be as large as that of a rod, ganglion cells have no outer segments with closely spaced disks or membrane lamellae, and the concentration of pigment is as much as 10^4 times lower than in a rod or cone (Do and Yau, 2010). The probability of absorption of a photon is therefore considerably lower. Although much more light is required to stimulate the ipRGCs, these cells can nevertheless produce maintained responses to continuous light spanning intensities from moonlight to full daylight (Figure 9.26B). At dimmer intensities, ipRGCs receive synaptic input from other retinal cells which can produce depolarization and action potentials. Together, the combined synaptic and intrinsic responses of the ipRGCs can cover the whole of the operating range of vision (Hattar et al., 2003; Guler et al., 2008). These cells therefore comprise an ideal system for adjusting the diameter of the pupil and setting the circadian clock according to the brightness of ambient illumination. They also seem to contribute to other visual tasks, including a primitive form of pattern vision (see for example Ecker et al., 2010; Stabio et al., 2018). Moreover, bright light makes us more cheerful, and the lack of light during the long winter months can trigger seasonal affective disorder (SAD). The ipRGCs may have a role in these phenomena as well (see Lazzerini Ospri et al., 2017).

Summary

Photoreceptors respond to the part of the electromagnetic spectrum we call light. They do this with a visual pigment, consisting of a chromophore covalently attached to a protein called opsin. From archaeobacteria to man, the mechanism of photon absorption is remarkably conserved. The chromophore is a derivative of vitamin A called retinal, and the absorption of light produces an isomerization: in bacteria, all-*trans* retinal is converted to 13-*cis*, whereas in animals 11-*cis* is converted to all-*trans*. For most visual pigments, the chromophore in the dark is protonated and forms a salt bridge with an adjacent, negatively charged amino acid. Isomerization produces a change in the shape of the chromophore within the opsin binding pocket, which breaks the salt bridge, triggering a change in conformation of rhodopsin to an active form that initiates the sensory cascade.

In bacteria, phototransduction is much like chemotaxis. Light produces an alteration of the concentration of phosphorylated CheY protein, which acts as a second messenger controlling the flagellar motor. In all animals, the visual pigment activates a heterotrimeric G protein and triggers a metabotropic cascade. Several different G-protein families and transduction cascades have been implicated in phototransduction, sometimes even in different cells of the same organism. The most thoroughly studied cascades are those that produce the depolarizing responses of the arthropods *Limulus* and *Drosophila*, and those producing the hyperpolarizations of vertebrate rods and cones.

In both *Limulus* and *Drosophila*, the photopigment is contained within numerous microvilli, collectively referred to as a rhabdomere, which greatly increase the surface area of the plasma membrane. Rh* activates a trimeric G protein with a $G\alpha_q$ subunit to produce $G\alpha_q \bullet GTP$. The $G\alpha_q \bullet GTP$ then stimulates a PLC, generating the two second messengers IP_3 and DAG. Despite many years of intense effort, it is still not clear which if either of these second messengers is directly responsible for gating the opening of the ion channels. The channels in *Drosophila* are the founding members of the *trp* family of proteins, of which two different forms are expressed: TRP and TRPL. In *Limulus*, on the other hand, the channels have been proposed to be similar to those in vertebrates, gated by cyclic nucleotides. Light produces a large increase in intracellular Ca^{2+}, in *Drosophila* mostly the result of Ca^{2+} entering the cell through the light-dependent channels, and in *Limulus* from IP_3-gated Ca^{2+} release. In both species, the increase in Ca^{2+} plays an important role in activation and modulation of the transduction cascade. Several different components of transduction are affected by Ca^{2+}, but it is not yet known exactly how the gain and sensitivity are regulated.

In the rods and cones of vertebrates, transduction occurs in a part of the cell called the outer segment, which contains the photopigment, transduction enzymes, and channels necessary for producing the light response. Rh* excites a G protein called transducin, which is a member of the α_i/α_o family and is coupled to PDE6. The $T\alpha$-GTP binds to the PDE6 γ inhibitory subunits, relieving inhibition and stimulating the PDE6 to hydrolyze cGMP. Rod and cone outer segments contain cation-permeable channels gated by cyclic nucleotides, which are open in darkness and allow a large influx of both Na^+ and Ca^{2+}. The decrease in cGMP concentration caused by PDE6 activation leads to closing of the channels, reduction in ion influx, and hyperpolarization of the membrane potential. The transduction cascade is turned off by the quenching of Rh*, produced by C-terminal phosphorylation of serines and threonines by rhodopsin kinase, followed by the binding of arrestin. The $T\alpha$-GTP is quenched when GTP is hydrolyzed to GDP, and the rate of this reaction is greatly accelerated by the GAP-complex proteins RGS9, $G\beta5L$, and R9AP together with the inhibitory γ subunit of PDE6. The decay of PDE6 activity is the rate-limiting step of inactivation, at least for mammalian rods.

The Ca^{2+} coming into the rod through the cyclic-nucleotide-gated channels in darkness is extruded by a Na^+/K^+-Ca^{2+} transporter, which exploits the energy of both the Na^+ and K^+ ion concentration gradients. Light closes the cyclic-nucleotide-gated channels, and continued extrusion by the transporter produces a decrease in intracellular Ca^{2+} concentration. The change in Ca^{2+} is thought to have several effects on the transduction cascade, but the most important seems to be the modulation of the rate of the guanylyl cyclase via small-molecular-weight binding proteins called GCAPs. As the Ca^{2+} concentration decreases in the light, Ca^{2+} comes off the GCAP proteins and cyclase activity is increased. The increase in cyclase in turn increases the cGMP concentration and re-opens a fraction of the channels, speeding the return of the photoreceptor current back toward its dark level. Ca^{2+} also plays an important role in light adaptation. If changes in outer segment Ca^{2+} concentration are prevented, adaptation is significantly impaired. The Ca^{2+} regulation of cyclase makes an important contribution to adaptation by preventing saturation of the photoreceptor response, but Ca^{2+} is also thought to regulate the cascade by additional mechanisms that have not as yet been clarified.

After bright light exposure, photoreceptors slowly recover sensitivity. The retinal isomerized by light from 11-*cis* to all-*trans* must be re-isomerized back to 11-*cis*. For microvillar photoreceptors, including those of arthropods, the chromophore remains attached to the protein and is re-isomerized by light. In ciliary rods and cones, the all-*trans* retinal is released from the photopigment, converted to all-*trans* retinol, and transported to an adjacent cell layer, either the retinal pigment epithelium or Müller glial cells. There the all-*trans* retinol is reconverted either enzymatically or by light exposure to an 11-*cis* isomer, which can diffuse or be transported back to the photoreceptor to recombine with opsin and regenerate the rod or cone visual pigment. During this process, the sensitivity of the photoreceptor is depressed as if it were illuminated by an equivalent background light. In molecular terms, this equivalent background is produced by low-level stimulation of the transduction cascade by intermediates of bleaching including opsin itself, which can activate the cascade, though with much less effectiveness than Rh*.

In addition to the photoreceptors, the retina of vertebrates contains a small fraction of ganglion cells that express a visual pigment and can respond directly to light. These cells, called intrinsically photosensitive retinal ganglion cells (ipRGCs), contain melanopsin, a visual pigment similar in structure to arthropod rhodopsin. Although much less sensitive than either rods or cones, the ipRGCs can respond to light with maintained firing of action potentials. The ipRGCs also receive synaptic input from other retina cells, and these signals, together with the intrinsic photosensitivity, enable the ipRGCs to respond to illumination over an extended range. They project to a variety of targets in the CNS to help control many tasks not requiring image formation, including the control of pupillary diameter and circadian rhythm. They may in addition have a limited role in visual detection and the modification of affect or mood.

CHAPTER 10

Extra sensory receptors

In addition to the five major senses of touch, hearing, smell, taste, and sight, there are modalities of sensation, some present in most animals but others unique to certain species. These include a sense called proprioception, which provides an awareness of the position of the body in space. Proprioception is a complicated sense, but one important contribution comes from stretch receptors and tendon organs associated with our muscles, which indicate the position and orientation of our limbs. I shall not discuss the physiology of these receptor organs in detail, because there is every reason to believe that mammalian stretch receptors and tendon organs are mechanoreceptors using mechanisms of transduction similar to those of touch receptors in the skin (see for example Woo et al., 2015; Chesler et al., 2016). Within the viscera of our body there are many kinds of enteroreceptors—sensory receptors that detect changes in the chemical composition of internal fluids or alteration in the shape of organs that mediate, for example, our sensations of hunger and thirst. The little that we know about these receptors suggests that they function much like the chemoreceptors and mechanoreceptors I have previously described in Chapters 5–8, though the receptor molecules and channels and perhaps even the details of the sensory cascades may be somewhat different.

Of greater interest for our purposes are three kinds of sensory receptors whose mechanisms of transduction seem to be novel. It is these receptors that are the subject of this last chapter. They include, first, thermoreceptors which sense changes in temperature. Like other vertebrates, we have them in our skin, but some beetles, snakes, and bats have them in specialized organs that detect infrared radiation and allow these animals literally to see in the dark. Electroreceptors in elasmobranchs and certain fresh-water fish sense electric fields in water. They are used for locating prey, avoiding enemies, and communicating with other members of the same species, particularly in streams or lakes that are turbid and have poor visibility. Finally, magnetoreceptors are cells that are presumed to sense the direction of the magnetic field. Many years of patient behavioral experimentation have shown that variations in the strength or direction of the Earth's magnetic field can be detected by many species and provide necessary aids in navigation, particularly for migrating birds. The cells that perform this detection and the mechanisms they use still remain to be discovered.

Thermoreception

The mammalian skin contains localized clusters of free nerve endings, which mediate our senses of hot and cold (see Hensel et al., 1974; Jänig, 2018). These endings can be identified by placing a small probe against the skin and stimulating with an electrical current, which produces a distinct sensation of warmth or cooling but only when the electrode is carefully positioned in a well-defined spot no more than 1 mm in diameter (see Hensel, 1973; Darian-Smith, 1984b; Willis and Coggeshall, 1991). The skin has more cold spots than warm spots; on the human nose, for example, there are eight to ten cold spots and only one warm spot per square centimeter of tissue. Each spot is apparently formed by the nerve terminals of a single sensory axon coming from one

Sensory Transduction. Second Edition. Gordon L. Fain, Oxford University Press (2020). © Gordon L. Fain 2020.
DOI: 10.1093/oso/9780198835028.001.0001

of the dorsal root ganglia, which I earlier described as the source of touch receptors innervating the skin (see Chapter 5). Cold sensitivity is apparently present in DRG cells having small, mostly myelinated axons which each extend a number of fine terminals into the basal lamina of the cutaneous epithelium (Hensel et al., 1974; Jänig, 2018). These DRG cells resemble touch receptors except that they lack any external specialization at the ends of their terminals. Warm receptors are formed predominantly from unmyelinated axons that terminate deeper in the skin and also elaborate free nerve endings.

The action potentials of thermoreceptors can be recorded with extracellular electrodes from the sensory nerves coming from the dorsal root ganglia (see Hensel, 1973; Darian-Smith and Johnson, 1977; Darian-Smith, 1984b; Willis and Coggeshall, 1991). When recordings are made from the median nerve of the macaque monkey coming from the wrist, a substantial fraction of unmyelinated fibers are warm receptors, and about one-third of all the myelinated fibers are cold receptors (Darian-Smith, 1984b). For a warm receptor, increases in temperature produce an increase in firing that adapts to a maintained, static spike frequency roughly proportional to the value of the ambient temperature (Figure 10.1A). The initial rapid increase in firing produces a signal that indicates a change in temperature, while the maintained firing rate is thought to be responsible for our perception of absolute temperature.

For a cold-sensitive fiber, decreases in temperature produce an increase in spike firing (Figure 10.1B), followed again by adaptation to a maintained frequency. This maintained discharge is formed by bursts of a number of action potentials in rapid succession, followed by a pause in firing (see for example Viana et al., 2002), and both the delay between bursts and the number of spikes per burst increase with decreasing temperature (Bade et al., 1979; Braun et al., 1980). This bursting of action potentials may be the feature of the response that helps us decide whether a light jacket will do or whether it is time to take the winter coat out of the closet.

The fine terminals of thermoreceptors are embedded in the skin and practically impossible to dissect free for intracellular or patch-clamp recording. It is nevertheless possible to record from the cell bodies of the fibers in preparations of single cells dissociated

from dorsal root ganglia, particularly from the more numerous cold-sensitive cells. Often the trigeminal ganglion is used—this is the nucleus that provides the sensory input to the face (Reid and Flonta, 2001; McKemy et al., 2002; Okazawa et al., 2002; Viana et al., 2002).

The voltage recording in Figure 10.2A is from a cold-sensitive cell dissociated from the mouse trigeminal ganglion. Cooling produces a depolarization and action potentials that adapt and then settle into a characteristic pattern of bursts like the extracellularly recorded spikes in Figure 10.1B. In voltage-clamped cells (Figure 10.2B), a decrease in temperature generates an inward current that rapidly adapts in normal solution but shows little adaptation when extracellular Ca^{2+} is removed from the Ringer. Similar responses are produced by exposing cells to menthol, which also stimulates cold receptors. The current in Figure 10.2B is caused by an increase in conductance, due to the opening of channels that are rather non-selectively permeable to monovalent and divalent cations, including Ca^{2+} (Reid and Flonta, 2001; McKemy et al., 2002; Okazawa et al., 2002). Single-channel responses can be recorded from excised patches (Figure 10.2C), suggesting that the gating of the channel does not require second messengers or other enzymes or metabolites. That is, the temperature sensitivity is apparently intrinsic to the channel protein itself, much as we saw earlier for the heat-sensitive TRPV1 protein (Figure 3.6B).

McKemy and colleagues (2002) took advantage of the menthol sensitivity of cold-sensitive cells to screen an expression library from the trigeminal ganglion. In this way, they identified a cDNA encoding a cold-sensitive channel protein (see also Peier et al., 2002). The protein they identified is now called TRPM8 and is another member of the TRP family (see Figures 3.6, 5.10D, and 9.11). When the gene for this protein is expressed in *Xenopus* oocytes or in a cell line, it confers all of the properties of the cold receptor including sensitivity to menthol (Figure 10.3). Decreases in temperature produce an increase in inward current, caused by an increase in permeability to cations including Na^+. This evidence substantiates our interpretation of the single-channel recordings from DRG cells in Figure 10.2C that cold sensation is ionotropic, mediated by proteins whose probability of channel opening is

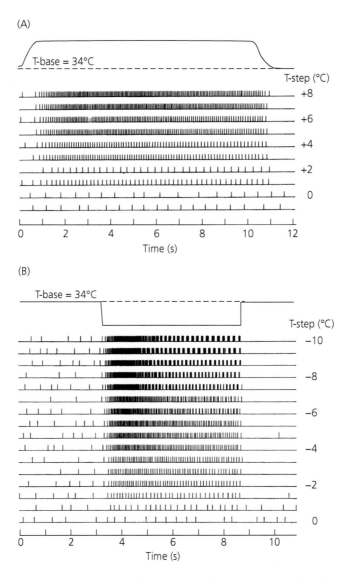

Figure 10.1 Temperature receptors in the hand of a monkey. Extracellular recording of action potentials from (A) a warm-sensitive neuron and (B) a cold-sensitive neuron from the median nerve of the macaque monkey. T-base, Ambient temperature from which temperature change was made; T-step, sign and amplitude of change of temperature. Uppermost traces in both parts of the figure give time courses of temperature changes. Dashed lines indicate ambient temperature. (From Darian-Smith and Johnson, 1977.)

altered specifically by a change in temperature without the necessity of a second messenger or sensory cascade.

When the gene for TRPM8 is deleted, mouse sensory neurons have greatly reduced sensitivity to moderate cold; moreover, *TRPM8–/–* mice show marked behavioral deficits in temperature prefer-ence and cold avoidance (Bautista et al., 2007; Colburn et al., 2007; Dhaka et al., 2007). This evidence indicates that much of our sense of cooling derives from the gating of TRPM8 channels, though other channel proteins seem also to contribute especially in the detection of very low, noxious temperatures (see McKemy, 2013).

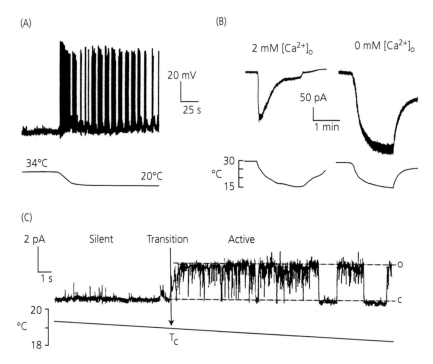

Figure 10.2 Responses of dissociated sensory cells to cooling. (A) Whole-cell patch recording of membrane voltage from a mouse trigeminal neuron. The resting potential of the cell was –65 mV. The time course of the change in temperature is indicated by trace below. (B) Whole-cell patch recording of membrane current of a rat dorsal-root ganglion cell in normal and zero Ca^{2+} solution. Time courses of changes in temperature are indicated by traces below. Holding potential was –60 mV. (C) Excised inside-out patch recording from a cell as in (B). Channels opened at the critical temperature (T_c) of 19°C. (A from Viana et al., 2002; B and C from Okazawa et al., 2002.)

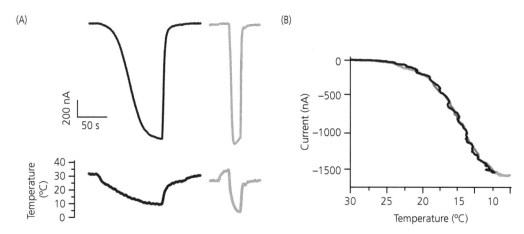

Figure 10.3 Recording from expressed TRPM8 channel. The gene for TRPM8 was expressed in a *Xenopus* oocyte, and responses were recorded with a two-electrode voltage clamp as in Figure 1.9A. (A) Decreases in the temperature of the solution perfusing the oocyte produced an inward current. Cooling ramps (bottom) were applied at two different rates (0.2°C s⁻¹, black lines; 1°C s⁻¹, blue lines). (B) Amplitude of current as a function of temperature. (From McKemy et al., 2002.)

TRP channels may also be responsible for the detection of heat. We already saw in Chapter 3 that TRPV1 channels can signal elevations of temperature by a mechanism that is ionotropic: even purified protein incorporated into artificial lipid membrane can be gated by temperature increases (see Figure 3.6B and Cao et al., 2013b). TRPV1 channels in rodents and probably in most mammals respond to high, noxious temperatures to signal pain. Several additional TRP channels have been identified that are sensitive to heat, including TRPV2, TRPV3, TRPV4, and TRPM3 (see Vriens et al., 2014; Palkar et al., 2015). A further channel, TRPM2, is widely expressed throughout the body and activated by oxidative stress (see Kashio and Tominaga, 2017), but it is also gated by temperature (see for example Song et al., 2016). Deletions of the TRPM2 gene can affect the tendency of mice to avoid heat that is elevated but below the pain threshold (Tan and McNaughton, 2016), suggesting that this channel may have a role in detecting moderate warmth like that of TRPM8 for moderate cold. Other channel types, including a Ca^{2+}-activated Cl^- channel (see Chapter 7), two-pore K^+ channels (see Chapter 3), and Orai channels (see Chapter 4) have been proposed as candidate receptors either for painful noxious heat or for less extreme temperature elevation, but their roles remain uncertain and controversial.

It is still unclear how increases or decreases in temperature can gate channel opening. The opening of the channel pore can be thought of as a chemical reaction, converting the closed conformation of the pore to the open conformation. The free energy (ΔG) of this reaction depends upon the change in enthalpy and entropy by the familiar equation

$$\Delta G = \Delta H - T\Delta S \qquad (10.1)$$

Like any chemical reaction, channel opening will proceed spontaneously only when ΔG is negative.

Equation (10.1) shows that the value and even sign of ΔG can change with temperature as $T\Delta S$ changes. The value of ΔS gives the change in entropy associated with the temperature-dependent change in channel conformation and is a measure of the change in disorder of the protein. Single-channel measurements from TRPV1 and TRPM8 channels indicate that channel gating is accompanied by a positive change in entropy for TRPV1 during warming but a negative change for TRPM8 during cooling (Yang et al., 2010). The result is that either warming the TRPV1 channel (which increases both T and ΔS) or cooling the TRPM8 channel (which decreases both T and ΔS) will tend to increase $T\Delta S$ and decrease ΔG to favor the reaction of channel opening. The changes in $T\Delta S$ are accompanied by changes in enthalpy (ΔH), as intramolecular bonds within the protein are formed or broken (see for example Brauchi et al., 2004).

The question now becomes, which part or parts of the channel undergo these changes in conformation? Meticulous investigation particularly of TRPV1 has shown that several protein domains are essential, including regions around the ion channel (Grandl et al., 2010; Yang et al., 2010) and near the amino terminus (Yao et al., 2011; see Zheng and Ma, 2014). It is possible, however, that channel opening may not depend upon clearly defined protein domains as in voltage-gated or ligand-gated channels, but may instead involve a large part of the channel structure (see Vriens et al., 2014). A fuller understanding may have to await the challenging experiment of determining the detailed conformation of the whole of the channel at different ambient temperatures.

Seeing in the dark: heat receptors as detectors of infrared

Several species have specialized thermoreceptors for sensing infrared irradiation (IR) (see Campbell et al., 2002). Adult beetles of the genus *Melanophila* congregate in huge numbers at forest fires and occasionally also at backyard barbeques. They mate even as the fire is still burning, and their larvae—which cannot withstand the defense mechanisms of living trees, take up residence in the dead wood. The beetles can detect fire from a distance as great as 80 km (50 miles) (see Hart, 1998). They do this with pit organs on either side of their bodies just underneath their middle legs. These organs are normally covered when the beetle is resting or walking but become exposed during flight (Vondran et al., 1995). They consist of a wax-filled cavity containing fifty to a hundred sensilla, similar in anatomy to the sensilla of mechanoreceptors (Figure 5.9) or chemoreceptors (Figures 2.8, 7.5, and 7.6) but each with a single receptor cell whose dendrite is surrounded by a bulbous sphere of cuticle. Stimulation of these organs with IR like that emitted by a forest fire (at wavelengths of 2–4 µm) produces a behavioral response (antenna twitching) and an increase in action potentials from receptor cells in the sensilla (Schmitz and Bleckmann, 1998; Schmitz et al., 2000). The mechanism of transduction is unknown.

Some vertebrate predators also have IR detectors, apparently for sensing the heat emitted from the body of their victims. Vampire bats have pits on the front of their faces adjacent to the nose (Kürten and Schmidt, 1981), which detect IR with modified TRPV1 proteins (Gracheva et al., 2011). They use these organs to locate patches of skin with blood vessels close to the surface. Snakes in the family *Boidae* (including pythons and boa constrictors) have a series of scales along their lips each with a small pit in its center (see Barrett et al., 1970). These pits are heavily innervated by heat-sensitive thermoreceptors, whose unmyelinated terminals break up into small filaments that lie just underneath the surface of the skin (Warren and Proske, 1968). The physiology of the terminals resembles that of heat receptors found elsewhere in vertebrate skin. They respond to any elevation of temperature including that produced by a warm object brought near the head of the snake, such as an unsuspecting mouse or even the hand of the experimenter.

The extracellular spike recordings in Figure 10.4 were made from the Australian python *Morelia spilotes* by dissecting single axons from the nerves coming from the labial scales (Warren and Proske, 1968). The fibers generally fire spontaneously, but the firing rate is greatly increased when IR is directed onto the scales (Figure 10.4A). Conversely, the cell is inhibited when a cold object is placed in its vicinity (Figure 10.4B). These receptors appear to enable the snakes to detect prey even in darkness, because blindfolded animals can strike at warm objects but not when the pits of the labial scales are covered (Noble and Schmidt, 1937).

The most carefully studied of the infrared sensors are the pit organs of another group of snakes called the Crotalinae or pit vipers, which include some of the most feared reptiles in North America—the water moccasin, the copperhead, and western rattlesnakes of the genus *Crotalus* (Barrett et al., 1970; Terashima and Goris, 1975; Newman and Hartline, 1982). These animals have two prominent pits on either side of the face between the eyes and mouth, each about 1–2 mm in diameter and 5 mm deep (Figure 10.5). In the middle of the depth of the pit there is a sensory membrane only 10-15 µm in thickness. Nearly half of the width of this membrane is occupied by a dense plexus of the terminals of from 6000 to 7000 nerve fibers, which appear to be a nearly uniform population of heat-sensing thermoreceptors.

The physiology of these thermoreceptors has been studied in some detail (Bullock and Diecke, 1956; Terashima et al., 1968; Desalvo and Hartline,

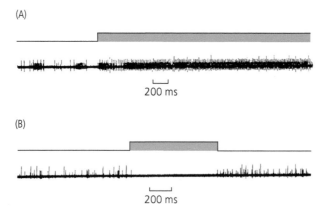

(A)

200 ms

(B)

200 ms

Figure 10.4 Extracellular recording of action potentials from heat-sensitive fibers of facial pits in the python *Morelia spilotes*. Responses to warming produced by the light of flashlight (A) and cooling produced by placing a cold object in front of the facial pits (B). The time course of the stimuli is indicated by blue bars. Contrast of records has been reversed. (From Warren and Proske, 1968.)

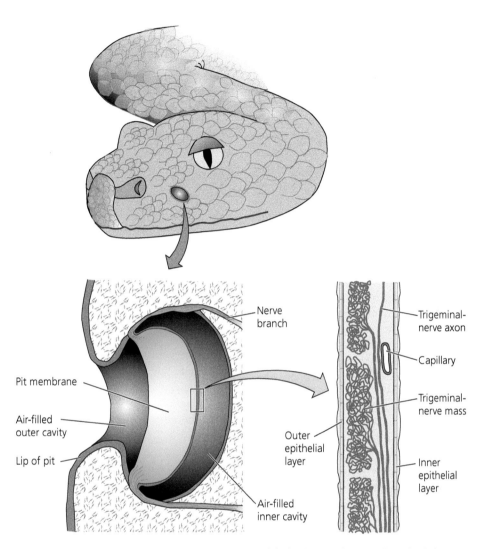

Figure 10.5 Rattlesnakes have two prominent pit organs, one on either side of the face between the eyes and mouth, which contain a sensory (pit) membrane with a dense plexus of terminals from a nearly uniform population of heat-sensitive thermoreceptors from the trigeminal nerve. (After Newman and Hartline, 1982.)

1978; Terashima and Liang, 1991). Like warm receptors in the skin (Figure 10.1A), they can be stimulated by an elevation in temperature, produced for example by flowing heated water across the pit membrane. Like the thermoreceptors of pythons (Figure 10.4), they can also be stimulated by the direct illumination of the pit opening with IR or any source of heat brought near the pit. Stimulation produces a rapid increase in axon firing, which then slowly adapts. As in pythons, most of the heat-

sensitive fibers are spontaneously active and inhibited by cooling. Most also respond to mechanical stimulation.

The response to temperature is extraordinarily sensitive: snakes can detect a change in temperature of the order of 0.003°C, corresponding to a thermal radiation of about 10^{-10} joules falling on the terminals of a single nerve fiber (Bullock and Diecke, 1956). This sensitivity reflects in part the properties of the thermoreceptive channels of the nerve terminals, which are TRPA1 channels acutely sensitive to warmth (Gracheva et al., 2010). The high sensitivity is also a consequence of the special anatomy of the pit organ. The thinness of the sensory membrane has the consequence that the nerve terminals are very close to the surface of the epithelium, in contrast to those in mammalian skin which are generally buried something like 300 μm into the epithelium. The IR falling on the pit organ needs to heat only the very thin sensory membrane, and little energy is lost to surrounding tissue.

The sensitivity of the thermoreceptors in the pit organ, though impressive, is still much less than that of an arthropod or vertebrate photoreceptor, which, as we have seen (Figure 2.13), can detect single photons of light. An energy of 10^{-10} joules falling on the nerve endings of a single pit viper thermoreceptor corresponds to something like 10^9–10^{10} photons of IR. Visual receptors are so much more sensitive because they use a mechanism of detection that is based on a pigment and a photochemical reaction. A mechanism of this kind is not possible for receptors responding to IR, because the energy of a photon of IR is not great enough to produce a chemical reaction with sufficient detectability (Luo et al., 2011).

It is quite important that the diameter of the opening of the pit organ be smaller than that of the sensory membrane. This difference has the effect that the pit organ can function as a crude pinhole camera: radiation coming from different sources will strike the sensory membrane in different places (Figure 10.6A). Since each thermoreceptor terminates in a different part of the sensory membrane, it is sensitive to a different region of space (Desalvo and Hartline, 1978). The snake can use the population of fibers to locate prey in complete darkness, which it does with frightening accuracy. In the experiment of Figure 10.6B (Newman and Hartline, 1982), the precision of detection was estimated by placing a hot soldering iron just beyond the range of a blind-folded snake, and then giving the animal a mild electric shock to provoke a strike. The accuracy of aim was measured by filming the snake with a video camera, and each blue dot gives the result of a separate trial. The figure provides an approximate indication of the error of the strike, which was rarely more than 5°. For those of us who enjoy hiking in the desert of Southern California, the moral is clear: never go out after darkness!

Electroreception

Sensory receptors for detecting electric fields are found in many aquatic vertebrates, including sharks and rays, some teleost fishes, the electric catfish and electric eel, many urodelian amphibians (salamanders and axolotls), and a variety of other species, including the duck-billed platypus and dolphins (Baker et al., 2013; Czech-Damal et al., 2013). The sensitivity in some species is almost beyond belief: the ray *Raja clavata* can respond to an electric field as small as 0.01–0.02 μV cm^{-1} (Kalmijn, 1966). This is the value produced by the voltage of a flashlight battery dropping over a distance of 1000 km, about that between San Francisco and Portland, Oregon (or Dover to Orkney). It accomplishes this task with an extensive network of ampullary organs (Figure 10.7A), consisting of canals typically 1 mm in diameter and as long as 15–20 cm, distributed over the body and opening directly into the sea water (see Murray, 1974; Szabo, 1974). The canals are filled with a jelly-like mucopolysaccharide and terminate within the animal in only a few places, called capsules.

Sharks and rays live in sea water containing a high concentration of dissolved ions, and skin resistance is low. This circumstance has the consequence that the electric field in the water is not much distorted by the body of the animal. The resistance of the wall lining the canal is quite high (Waltman, 1966), so that little voltage is lost by leakage through the canal wall. The animal can then compare the voltage at the opening of the canal with the voltage at the canal terminus, within the body interior. If the canal and the electric field are oriented in the same direction, the voltage simulating the receptor at the end of the canal can be calculated with little error by multiplying the electric field by the canal length. For a given electric field (in volts per centimeter), the longer the canal (in centimeters), the greater the voltage difference stimulating the receptor. Since each skate has many canals of different lengths pointing in different directions, it can use this diversity to detect sources of voltage of varying amplitude in different locations.

In the capsule, each canal swells up to form an ampulla, consisting of a series of small sacs called *alveoli* (Figure 10.7B). The body wall of the alveolus

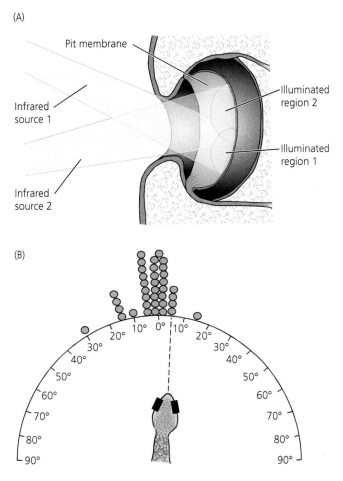

Figure 10.6 Infrared "vision:" optics of the pit organ and spatial accuracy of detection. (A) The pit organ of a snake illuminated from two infrared sources in different positions, showing the "pinhole camera" focusing of the image. (B) Accuracy of detection of striking to a heat source (soldering iron) by a blind-folded rattlesnake. (After Newman and Hartline, 1982.)

is an epithelium made of a single layer of receptor cells and supporting cells (Figure 10.7C), connected to one another on their luminal surface by an extensive network of tight junctions which form a high-resistance barrier and force the electric field to drop across the membranes of the receptors.

The receptors are thought to have evolved from hair cells of the lateral-line system (see Chapter 6). They have a single cilium which projects into the lumen of the alveolus, analogous to a hair cell kinocilium (Waltman, 1966). Each receptor cell makes four to five synapses with the five to ten nerve fibers that innervate the several thousand receptors of each ampulla, and these synapses have distinctive synaptic ribbons and postsynaptic densities (see

Figure 2.12B). As we shall see, synaptic transmission in electroreceptors is quite sensitive, and it is likely that these specialized synaptic structures are in large part responsible. The transmitter is excitatory and probably glutamate (Bennett and Obara, 1984).

Ampullary receptors with a similar morphology are found in many other species. Those of the freshwater fish *Eigenmannia* are typical (Figure 10.8A). There are many fewer sensory cells than in skate, and each ampulla is innervated by only a single nerve fiber. The canal is also much shorter. Freshwater fish live in a medium with a low concentration of dissolved ions and therefore a low tonicity. To keep from losing ions and gaining water, the fish

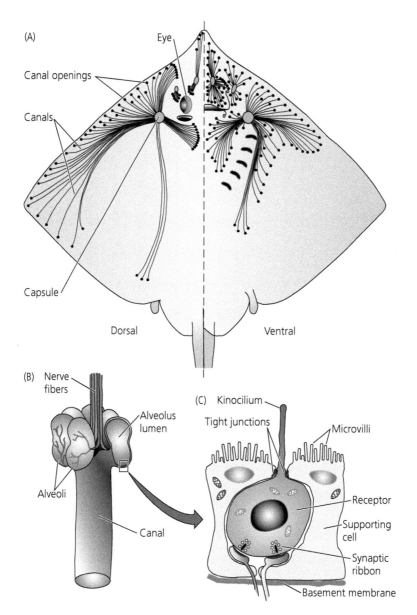

Figure 10.7 Ampullary electroreceptors of the skate. (A) Canal system of the skate *Raja clavate*. (B) End of canal with ampullary swelling containing electroreceptors. (C) Epithelium of an alveolus showing an electroreceptor with a single kinocilium and synaptic ribbons. (After Waltman, 1966; Murray, 1974; Clusin and Bennett, 1977a.)

are provided with a skin of high resistance. The resistance *inside* the fish is, however, comparatively low, so that the potential of the inside of the fish is everywhere nearly the same. It therefore makes no difference how long the canal is, provided it is long enough to penetrate the high resistance of the skin.

Shorter canals are advantageous because they take up less space and there is less chance for voltage to be lost by leakage across the canal wall (see Moller, 1995).

In fresh-water fish, the physiology of ampullary receptors is particularly simple (see Bennett, 1968,

(A)

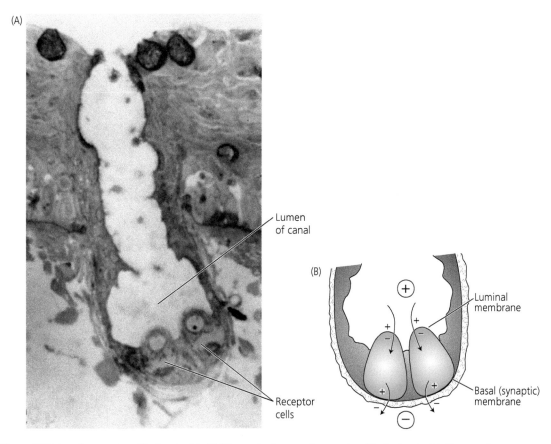

Lumen of canal

(B)

Luminal membrane

Basal (synaptic) membrane

Receptor cells

Figure 10.8 Ampullary receptors of fresh-water fish. (A) Micrograph of an ampullary receptor of the fresh-water fish *Eigenmannia*. Magnification 810×. (B) Flow of current across receptors produced by an external positive voltage. See text. (A from Bennett, 1971b.)

1971b; Bennett and Obara, 1984; Zakon, 1984). When the voltage outside the fish is more positive than inside, this voltage difference falls across the membranes of the receptor cell, as illustrated in Figure 10.8B. Because current always flows from positive to negative, it will flow through the receptor *inward* across the passive resistance of the luminal membrane and *outward* across the passive resistance of the basal (synaptic) membrane. By Ohm's law, current flowing inward across the resistance of the membrane produces an inside-negative change in membrane potential, so the luminal membrane hyperpolarizes. Current flowing outward across the basal membrane, on the other hand, depolarizes. The receptor is a secondary sense cell and does not normally produce action potentials, but depolarization of the basal membrane is sufficient to increase the

probability of opening of voltage-gated Ca^{2+} channels at the synapse, letting Ca^{2+} into the cell and increasing the release of an excitatory transmitter. The transmitter then depolarizes the postsynaptic nerve fiber, increasing the rate of spike firing.

A recording from the ampullary electroreceptor of the fresh-water fish *Gymnotus carapo* is shown in Figure 10.9. The insets give actual traces, with spike discharges from the nerve illustrated above and the stimulus waveform below. In the absence of any stimulation (upper trace), there was a resting discharge from the nerve, produced by a tonic release of synaptic transmitter. When the outside of the canal was made positive (middle trace), the spike frequency increased and then slowly adapted. Cessation of stimulation produced a brief inhibition. Negative stimuli, on the other hand, decreased spike firing (lower trace), but this response also adapted. Cessation of the negative stimulus produced a brief increase in excitation, which eventually settled down to the frequency of spontaneous activity.

The graph in Figure 10.9 shows that the rate of spike firing can be significantly altered by a stimulus voltage of only a fraction of a millivolt. Synaptic transmission must be rather sensitive for a voltage of this magnitude to release enough synaptic transmitter to change the rate of spike firing in the nerve. As impressive as this may seem, the sensitivity of elasmobranch electroreceptors is even greater. For the ampullary receptors in the skate *Raja* (Figure 10.7), a clear change in postsynaptic voltage can be produced by a stimulus as small as 5 μV (Murray, 1965; Clusin and Bennett, 1979a). These electroreceptors differ from the ampullary receptors of teleost fish in that the polarity of stimulation is reversed: negative voltages produce excitation, and positive voltages inhibition (Murray, 1965). The direction of current flow during an excitatory response is the opposite of that shown for a fresh-water fish in Figure 10.8B, depolarizing the luminal membrane but *hyperpolarizing* the synaptic membrane. A current of this sign should decrease the release of transmitter. Nevertheless, it is current flowing across the receptor membrane in this direction that normally excites the postsynaptic nerve.

One possible explanation is that the transmitter is inhibitory, but this seems not to be the case (see Bennett and Obara, 1984). Every indication is that the transmitter is glutamate, just as in teleost fishes (Steinbach, 1974). We could also imagine that hyperpolarizing stimuli somehow trigger the release of transmitter, but release at this synapse is dependent on Ca^{2+} (Steinbach, 1974) and is mediated by voltage-gated Ca^{2+} channels (Bellono et al., 2017), which open to depolarization. The correct explanation is altogether different. First proposed by Murray (1965) and strongly supported by the work of Obara and Bennett (1972), it was definitively demonstrated by Clusin and Bennett who, in an elegant series of papers (1977a, 1977b, 1979a, 1979b), showed why skate electroreceptors behave in this peculiar fashion.

When the arrows in Figure 10.8B are reversed, the currents produced by negative stimuli can be seen to depolarize the luminal membrane. Clusin and Bennett showed that this membrane contains voltage-gated Ca^{2+} channels which, in the isolated ampulla, produce Ca^{2+}-dependent action potentials (Figure 10.10A) whose amplitude and duration depend upon the strength of stimulation (Clusin

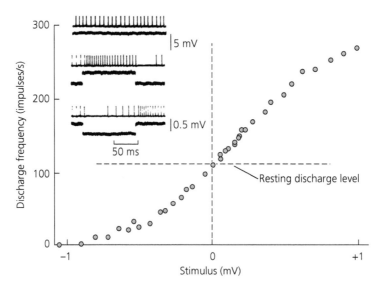

Figure 10.9 Electrical recording from nerve fibers of an ampullary receptor of the fish *Gymnotus carapo*. Action-potential (discharge) frequency of a nerve has been plotted against the amplitude of the voltage stimulus applied to the top of the canal at the receptor opening. Insets show sample recording for (from top to bottom) no stimulation (spontaneous discharge), positive voltage stimulation, and negative voltage stimulation. (After Bennett, 1968.)

and Bennett, 1977a, 1977b). These regenerative events spread to the basal membrane and trigger the release of synaptic transmitter.

If recordings are made instead from the ampulla so as to mimic the conditions that would occur in an animal in its native environment, simulating the low resistance of the animal's skin and high conductance of sea water, the Ca^{2+} channels are found to be poised very near threshold and spontaneously active. They generate spontaneous oscillations (Figure 10.10B), whose behavior can be altered by very small changes in voltage. Stimulation with a lumen-negative voltage of only a few microvolts depolarizes the luminal membrane, producing a small increase in the oscillations (upper traces) and generating a clear response from the nerve, whose action potentials had been blocked in this experiment with the sodium-channel blocker tetrodotoxin (TTX). A lumen-positive voltage (lower traces) decreases the amplitude of the oscillations and hyperpolarizes the nerve.

The Ca^{2+} channels responsible for these oscillations have unusual properties (Bellono et al., 2017). The skate channel is related to a mammalian channel called $Ca_V1.3$, but the skate channel has a series of charged amino acids near the intracellular membrane surface. This sequence renders the probability of channel opening exquisitely sensitive to membrane voltage. Furthermore, skate electroreceptors express a K^+ conductance activated by Ca^{2+} (King et al., 2016; Bellono et al., 2017). The oscillations are produced by the voltage-dependent opening of the Ca^{2+} channels, followed by entry of Ca^{2+} and opening of Ca^{2+}-activated K^+ channels, resulting in alternating depolarization and hyperpolarization. This mechanism resembles the one responsible for electrical resonance in turtle hair cells (Figure 6.22). In sharks, the electroreceptors have similar Ca^{2+} channels but different K^+ channels gated by voltage instead of Ca^{2+} entry, and instead of oscillations the electroreceptors of sharks produce larger, Ca^{2+}-dependent action potentials (Bellono et al., 2018). Both the oscillations and action potentials spread from the luminal membrane to the basal membrane of the receptor and have the same result of increasing release of transmitter onto the nerve, signaling the presence of a voltage in the surrounding water.

The very high sensitivity of ampullary receptors is exploited by elasmobranches to detect electric fields produced, for example, by gill and muscle movements of other fish. A shark can locate a flatfish hidden beneath the sand even if the fish is completely enclosed in agar so that it emits no odor (Kalmijn, 1971). The shark makes well-aimed turns

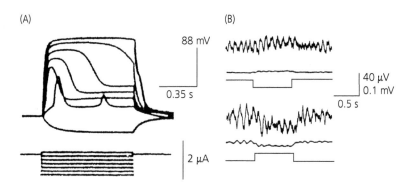

Figure 10.10 Electrical excitability of skate electroreceptor. (A) Ca^{2+}-dependent action potentials produced by the luminal membrane during application of current to the ampullary canal. Upper traces, responses of luminal membrane to applied current. Lower traces, amplitude and timing of stimuli with excitatory, lumen-negative stimuli plotted downward. Lumen-negative stimuli cause the luminal membrane to depolarize, and action potentials became increasingly prolonged as the amplitude of the current was increased. (B) Recordings made with the resistance across the receptor epithelium reduced by shorting with a salt bridge (tube containing salt in agar connecting the two solutions). Upper and lower panels each have two sets of three traces. The upper trace in each set shows voltage oscillations produced by voltage-gated Ca^{2+} channels in the luminal membrane (calibration, 40 μV). The middle traces show the postsynaptic potential of nerve (calibration, 0.1 mV). The lower trace in each set gives the stimuli applied to the canal, which were lumen-negative 5 μV (upper set of traces) and lumen-positive 10 μV (lower set of traces). (A from Clusin and Bennett, 1977a; B from Clusin and Bennett, 1979a.)

about the position where the fish is buried and may even try to dig about in the sand. If, however, the flatfish is covered in an electrically insulating film, the shark ignores it completely.

Tuberous receptors and electrolocation

Ampullary receptors are sometimes called tonic electroreceptors, because they give relatively sustained responses to steady changes in voltage and adapt rather slowly (Figure 10.9). A different sort of electroreceptive organ called a tuberous organ is found in two groups of fresh-water teleost fish and apparently *only* in these fish: the African mormyrids (of the order Mormyriformes) and the South-American gymnotids or knifefish (of the order Gymnotiformes) (see Moller, 1995). The receptors of these organs face a lumen not directly connected to the exterior by an obvious canal (Bennett, 1971b; Szabo, 1974; Zakon, 1984). Instead, the lumen is covered by loosely packed layers of epithelial cells capped by epidermis, probably providing a pathway of low resistance to the exterior (Figure 10.11). There is a variable—but generally small—number of receptor cells, often with a large percentage of their plasma membrane projecting into the lumen (Szamier and Wachtel, 1970). These cells are held together at their bases by a network of tight junctions (see Bennett, 1971b), which probably act to increase the resistance to current flow between the cells, again forcing most of the current to pass across the membranes of the receptors. The receptors lack cilia but often have large numbers of microvilli, increasing the surface area of the luminal membrane and so decreasing its resistance and increasing its capacitance. The basal membrane contains synaptic ribbons like those of ampullary receptors, and all the cells of an organ appear to synapse onto a single afferent nerve fiber.

Gymnotid and mormyrid fish live in muddy water or are nocturnal, and they use tuberous organs to detect an electric field that they themselves produce. All of these fish have electric organs which emit brief voltage pulses or sinusoidal waves of voltage of a few hundred millivolts to a few volts in amplitude, at frequencies ranging from a few per second to 1000–1500 Hz. They use these voltage pulses together with their tuberous organs in a special form

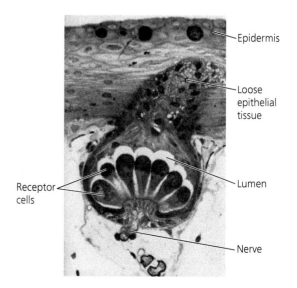

Figure 10.11 Anatomy of the tuberous organ of gymnotid fish *Hypopomus*. Light micrograph of the tuberous organ. Magnification 775×. (From Szamier and Wachtel, 1970.)

of sensory detection called *electrolocation* (Heiligenberg, 1977; Moller, 1995; von der Emde, 2001).

Electrolocation was first discovered by H. W. Lissmann, who was visiting London Zoo and noticed that the mormyrid *Gymnarchus niloticus* could avoid objects even when swimming backward. Lissmann was given a single live *Gymnarchus* as a wedding present (see Moller, 1995) and was able to get it to live in his laboratory aquarium long enough to confirm that it has an extraordinary ability to avoid objects even when it cannot see them. He then showed that the fish emits a surprisingly regular oscillating wave of voltage at a frequency of 250–300 Hz (Lissmann, 1951). It could clearly detect its own voltage, because when a signal like the one it emitted was fed back into the aquarium, the fish could locate the stimulating electrodes and would attack them.

The electric organs of *Gymnarchus*, like those of all mormyrid fishes, are composed of many single cells (called *electrocytes*) placed in close apposition (see Bennett, 1971a; Bass, 1984). Similar organs are found in gymnotids—in *Gymnotus* they consist of four columns of drum-shaped cells 300 μm thick and 1 mm in diameter, running down both sides of the fish from just behind the chin to the tip of the tail

(Schwartz et al., 1975). The columns of electrocytes are shown in an unstained preparation in Figure 10.12. On the right, single electrocytes have been outlined in each of the four columns. These cells are modified muscle cells and are excited by cholinergic input coming from the spinal cord. The depolarization of the electrocytes by acetylcholine (ACh) can be magnified by the generation of action potentials on one or both membrane surfaces of the electrocyte (see for example Bennett and Grundfest, 1959). Since electrocytes are placed end to end down the body of the fish and are surrounded by insulating tissue, their voltages sum like batteries placed in series and produce an electric field around the fish. *Gymnotus* is said to be weakly electric, because the amplitude of the voltage is no more than a fraction of a volt in water. Some animals such as electric rays (*Torpedo*) have electric organs of similar morphology but which are much larger and produce voltages great enough to be used in defense and to stun prey. The electric organ of *Gymnotus* is more modest but produces a voltage sufficiently large for electrolocation.

The principle of electrolocation is illustrated in Figure 10.13 (Heiligenberg, 1977; Moller, 1995). Current generated by the electric organ leaves the fish from the front of the animal at the anterior end of the electric organ and then returns to the posterior end of the organ at the animal's tail; or just the reverse, from the tail to the head, depending upon the orientation and electrical properties of the electrocytes. This current produces a three-dimensional gradient of voltage in the water, which can be detected by the animal (Figure 10.13A). The voltage decays rapidly with distance and is useful for electrolocation only within one to two body lengths of the fish, but this is enough to assist in schooling and in the maintenance of position above the river floor or at a fixed distance from nearby obstacles.

The lines of current flow produced by discharge of the electric organ are disturbed by any insulator (e.g. rock) or conductor (e.g. worm or water plant) located in the vicinity of the fish. Actual measurements of the electric field are given in Figure 10.13B for the gymnotid *Eigenmannia* (Heiligenberg, 1973). The lines in the figure are those of equal potential and map out the electric field, which is distorted in different ways by a Plexiglass insulator or an aluminum rod. These changes in voltage are detected by the tuberous organs and can be used by the fish quite accurately to sense objects in the environment, even in complete darkness (von der Emde et al., 1998). Keep in mind that voltage pulses

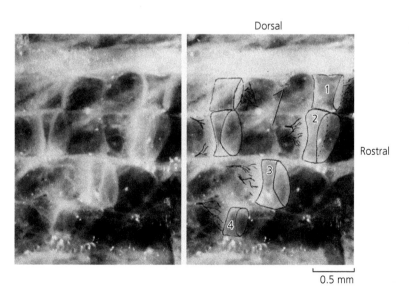

Figure 10.12 The electric organ of *Gymnotus*. Left, photograph of unstained preparation of the organ from the tail of the fish. Right, the same photograph but with individual cells outlined. The main pattern of innervation is also indicated. Representative electrocytes have been numbered in each of the four columns of cells. (From Bennett, 1971a.)

(A)

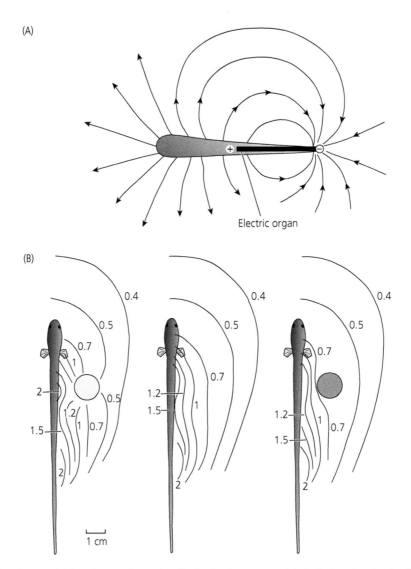

Electric organ

(B)

1 cm

Figure 10.13 Electrolocation. (A) Lines of current flow produced by the electric organ around the body of an electrolocating fish. (B) Lines of equal potential measured in water during discharge of electric organs for the gymnotid *Eigenmannia*. Left, with insulator (plexiglass rod); middle, undisturbed fish; right, with conductor (aluminum rod). (After Heiligenberg, 1977.)

are produced many times a second and are quite brief. In *Gymnotus*, for example, the electric organ discharges fifty times a second in an undisturbed fish (though 200 times a second when the animal is feeding), producing an electric field that lasts less than a millisecond. Since the electroreceptors in the tuberous organs also respond transiently, the two systems working together can provide information at rather high temporal resolution.

The responses of tuberous electroreceptors vary considerably from species to species (Bennett, 1971b; Szabo and Fessard, 1974; Zakon, 1984). All give one or a few spikes even to maintained stimulation and are therefore often called phasic electroreceptors. Most fish seem to have two kinds. One is called a rapid timing unit, or a T unit; or, in mormyrids, a K unit. It responds generally with only a single spike. In the CNS of the fish, the signals from these receptors are inhibited during the discharge of the animal's own electric organ (Zipser and Bennett, 1976), indicating that T and K units are probably used in detecting the electrical discharges of other fish rather than in electrolocation. A second kind of tuberous electroreceptor is called an amplitude-modulated unit, or a P unit; or, in mormyrids, a D unit. It gives a burst of several spikes whose number depends upon the strength of the detected stimulus.

An example of a recording from an amplitude-modulated unit of the electric eel *Electrophorus* is given in Figure 10.14 (from Hagiwara et al., 1965). Electric eels are not true eels but rather gymnotids. *Electrophorus* is a strongly electric fish and can on occasion produce signals as large as 500 volts to shock predators or prey, but like other members of its order it can also generate the much smaller voltages used in electrolocation. The uppermost record in Figure 10.14 shows the control response of a nerve fiber coming from a single tuberous organ to a low-voltage discharge of the eel's own electric organ. The response consists of a brief burst of one to fifteen impulses, with the number of spikes growing in a graded fashion with increasing intensity of the electric field. A metal plate was placed in the water next to the eel at different positions down the length of the animal's body, and the response was measured from this same tuberous organ at each of these positions. The number next to each trace gives the position of the center of the metal plate according to the scale shown next to the animal. The number of spikes changed as the position of the plate was altered. Similar results have been obtained from several other species, both gymnotids and mormyrids (see Bennett, 1971b; Moller, 1995). The change in spike number apparently provides the signal the animal uses during electrolocation to sense the position and size of the object.

Electrolocation and electrical communication are highly developed among gymnotid and mormyrid fishes (see Heiligenberg, 1977; Moller, 1995). Electrical signals are used in a variety of social interactions, including species and gender recognition, aggression, and territoriality. In some species, dominant animals emit pulses at a frequency higher or lower than those of other animals, and the frequency of discharge of males and females may also

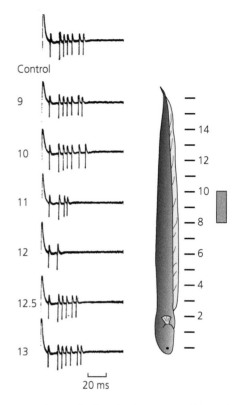

Figure 10.14 Recording from the tuberous organ of *Electrophorus* during electrolocation. A metal bar was placed in the water next to the fish at different positions indicated with numbers next to the drawing of the animal (right). The bar is shown with its middle at position 9. Recordings show the change in the number of spikes recorded from the electroreceptor produced during the discharge of the animal's electric organ as the position of the middle of the bar was changed. Control, recording with bar absent. (After Hagiwara et al., 1965.)

differ. The frequency can be changed to avoid confusion with the discharges of other, nearby fish, in a maneuver called *jamming avoidance*. These adaptations allow electrolocating fish to live and breed in rivers such as the Amazon, which is so turbid that no light descends below 7 m of depth.

Magnetoreception

The clearest example of an animate response to the Earth's magnetic field is the magnetotaxis of sedimentary bacteria. Magnetotaxis was first discovered by Richard Blakemore (1975), who was isolating bacteria from marsh sediment and observing them in a light microscope. He noticed that many of the bacteria persistently accumulated in one region of a drop of muddy water on his microscope slide. After moving the light to various places and testing other possible explanations without success, he approached the slide with a small magnet. The bacteria immediately changed direction. A few simple tests then demonstrated that the bacteria in his sample consistently swam toward magnetic north.

The reason for this preferential orientation became apparent when Blakemore examined the bacteria at low power in the electron microscope (see also Blakemore and Frankel, 1981; Blakemore, 1982). The north-seeking bacteria all contained one or more chains of electron-dense particles, each roughly cuboidal or octahedral in shape, about 50 nm in width, and surrounded by a thin membrane (Figure 10.15A). These particles consist almost entirely of magnetite (Frankel et al., 1979), in small enough crystals to form single domains of uniform orientation. The particles align with one another and together form a small permanent magnetic dipole, like a compass needle. The amount of magnetite is sufficiently

large to overcome Brownian motion and bias the orientation of the bacteria toward magnetic north. The bacteria synthesize the magnetite from iron they pick up from the surrounding medium, and if they are grown in the absence of iron, they do not form particles of magnetite and are not magnetotactic.

The key to understanding the selective advantage produced by the magnetite is to realize that the magnetic field generated by the Earth has a vertical as well as a horizontal component. The field is perpendicular to the surface of the Earth at the poles, pointing upward at the South Pole and downward at the North, and it is parallel to the Earth at the equator (Figure 10.16A). At intermediate longitudes, the vectors of the field point inward or outward at angles of inclination, which change in steepness with distance from either pole (Figure 10.16B). An ordinary compass is constrained by its metal frame to rotate only in the horizontal plane, but an inclination compass with a needle suspended at its middle and free to rotate over 360° will point downward in Massachusetts, where Blakemore collected his bacteria. The magnetite in the bacterium therefore

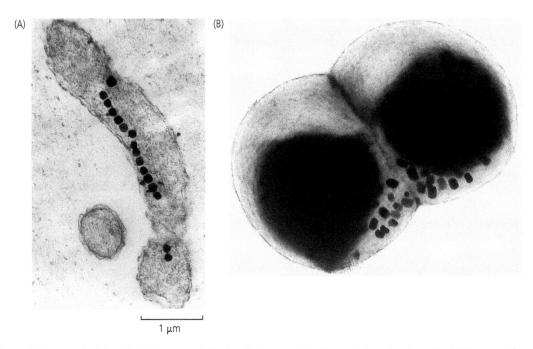

(A) (B)

1 μm

Figure 10.15 Magnetite in bacteria. (A) Magnetotactic flagellated bacterium collected from a Durham, New Hampshire (USA) water-purification plant. Scale bar, 1 μm. (B) Magnetotactic bacteriococcus in the process of undergoing cell division. (A courtesy of R. B. Frankel; B from Frankel and Blakemore, 1989.)

(A)

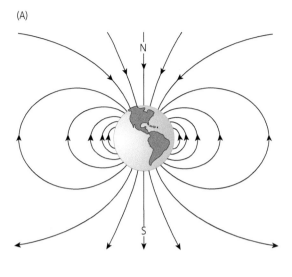

(B)

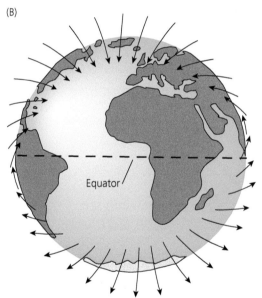

Equator

Figure 10.16 Magnetic field of Earth. (A) Orientation of magnetic field lines around Earth. (B) Orientation of the vectors of the magnetic field near the surface of the Earth. (After Able, 1994; Hughes, 1999; Wiltschko and Wiltschko, 2001.)

acts as a tiny inclination compass which orients the organism so that it swims downward into the sediment, into regions of low oxygen tension where the bacteria grow best.

Magnetotactic bacteria have also been isolated in sediments in South America and New Zealand. Here the magnetite is oriented with its south pole at

the front of the organism, so that the bacteria again swim downward. The orientation of the magnetite is apparently determined at cell division, since the cleavage plane usually passes down the middle of the organism, dividing the chain of magnetite into two (Figure 10.15B). As the bacterium synthesizes additional particles of magnetite, these particles become oriented in the same direction as the rest of the chain. Occasionally, the cleavage plane does not include any of the particles of magnetite, and one of the resulting bacteria must synthesize its whole compass needle *de novo*. In this case, there is an equal probability that the magnetite will form with the north or south pole at the front of the organism. As a consequence, a population of bacteria will always contain a few individuals pointing in the wrong direction. If a colony of predominantly north-seeking bacteria is placed in an artificial magnetic field of reversed orientation, the colony gradually changes to south-seeking, just as a result of natural selection. This flexibility may be important in the life of the colony, because the direction of the magnetic field changes periodically, reversing north and south poles. This reversal happened most recently less than a million years ago.

Magnetoreception in migrating birds

The discovery of magnetotaxis in bacteria provided the first universally accepted evidence that living organisms can respond to the Earth's magnetic field. The mechanism of this response, however, differs fundamentally from chemotaxis in bacteria (see Chapter 7), where receptors activate a transduction cascade. The magnetite serves only to orient the bacterium and does this perfectly well in a dead organism. A description of magnetotaxis might even seem out of place in a book on sensory transduction, were it not for the discovery that higher animals are also sensitive to a magnetic field (see Clites and Pierce, 2017; Mouritsen, 2018).

The first and still the clearest evidence for magnetoreception in vertebrates has been obtained for migrating songbirds. Robins, warblers, and indigo buntings, which fly between their summer and winter breeding and feeding grounds, all have been shown to orient in a magnetic field. Many of these experiments were done with a funnel cage (Figure 10.17A), based on a design by Emlen and Emlen (1966). Migrating birds at the time of departure become agitated and exhibit *Zugunruhe* (from the German *Zug* which can mean migration, and *Unruhe*, restlessness). They attempt to climb up the cage, and if their feet are marked with ink (or the side of the cage is lined with typewriter correction paper), the marks they make as they attempt to ascend can be recorded (Figure 10.17B). These marks indicate which direction the animals are choosing as they attempt to leave the cage, giving an indication of the direction of their intended migration. This technique was first used in a collaboration between the Emlen laboratory and Wolfgang and Roswitha Wiltschko, who had already obtained evidence that the direction of

Zugunruhe could be altered by a magnetic field (Wiltschko and Wiltschko, 1972). With the Emlen laboratory (Emlen et al., 1976), the Wiltschkos provided clear evidence that the direction of markings in the funnel cage could be changed by an externally applied field produced by an electromagnet but only during the season when the birds were beginning their migration.

The Wiltschkos then continued and greatly extended these results (see Wiltschko and Wiltschko, 1995, 2005). Figure 10.18 illustrates one of their most important findings (Wiltschko and Wiltschko, 1972). Each triangle gives the mean heading of an individual bird, and the arrow is a vector giving the grand mean of orientation for all the animals, with

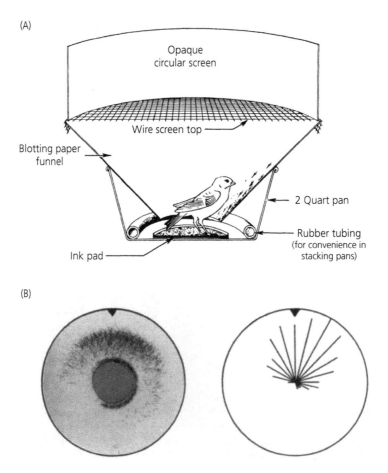

Figure 10.17 Measuring *Zugunruhe* with an Emlen funnel cage. (A) Design of cage. The bird stands in an ink pad. When it attempts to fly out of the cage, it leaves its footprints on blotting paper lining the side of the cage. (B) Sample footprint record for an indigo bunting tested in the spring (of 1965) under an artificial planetarium sky set for the condition of the sky in Ithaca, New York, where recordings were made. Hops by bird show preference of orientation. (From Emlen and Emlen, 1966.)

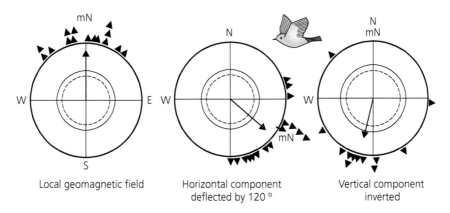

Figure 10.18 Migrating birds have an inclination compass. Orientation behavior of migrating European robins during the spring, tested in funnel cages with the local magnetic field (left) and in two artificial fields (middle and right) produced by an electromagnet (Helmholtz coil). mN, Magnetic north. The triangles at the periphery of the circles mark the mean headings of individual birds. The arrows show the grand mean vectors for all of the birds, with the angle of the vector indicating the mean orientation of the birds and the length of the vector indicating significance. The two inner circles are the borders for 5 percent (dashed) and 1 percent statistical significance. Birds reversed direction when the vertical component of the magnetic field was inverted, even though the horizontal component (and the direction indicated by a compass) was unchanged. (From Wiltschko and Wiltschko, 2005.)

its length indicating the statistical significance of the measurements. In the experiments to the left, no magnetic field was applied, and the birds headed in a direction close to magnetic north (mN). When the horizontal component of the magnetic field was altered, the birds responded by attempting to fly off to the southeast, in approximate correspondence to the altered field. The direction was also altered when the vertical component of the field was inverted (diagram on right), nearly reversing the orientation of the birds to magnetic south. This result is significant, because a reading from a pocket compass would not be affected if only the vertical component of the field were changed. A compass needle is constrained by its mounting to measure only the horizontal component, as I have said. The birds, on the other hand, can respond to both horizontal and vertical components, showing that they are somehow provided with an inclination compass to determine the direction of their flight.

How do birds sense magnetic inclination? One possibility is that they use magnetite like magneto-tactic bacteria. There have been numerous reports of magnetite particles in the head or beak of birds (see Shaw et al., 2015), but these findings have now been called into question. The iron deposits in the bird beak seem to be mostly located in macrophages rather than in neurons (Treiber et al., 2012, 2013).

Iron accumulations can also occur extracellularly, perhaps as a result of contamination during tissue preparation (Edelman et al., 2015). It remains possible that magnetite will eventually be discovered somewhere in the bird nervous system, but no compelling evidence has yet been obtained.

A second theory is biochemical: a magnetic field is postulated to affect the rate of some reaction that eventually gates the opening or closing of ion channels. Although this theory might seem implausible because the magnetic field of the Earth is so weak, even a weak field can alter the rate of a reaction if that reaction proceeds through the formation of a pair of free radicals (see McFadden and Al-Khalili, 2014; Hore and Mouritsen, 2016). Attention has focused in particular on cryptochromes, which are a diverse group of proteins known to be important in the control of circadian rhythm (see Michael et al., 2017). The cryptochromes have been suggested as magnetoreceptors because certain of the cryptochromes can bind to flavin adenine dinucleotide (FAD). The magnetic sense of birds seems to require light particularly at shorter wavelengths (Wiltschko et al., 1993), and short-wavelength light can be absorbed by FAD in association with cryptochrome to produce a pair of free radicals. A magnetic field could then interact with the spin of the free radicals to alter the rate of reduction of FAD to FADH.

Moreover, cryptochromes are present in the nervous system and, in particular, in the outer segments of cone photoreceptors of the bird retina (see Figure 10.19 and Nießner et al., 2011; Gunther et al., 2018).

This attractive hypothesis has received considerable attention and stimulated many recent studies (summarized in Hore and Mouritsen, 2016; Mouritsen, 2018), but there are problems. Because birds have an inclination compass, the cryptochrome and FAD would have to be oriented in some way. This orientation would need to be different in different groups of cells so that the bird could determine the angle of magnetic inclination. Cryptochrome is,

however, a soluble protein. It could in theory be bound to some other protein, but what protein would this be? The outer segments of photoreceptors seem unlikely locations for magnetic field analyzers, because most of the proteins to which the cryptochrome could conceivably bind are relatively mobile and would not be able to hold the cryptochrome in a fixed orientation (see Chapter 9). The proteins at the edges of the outer segments do seem rather more solidly held in place, but the cryptochrome in Figure 10.19 seems distributed all across the outer segment and is not particularly concentrated at the edges.

What would the FADH do? There is no evidence of an effect of FAD/FADH on cyclic-nucleotide-gated channels of photoreceptors or indeed any other channel type. Bird photoreceptors are particularly problematic for the cryptochrome hypothesis because most avian cones have oil droplets in their inner segments, which have a high concentration of carotenoid, filtering out the light reaching the outer segment in just the region of wavelengths where FAD has been postulated to absorb it (Hart, 2001; Toomey et al., 2015).

None of these objections is fatal to the radical-pair hypothesis or, more generally, to some biochemical explanation of the sensitivity of migrating birds to the inclination of the magnetic field. The behavioral data are so compelling, and the mechanism of detection so specific, that the elucidation of transduction should have been easy. Instead, it has turned out to be unexpectedly difficult. It is perhaps fitting that I end the book with this one sensory modality whose elucidation has so far escaped all of the best efforts of the scientific community. Gene cloning, patch-clamp recording, and techniques of imaging and structural determination have been instrumental in revealing the basic mechanisms of transduction in nearly every other modality. Perhaps the persistent application of these and other evolving methods will eventually unravel this last and most mysterious of the senses.

Summary

In addition to the familiar five senses, many animals have sensory modalities whose mechanisms of transduction appear to be novel. Thermoreceptors respond to a change in temperature. In the mammalian skin there are two kinds, excited by either

Figure 10.19 Cryptochrome in the outer segment of a bird cone. Immunolabeling of cryptochrome 1a (black dots) in the outer segment of a robin cone. The cone was judged from its shape to contain a photopigment sensitive to short wavelengths in the ultraviolet and violet part of the spectrum. (From Nießner et al., 2011.)

0.2 μm

warming or cooling. Both respond to temperature change with a transient increase in spike firing, followed by adaptation to a steady or bursting discharge rate that appears to be responsible for our sense of absolute temperature.

Recordings from cells dissociated from dorsal root or trigeminal ganglia have demonstrated populations of cells responding to both cooling and warming. For cold-sensitive cells, the response is produced by the opening of channels permeable to cations, including Ca^{2+}. The response can be recorded even from isolated channels in excised patches, showing that temperature change gates channels directly without the requirement for a metabotropic cascade. Several receptor proteins have now been identified, which are thought to mediate our senses of moderate and painful cooling and warmth. The clearest examples are members of the TRP family of membrane channels which signal by temperature-dependent changes in protein conformation.

Some animals have warm receptors used for the detection of infrared irradiation in specialized pit organs. In rattlesnakes, the axon terminals of warm receptors express a temperature-dependent TRP channel and are densely packed within a thin sensory epithelium. They can respond to changes in temperature of a small fraction of a degree. The pit organs are used by the animal to locate sources of warmth and provide a frighteningly accurate system of detection for locating warm-blooded prey even in complete darkness.

Electroreceptors responding to external electric fields have been described in many aquatic vertebrates. Some animals have ampullary organs with receptor cells lying at the end of closed canals, which can directly sense an electric field in the surrounding water. In elasmobranchs (sharks and skates), the canals can be quite long and the ampullary receptors can be incredibly sensitive. Electrophysiological recording has shown that part of the reason for this sensitivity is a mechanism of amplification produced by a combination of voltage-gated Ca^{2+} channels and Ca^{2+}-activated or voltage-gated K^+ channels. In skate electroreceptors, the Ca^{2+} and K^+ channels in concert produce spontaneous oscillations of membrane potential near threshold. External voltage differences magnify these oscillations, which propagate to the presynaptic membrane and alter afferent nerve firing. Elasmobranchs can use ampullary receptors to detect prey hidden in the sea bottom from small electric fields produced by gill and muscle movement.

Two groups of fish, the African mormyrids and the South-American gymnotids, have electroreceptors in tuberous organs that respond to voltage fields produced by the fish themselves. These fishes have electric organs running down the length of the animal that act like batteries in series and produce a flow of current in the surrounding water. This current generates an electric field whose distribution is altered by adjacent conductors and insulators. The distortions of the field are detected by the receptors of the tuberous organs with high temporal resolution in a process called electrolocation, which the animals use in orientation and schooling, as well as for species recognition and territoriality.

Just as some species can sense an electric field, there is increasing evidence that some are also capable of detecting a magnetic field. Sedimentary bacteria synthesize magnetite and orient in a magnetic field much like a compass needle. This mechanism ensures that the bacterium swims downward into the sediment where it grows best. Some species of birds can detect the inclination of the magnetic field and use this information to determine the direction of their migration. They may also use accumulations of magnetite as tiny magnets somewhere in their nervous system, but the evidence has so far been inconclusive. Some other biochemical pathway may signal the orientation of the magnetic field, perhaps by the formation of pairs of free radicals and the alteration of the rate of a chemical reaction. Some evidence seems to support a role for proteins called cryptochromes, but no clear support for this mechanism has yet been obtained. Of all the senses we have described, magnetoreception continues to be the least well understood. The mechanism by which a magnetic field produces a neural signal remains an enigma.

References

Able KP (1994) Magnetic orientation and magnetoreception in birds. Prog Neurobiol 42:449–73.

Abraira VE, Ginty DD (2013) The sensory neurons of touch. Neuron 79:618–39.

Abuin L, Bargeton B, Ulbrich MH, ti Isacoff EY, Kellenberger S, Benton R (2011) Functional architecture of olfactory ionotropic glutamate receptors. Neuron 69:44–60.

Acharya JK, Jalink K, Hardy RW, Hartenstein V, Zuker CS (1997) InsP3 receptor is essential for growth and differentiation but not for vision in *Drosophila*. Neuron 18:881–7.

Adler E, Hoon MA, Mueller KL, Chandrashekar J, Ryba NJ, Zuker CS (2000) A novel family of mammalian taste receptors. Cell 100:693–702.

Adler J (1966) Chemotaxis in bacteria. Science 153:708–16.

Adler J (1975) Chemotaxis in bacteria. Annu Rev Biochem 44:341–56.

Adrian ED (1928) The Basis of Sensation. London: Christopers.

Adrian ED (1931) The Mechanism of Nervous Action. Philadelphia: University of Pennsylvania Press.

Adrian ED (1947) The Physical Background of Perception. Oxford: Clarendon.

Akabas MH, Dodd J, Al-Awqati Q (1988) A bitter substance induces a rise in intracellular calcium in a subpopulation of rat taste cells. Science 242:1047–50.

Alam A, Jiang Y (2011) Structural studies of ion selectivity in tetrameric cation channels. J Gen Physiol 137:397–403.

Albert JT, Göpfert MC (2015) Hearing in *Drosophila*. Curr Opin Neurobiol 34:79–85.

Albert JT, Nadrowski B, Gopfert MC (2007) Mechanical signatures of transducer gating in the *Drosophila* ear. Curr Biol 17:1000–1006.

Alloway PG, Dolph PJ (1999) A role for the light-dependent phosphorylation of visual arrestin. Proc Natl Acad Sci U S A 96:6072–7.

Amjad A, Hernandez-Clavijo A, Pifferi S, Maurya DK, Boccaccio A, Franzot J, Rock J, Menini A (2015) Conditional knockout of TMEM16A/anoctamin1 abolishes the calcium-activated chloride current in mouse vomeronasal sensory neurons. J Gen Physiol 145:285–301.

Angueyra JM, Pulido C, Malagon G, Nasi E, Gomez Mdel P (2012) Melanopsin-expressing amphioxus photoreceptors transduce light via a phospholipase C signaling cascade. PLoS One 7:e29813.

Araneda RC, Kini AD, Firestein S (2000) The molecular receptive range of an odorant receptor. Nature Neuroscience 3:1248–55.

Armitage JP (1999) Bacterial tactic responses. Adv Microb Physiol 41:229–89.

Arnadottir J, O'Hagan R, Chen Y, Goodman MB, Chalfie M (2011) The DEG/ENaC protein MEC-10 regulates the transduction channel complex in *Caenorhabditis elegans* touch receptor neurons. J Neurosci 31:12695–704.

Arshavsky V, Bownds MD (1992) Regulation of deactivation of photoreceptor G protein by its target enzyme and cGMP. Nature 357:416–17.

Art JJ, Fettiplace R (1987) Variation of membrane properties in hair cells isolated from the turtle cochlea. J Physiol 385:207–42.

Art JJ, Wu YC, Fettiplace R (1995) The calcium-activated potassium channels of turtle hair cells. J Gen Physiol 105:49–72.

Artamonov ID, Zolotarev AS, Kostina MB, Khoroshilova NI, Feigina MI (1983) Primary structure of rhodopsin. I. Cyanogen bromide degradation peptides (in Russian). Bioorg Khim 9:1301–16.

Ashmore J (2018) Outer hair cells and electromotility. Cold Spring Harb Perspect Med Sep 4. pii:a033522.

Ashmore JF (1987) A fast motile response in guinea-pig outer hair cells: the cellular basis of the cochlear amplifier. J Physiol 388:323–47.

Ashmore JF (1992) Mammalian hearing and the cellular mechanisms of the cochlear amplifier. Soc Gen Physiol Ser 47:395–412.

Ashmore JF (2008) Cochlear outer hair cell motility. Physiol Rev 88:173–210.

Assad JA, Corey DP (1992) An active motor model for adaptation by vertebrate hair cells. J Neurosci 12:3291–309.

Assad JA, Hacohen N, Corey DP (1989) Voltage dependence of adaptation and active bundle movement in bullfrog saccular hair cells. Proc Natl Acad Sci U S A 86:2918–22.

Assad JA, Shepherd GM, Corey DP (1991) Tip-link integrity and mechanical transduction in vertebrate hair cells. Neuron 7:985–94.

Asteriti S, Grillner S, Cangiano L (2015) A Cambrian origin for vertebrate rods. Elife 4.

Avraham KB, Hasson T, Steel KP, Kingsley DM, Russell LB, Mooseker MS, Copeland NG, Jenkins NA (1995) The mouse Snell's waltzer deafness gene encodes an unconventional myosin required for structural integrity of inner ear hair cells. Nat Genet 11:369–75.

Azevedo A, Doan T, Moaven H, Sokal I, Baameur F, Vishnivetskiy S, Homan K, Tesmer J, Gurevich V, Chen J, Rieke F (2015) C-terminal threonines and serines play distinct roles in the desensitization of rhodopsin, a G protein-coupled receptor. eLife 2015;4:e05981.

Bacigalupo J, Johnson EC, Vergara C, Lisman JE (1991) Light-dependent channels from excised patches of Limulus ventral photoreceptors are opened by cGMP. Proc Natl Acad Sci U S A 88:7938–42.

Bade H, Braun HA, Hensel H (1979) Parameters of the static burst discharge of lingual cold receptors in the cat. Pflugers Arch 382:1–5.

Bai L, Lehnert BP, Liu J, Neubarth NL, Dickendesher TL, Nwe PH, Cassidy C, Woodbury CJ, Ginty DD (2015) Genetic identification of an expansive mechanoreceptor sensitive to skin stroking. Cell 163:1783–95.

Baird IL (1960) A survey of the periotic labyrinth in some representative recent reptiles. Univ Kansas Sci Bull 41.

Baird RA (1994) Comparative transduction mechanisms of hair cells in the bullfrog utriculus. II. Sensitivity and response dynamics to hair bundle displacement. J Neurophysiol 71:685–705.

Bakalyar HA, Reed RR (1990) Identification of a specialized adenylyl cyclase that may mediate odorant detection. Science 250:1403–6.

Baker CV, Modrell MS, Gillis JA (2013) The evolution and development of vertebrate lateral line electroreceptors. J Exp Biol 216:2515–22.

Bandyopadhyay BC, Payne R (2004) Variants of TRP ion channel mRNA present in horseshoe crab ventral eye and brain. J Neurochem 91:825–35.

Barber VC, Evans EM, Land MF (1967) The fine structure of the eye of the mollusc Pecten maximus. Z Zellforsch Mikrosk Anat 76:25–312.

Barnes S, Hille B (1989) Ionic channels of the inner segment of tiger salamander cone photoreceptors. J Gen Physiol 94:719–43.

Barrett R, Maderson PFA, Meszler RM (1970) The pit organs of snakes. In: Biology of the Reptilia, Vol. 2, Morphology B (Gans C, ed), pp 277–314. New York: Academic Press.

Barretto RP, Gillis-Smith S, Chandrashekar J, Yarmolinsky DA, Schnitzer MJ, Ryba NJ, Zuker CS (2015) The neural representation of taste quality at the periphery. Nature 517:373–6.

Bass AH (1984) Electric organs revisited. In: Electroreception (Bullock TH, Heiligenberg W, eds), pp 13–70. New York: John Wiley and Sons.

Bautista DM, Siemens J, Glazer JM, Tsuruda PR, Basbaum AI, Stucky CL, Jordt SE, Julius D (2007) The menthol receptor TRPM8 is the principal detector of environmental cold. Nature 448:204–8.

Baylor DA, Hodgkin AL (1974) Changes in time scale and sensitivity in turtle photoreceptors. J Physiol 242:729–58.

Baylor DA, Nunn BJ (1986) Electrical properties of the light-sensitive conductance of rods of the salamander Ambystoma tigrinum. J Physiol 371:115–45.

Baylor DA, Lamb TD, Yau KW (1979a) Responses of retinal rods to single photons. J Physiol 288:613–34.

Baylor DA, Lamb TD, Yau KW (1979b) The membrane current of single rod outer segments. J Physiol 288:589–611.

Baylor DA, Matthews G, Yau KW (1980) Two components of electrical dark noise in toad retinal rod outer segments. J Physiol 309:591–621.

Baylor DA, Nunn BJ, Schnapf JL (1984) The photocurrent, noise and spectral sensitivity of rods of the monkey Macaca fascicularis. J Physiol 357:575–607.

Baylor DA, Nunn BJ, Schnapf JL (1987) Spectral sensitivity of cones of the monkey Macaca fascicularis. J Physiol (Lond) 390:145–60.

Beech DJ, Barnes S (1989) Characterization of a voltage-gated K+ channel that accelerates the rod response to dim light. Neuron 3:573–81.

Bellono NW, Leitch DB, Julius D (2017) Molecular basis of ancestral vertebrate electroreception. Nature 543:391–6.

Bellono NW, Leitch DB, Julius D (2018) Molecular tuning of electroreception in sharks and skates. Nature 558:122–6.

Belluscio L, Gold GH, Nemes A, Axel R (1998) Mice deficient in G(olf) are anosmic. Neuron 20:69–81.

Belluscio L, Koentges G, Axel R, Dulac C (1999) A map of pheromone receptor activation in the mammalian brain. Cell 97:209–20.

Bennett MVL (1968) Similarities between chemically and electrically mediated transmission. In: Physiological and Biochemical Aspects of Nervous Integration (Carlson FD, ed), pp 73–128. Englewood Cliffs, NJ: Prentice-Hall.

Bennett MVL (1971a) Electric organs. In: Fish Physiology, Vol. 5, Sensory Systems and Electric Organs (Hoar WS, Randall DS, eds), pp 347–491. New York: Academic Press.

Bennett MVL (1971b) Electroreception. In: Fish Physiology, Vol. 5, Sensory Systems and Electric Organs (Hoar WS, Randall DS, eds), pp 493–574. New York: Academic Press.

Bennett MVL, Grundfest H (1959) Electrophysiology of electric organ in Gymnotus carpo. J Gen Physiol 42:1067–104.

Bennett MVL, Obara S (1984) Ionic mechanisms and pharmacology of electroreceptors. In: Electroreception (Bullock TH, Heiligenberg W, eds), pp 157–81. New York: John Wiley and Sons.

Benton R, Sachse S, Michnick SW, Vosshall LB (2006) Atypical membrane topology and heteromeric function of *Drosophila* odorant receptors in vivo. PLoS Biol 4:e20.

Benton R, Vannice KS, Gomez-Diaz C, Vosshall LB (2009) Variant ionotropic glutamate receptors as chemosensory receptors in *Drosophila*. Cell 136:149–62.

Berg HC (1975) Chemotaxis in bacteria. Annu Rev Biophys Bioeng 4:119–36.

Berg HC (1988) A physicist looks at bacterial chemotaxis. Cold Spring Harb Symp Quant Biol 53:1–9.

Berghard A, Buck LB, Liman ER (1996) Evidence for distinct signaling mechanisms in two mammalian olfactory sense organs. Proc Natl Acad Sci U S A 93:2365–9.

Berry RM, Armitage JP (1999) The bacterial flagella motor. Adv Microb Physiol 41:291–337.

Berson DM, Dunn FA, Takao M (2002) Phototransduction by retinal ganglion cells that set the circadian clock. Science 295:1070–3.

Beurg M, Fettiplace R (2017) PIEZO2 as the anomalous mechanotransducer channel in auditory hair cells. J Physiol 595:7039–48.

Beurg M, Fettiplace R, Nam JH, Ricci AJ (2009) Localization of inner hair cell mechanotransducer channels using high-speed calcium imaging. Nat Neurosci 12:553–8.

Beurg M, Tan X, Fettiplace R (2013) A prestin motor in chicken auditory hair cells: active force generation in a nonmammalian species. Neuron 79:69–81.

Beurg M, Kim KX, Fettiplace R (2014) Conductance and block of hair-cell mechanotransducer channels in transmembrane channel-like protein mutants. J Gen Physiol 144:55–69.

Bhandawat V, Reisert J, Yau KW (2005) Elementary response of olfactory receptor neurons to odorants. Science 308:1931–4.

Billig GM, Pal B, Fidzinski P, Jentsch TJ (2011) Ca^{2+}-activated Cl^- currents are dispensable for olfaction. Nat Neurosci 14:763–9.

Birnbaumer L, Birnbaumer M (1994) G proteins in signal transduction. In: Handbook of Biomembranes (Shinitzky M, ed), pp 153–252. Rehovoth, Israel: Balaban Publishers.

Blakemore R (1975) Magnetotactic bacteria. Science 190:377–9.

Blakemore R, Frankel RB (1981) Magnetic navigation in bacteria. Scientif American 245(6):58–65.

Blakemore RP (1982) Magnetotactic bacteria. Annu Rev Microbiol 36:217–38.

Blest AD (1978) The rapid synthesis and destruction of photoreceptor membrane by a dinopid spider: a daily cycle. Proc R Soc London B 200: 463–83.

Blest AD (1988) The turnover of phototransductive membrane in compound eyes and ocelli. Adv Insect Physiol 20:1–53.

Bloomquist BT, Shortridge RD, Schneuwly S, Perdew M, Montell C, Steller H, Rubin G, Pak WL (1988) Isolation of a putative phospholipase C gene of *Drosophila*, norpA, and its role in phototransduction. Cell 54:723–33.

Boccaccio A, Lagostena L, Hagen V, Menini A (2006) Fast adaptation in mouse olfactory sensory neurons does not require the activity of phosphodiesterase. J Gen Physiol 128:171–84.

Boeckh J, Kaissling KE, Schneider D (1965) Insect olfactory receptors. Cold Spring Harb Symp Quant Biol 30:263–80.

Bok D (1993) The retinal pigment epithelium: a versatile partner in vision. J Cell Sci Suppl 17:189–95.

Bollepalli MK, Kuipers ME, Liu CH, Asteriti S, Hardie RC (2017) Phototransduction in *Drosophila* is compromised by Gal4 expression but not by InsP3 receptor knockdown or mutation. eNeuro 4(3). pii:ENEURO.0143-17.2017.

Bonigk W, Bradley J, Muller F, Sesti F, Boekhoff I, Ronnett GV, Kaupp UB, Frings S (1999) The native rat olfactory cyclic nucleotide-gated channel is composed of three distinct subunits. J Neurosci 19:5332–47.

Boschat C, Pelofi C, Randin O, Roppolo D, Luscher C, Broillet MC, Rodriguez I (2002) Pheromone detection mediated by a V1R vomeronasal receptor. Nat Neurosci 5:1261–2.

Bowmaker JK, Dartnall HJ, Mollon JD (1980) Microspectrophotometric demonstration of four classes of photoreceptor in an old world primate, Macaca fascicularis. J Physiol 298:131–43.

Bownds D, Dawes J, Miller J, Stahlman M (1972) Phosphorylation of frog photoreceptor membranes induced by light. Nat New Biol 237:125–7.

Bownds MD, Arshavsky VY (1995) What are the mechanisms of photoreceptor adaptation? Behav Brain Sci 18:415–24.

Bozza T, Feinstein P, Zheng C, Mombaerts P (2002) Odorant receptor expression defines functional units in the mouse olfactory system. J Neurosci 22:3033–43.

Brauchi S, Orio P, Latorre R (2004) Clues to understanding cold sensation: thermodynamics and electrophysiological analysis of the cold receptor TRPM8. Proc Natl Acad Sci U S A 101:15494–9.

Braun HA, Bade H, Hensel H (1980) Static and dynamic discharge patterns of bursting cold fibers related to hypothetical receptor mechanisms. Pflugers Arch 386:1–9.

Brehm P, Eckert R (1978) An electrophysiological study of the regulation of ciliary beating frequency in *Paramecium*. J Physiol 283:557–68.

Brehm P, Kullberg R, Moody-Corbett F (1984) Properties of non-junctional acetylcholine receptor channels on innervated muscle of *Xenopus laevis*. J Physiol 350:631–48.

Brindley GS (1960) Physiology of the retina and visual pathway. London: Edward Arnold.

Brini M, Carafoli E (2009) Calcium pumps in health and disease. Physiol Rev 89:1341–78.

Brini M, Carafoli E (2011) The plasma membrane Ca^{2+} ATPase and the plasma membrane sodium calcium exchanger cooperate in the regulation of cell calcium. Cold Spring Harb Perspect Biol 3(2). pii:a004168.

Brown AM (1993) Functional bases for interpreting amino acid sequences of voltage- dependent K+ channels. Annu Rev Biophys Biomol Struct 22:173–98.

Brown JE, Mote MI (1974) Ionic dependence of reversal voltage of the light response in *Limulus* ventral photoreceptors. J Gen Physiol 63:337–50.

Brown KT, Flaming DG (1977) New microelectrode techniques for intracellular work in small cells. Neuroscience 2:813–27.

Brown PK, Gibbons RI, Wald G (1963) The visual cells and visual pigments of the mudpuppy, *Necturus*. J Cell Biol 19:79–106.

Brownell WE, Bader CR, Bertrand D, de Ribaupierre Y (1985) Evoked mechanical responses of isolated cochlear outer hair cells. Science 227:194–6.

Brunet LJ, Gold GH, Ngai J (1996) General anosmia caused by a targeted disruption of the mouse olfactory cyclic nucleotide-gated cation channel. Neuron 17:681–93.

Brunner JD, Lim NK, Schenck S, Duerst A, Dutzler R (2014) X-ray structure of a calcium-activated TMEM16 lipid scramblase. Nature 516:207–12.

Buck L, Axel R (1991) A novel multigene family may encode odorant receptors: a molecular basis for odor recognition. Cell 65:175–87.

Buck LB (2000) The molecular architecture of odor and pheromone sensing in mammals. Cell 100:611–18.

Bullock TH, Diecke FPJ (1956) Properties of an infra-red receptor. J Physiol 134:47–87.

Bullock TH, Horridge GA (1965) Structure and Function in the Nervous System of Invertebrates. San Francisco: WH Freeman.

Burgoyne T, Meschede IP, Burden JJ, Bailly M, Seabra MC, Futter CE (2015) Rod disc renewal occurs by evagination of the ciliary plasma membrane that makes cadherin-based contacts with the inner segment. Proc Natl Acad Sci U S A 112:15922–7.

Burkhardt W, Braitenberg V (1976) Some peculiar synaptic complexes in the first visual ganglion of the fly, *Musca domestica*. Cell Tissue Res 173:287–308.

Burns M, Mendez A, Chen J, Baylor D (2002) Dynamics of cyclic GMP synthesis in retinal rods. Neuron 36:81.

Burns ME (2010) Deactivation mechanisms of rod phototransduction: the Cogan lecture. Invest Ophthalmol Vis Sci 51:1282–8.

Burns ME, Mendez A, Chen CK, Almuete A, Quillinan N, Simon MI, Baylor DA, Chen J (2006) Deactivation of phosphorylated and nonphosphorylated rhodopsin by arrestin splice variants. J Neurosci 26:1036–44.

Butterwick JA, Del Marmol J, Kim KH, Kahlson MA, Rogow JA, Walz T, Ruta V (2018) Cryo-EM structure of the insect olfactory receptor Orco. Nature 560:447–52.

Byk T, Bar-Yaacov M, Doza YN, Minke B, Selinger Z (1993) Regulatory arrestin cycle secures the fidelity and maintenance of the fly photoreceptor cell. Proc Natl Acad Sci U S A 90:1907–11.

Caicedo A, Kim KN, Roper SD (2002) Individual mouse taste cells respond to multiple chemical stimuli. J Physiol 544:501–9.

Cajal R y (1893/1973) The Vertebrate Retina, translation by Maguire D and Rodieck RW of La rétine des vertébrés, published in The Vertebrate Retina, Principles of Structure and Function. San Francisco: Freeman.

Cajal R y (1911/1998) Histology of the Nervous System of Man and Vertebrates, translation by Swanson N and Swanson LW of Histologie du systeme nerveux de l'homme et des vertebres. Oxford: Oxford University Press.

Calman BG, Chamberlain SC (1982) Distinct lobes of *Limulus* ventral photoreceptors. II. Structure and ultrastructure. J Gen Physiol 80:839–62.

Calvert PD, Ho TW, LeFebvre YM, Arshavsky VY (1998) Onset of feedback reactions underlying vertebrate rod photoreceptor light adaptation. J Gen Physiol 111:39–51.

Cameron P, Hiroi M, Ngai J, Scott K (2010) The molecular basis for water taste in *Drosophila*. Nature 465:91–5.

Campbell AL, Naik RR, Sowards L, Stone MO (2002) Biological infrared imaging and sensing. Micron 33:211–25.

Cao E, Liao M, Cheng Y, Julius D (2013a) TRPV1 structures in distinct conformations reveal activation mechanisms. Nature 504:113–18.

Cao E, Cordero-Morales JF, Liu B, Qin F, Julius D (2013b) TRPV1 channels are intrinsically heat sensitive and negatively regulated by phosphoinositide lipids. Neuron 77:667–79.

Carlson SD, Saint Marie RL, Chi C (1984) The photoreceptor cells. In: Insect Ultrastructure, Vol. 2 (King RC, Akai H, eds), pp 397–433. New York: Plenum.

Cattaert D, Le Ray D (2001) Adaptive motor control in crayfish. Prog Neurobiol 63:199–240.

Catterall WA (2012) Voltage-gated sodium channels at 60: structure, function, and pathophysiology. J Physiol 590:2577–89.

Cervetto L, Piccolino M (1974) Synaptic transmission between photoreceptors and horizontal cells in the turtle retina. Science 183:417–19.

Cervetto L, Lagnado L, Perry RJ, Robinson DW, McNaughton PA (1989) Extrusion of calcium from rod outer segments is driven by both sodium and potassium gradients. Nature 337:740–3.

Chaban B, Hughes HV, Beeby M (2015) The flagellum in bacterial pathogens: for motility and a whole lot more. Semin Cell Dev Biol 46:91–103.

Chabre M, Cone R, Saibil H (2003) Biophysics: is rhodopsin dimeric in native retinal rods? Nature 426:30–1; discussion 31.

Chalfie M, Au M (1989) Genetic control of differentiation of the *Caenorhabditis elegans* touch receptor neurons. Science 243:1027–33.

Chalfie M, Sulston J (1981) Developmental genetics of the mechanosensory neurons of *Caenorhabditis elegans*. Dev Biol 82:358–70.

Chalfie M, Thomson JN (1982) Structural and functional diversity in the neuronal microtubules of *Caenorhabditis elegans*. J Cell Biol 93:15–23.

Chamberlain SC, Barlow RB, Jr (1979) Light and efferent activity control rhabdom turnover in *Limulus* photoreceptors. Science 206:361–3.

Chambers MR, Andres KH, von Duering M, Iggo A (1972) The structure and function of the slowly adapting type II mechanoreceptor in hairy skin. Q J Exp Physiol Cogn Med Sci 57:417–45.

Chamero P, Marton TF, Logan DW, Flanagan K, Cruz JR, Saghatelian A, Cravatt BF, Stowers L (2007) Identification of protein pheromones that promote aggressive behaviour. Nature 450:899–902.

Chamero P, Weiss J, Alonso MT, Rodriguez-Prados M, Hisatsune C, Mikoshiba K, Leinders-Zufall T, Zufall F (2017) Type 3 inositol 1,4,5-trisphosphate receptor is dispensable for sensory activation of the mammalian vomeronasal organ. Sci Rep 7:10260.

Chandrashekar J, Mueller KL, Hoon MA, Adler E, Feng L, Guo W, Zuker CS, Ryba NJ (2000) T2Rs function as bitter taste receptors. Cell 100:703–11.

Chandrashekar J, Kuhn C, Oka Y, Yarmolinsky DA, Hummler E, Ryba NJ, Zuker CS (2010) The cells and peripheral representation of sodium taste in mice. Nature 464:297–301.

Chang RB, Waters H, Liman ER (2010) A proton current drives action potentials in genetically identified sour taste cells. Proc Natl Acad Sci U S A 107:22320–5.

Chaudhari N, Landin AM, Roper SD (2000) A metabotropic glutamate receptor variant functions as a taste receptor. Nat Neurosci 3:113–19.

Chen CK, Inglese J, Lefkowitz RJ, Hurley JB (1995a) Ca^{2+}-dependent interaction of recoverin with rhodopsin kinase. J Biol Chem 270:18060–6.

Chen CK, Burns ME, Spencer M, Niemi GA, Chen J, Hurley JB, Baylor DA, Simon MI (1999a) Abnormal photoresponses and light-induced apoptosis in rods lacking rhodopsin kinase. Proc Natl Acad Sci U S A 96:3718–22.

Chen CK, Burns ME, He W, Wensel TG, Baylor DA, Simon MI (2000) Slowed recovery of rod photoresponse in mice lacking the GTPase accelerating protein RGS9-1. Nature 403:557–60.

Chen CK, Woodruff ML, Chen FS, Chen D, Fain GL (2010a) Background light produces a recoverin-dependent modulation of activated-rhodopsin lifetime in mouse rods. J Neurosci 30:1213–20.

Chen FH, Baumann A, Payne R, Lisman JE (2001a) A cGMP-gated channel subunit in *Limulus* photoreceptors. Vis Neurosci 18:517–26.

Chen FH, Ukhanova M, Thomas D, Afshar G, Tanda S, Battelle BA, Payne R (1999b) Molecular cloning of a putative cyclic nucleotide-gated ion channel cDNA from *Limulus polyphemus*. J Neurochem 72:461–71.

Chen J, Makino CL, Peachey NS, Baylor DA, Simon MI (1995b) Mechanisms of rhodopsin inactivation in vivo as revealed by a COOH-terminal truncation mutant. Science 267:374–7.

Chen J, Woodruff ML, Wang T, Concepcion F, Tranchina D, Fain GL (2010b) Channel modulation and the mechanism of light adaptation in mouse rods. J Neurosci 30:16232–40.

Chen P, Hao W, Rife L, Wang XP, Shen D, Chen J, Ogden T, Van Boemel GB, Wu L, Yang M, Fong HK (2001b) A photic visual cycle of rhodopsin regeneration is dependent on Rgr. Nat Genet 28:256–60.

Chen TW, Wardill TJ, Sun Y, Pulver SR, Renninger SL, Baohan A, Schreiter ER, Kerr RA, Orger MB, Jayaraman V, Looger LL, Svoboda K, Kim DS (2013) Ultrasensitive fluorescent proteins for imaging neuronal activity. Nature 499:295–300.

Chen TY, Yau KW (1994) Direct modulation by Ca^{2+}-calmodulin of cyclic nucleotide-activated channel of rat olfactory receptor neurons. Nature 368:545–8.

Chen TY, Peng YW, Dhallan RS, Ahamed B, Reed RR, Yau KW (1993) A new subunit of the cyclic nucleotide-gated cation channel in retinal rods. Nature 362:764–7.

Chen Y, Sun XD, Herness S (1996) Characteristics of action potentials and their underlying outward currents in rat taste receptor cells. J Neurophysiol 75:820–31.

Chen Y, Bharill S, Isacoff EY, Chalfie M (2015) Subunit composition of a DEG/ENaC mechanosensory channel of *Caenorhabditis elegans*. Proc Natl Acad Sci U S A 112:11690–5.

Chen Y, Bharill S, Altun Z, O'Hagan R, Coblitz B, Isacoff EY, Chalfie M (2016) *Caenorhabditis elegans* paraoxonase-like proteins control the functional expression of DEG/ENaC mechanosensory proteins. Mol Biol Cell 27:1272–85.

Chen Z, Wang Q, Wang Z (2010c) The amiloride-sensitive epithelial Na^+ channel PPK28 is essential for *Drosophila* gustatory water reception. J Neurosci 30:6247–52.

Cheng LE, Song W, Looger LL, Jan LY, Jan YN (2010) The role of the TRP channel NompC in *Drosophila* larval and adult locomotion. Neuron 67:373–80.

Cherry JA, Davis RL (1995) A mouse homolog of dunce, a gene important for learning and memory in *Drosophila*, is preferentially expressed in olfactory receptor neurons. J Neurobiol 28:102–13.

Chesler AT, Szczot M, Bharucha-Goebel D, Ceko M, Donkervoort S, Laubacher C, Hayes LH, Alter K,

Zampieri C, Stanley C, Innes AM, Mah JK, Grosmann CM, Bradley N, Nguyen D, Foley AR, Le Pichon CE, Bonnemann CG (2016) The Role of PIEZO2 in human mechanosensation. N Engl J Med 375:1355–64.

Christensen AP, Corey DP (2007) TRP channels in mechanosensation: direct or indirect activation? Nat Rev Neurosci 8:510–21.

Chung YD, Zhu J, Han Y, Kernan MJ (2001) nompA encodes a PNS-specific, ZP domain protein required to connect mechanosensory dendrites to sensory structures. Neuron 29:415–28.

Chyb S, Raghu P, Hardie RC (1999) Polyunsaturated fatty acids activate the *Drosophila* light-sensitive channels TRP and TRPL. Nature 397:255–9.

Cideciyan AV, Zhao X, Nielsen L, Khani SC, Jacobson SG, Palczewski K (1998) Null mutation in the rhodopsin kinase gene slows recovery kinetics of rod and cone phototransduction in man. Proc Natl Acad Sci U S A 95:328–33.

Clapham DE, Neer EJ (1997) G protein beta gamma subunits. Annu Rev Pharmacol Toxicol 37:167–203.

Clapp TR, Yang R, Stoick CL, Kinnamon SC, Kinnamon JC (2004) Morphologic characterization of rat taste receptor cells that express components of the phospholipase C signaling pathway. J Comp Neurol 468:311–21.

Clapp TR, Medler KF, Damak S, Margolskee RF, Kinnamon SC (2006) Mouse taste cells with G protein-coupled taste receptors lack voltage-gated calcium channels and SNAP-25. BMC Biol 4:7.

Clark AW, Millecchia R, Mauro A (1969) The ventral photoreceptor cells of *Limulus*. I. The microanatomy. J Gen Physiol 54:289–309.

Clites BL, Pierce JT (2017) Identifying cellular and molecular mechanisms for magnetosensation. Annu Rev Neurosci 40:231–50.

Clusin WT, Bennett MVL (1977a) Calcium-activated conductance in skate electroreceptors. Current clamp experiments. J Gen Physiol 69:121–43.

Clusin WT, Bennett MVL (1977b) Calcium-activated conducctance in skate electroreceptors. Voltage clamp experiments. J Gen Physiol 69:145–82.

Clusin WT, Bennett MVL (1979a) The oscillatory responses of skate electroreceptors to small voltage stimuli. J Gen Physiol 73:685–702.

Clusin WT, Bennett MVL (1979b) The ionic basis of oscillatory responses of skate electroreceptors. J Gen Physiol 73:703–23.

Clyne PJ, Warr CG, Freeman MR, Lessing D, Kim J, Carlson JR (1999) A novel family of divergent seven-transmembrane proteins: candidate odorant receptors in *Drosophila*. Neuron 22:327–38.

Cobbs WH, Pugh EN, Jr (1987) Kinetics and components of the flash photocurrent of isolated retinal rods of the larval salamander, *Ambystoma tigrinum*. J Physiol 394:529–72.

Coggeshall RE (1969) A fine structure analysis of the statocyst in *Aplysia californica*. J Morphol 127:113–32.

Cohen MJ (1955) The function of receptors in the statocyst of the lobster *Homarus americanus*. J Physiol 130:9–34.

Cohen MJ (1960) The response patterns of single receptors in the crustacean statocyst. Proc R Soc London B 152:30–44.

Colburn RW, Lubin ML, Stone DJ, Jr, Wang Y, Lawrence D, D'Andrea MR, Brandt MR, Liu Y, Flores CM, Qin N (2007) Attenuated cold sensitivity in TRPM8 null mice. Neuron 54:379–86.

Cook B, Bar-Yaacov M, Cohen Ben-Ami H, Goldstein RE, Paroush Z, Selinger Z, Minke B (2000) Phospholipase C and termination of G-protein-mediated signalling in vivo. Nat Cell Biol 2:296–301.

Cook NJ, Hanke W, Kaupp UB (1987) Identification, purification, and functional reconstitution of the cyclic GMP-dependent channel from rod photoreceptors. Proc Natl Acad Sci U S A 84:585–9.

Copenhagen DR, Jahr CE (1989) Release of endogenous excitatory amino acids from turtle photoreceptors. Nature 341:536–9.

Corey DP, Hudspeth AJ (1979a) Ionic basis of the receptor potential in a vertebrate hair cell. Nature 281:675–7.

Corey DP, Hudspeth AJ (1979b) Response latency of vertebrate hair cells. Biophys J 26:499–506.

Corey DP, Hudspeth AJ (1983) Analysis of the microphonic potential of the bullfrog's sacculus. J Neurosci 3:942–61.

Corey DP, Akyuz N, Holt JR (2018) Function and dysfunction of tmc channels in inner ear hair cells. Cold Spring Harb Perspect Med Oct 5. pii:a033506.

Corns LF, Johnson SL, Kros CJ, Marcotti W (2014) Calcium entry into stereocilia drives adaptation of the mechano-electrical transducer current of mammalian cochlear hair cells. Proc Natl Acad Sci U S A 111:14918–23.

Cornwall MC, Gorman AL (1983a) The cation selectivity and voltage dependence of the light-activated potassium conductance in scallop distal photoreceptor. J Physiol 340:287–305.

Cornwall MC, Gorman AL (1983b) Colour dependence of the early receptor potential and late receptor potential in scallop distal photoreceptor. J Physiol 340:307–34.

Cornwall MC, Fain GL (1994) Bleached pigment activates transduction in isolated rods of the salamander retina. J Physiol 480:261–79.

Coste B (2011) Détecter la pression? Identification de deux protéines activées par les forces mécaniques. Med Sci (Paris) 27:17–19.

Coste B, Mathur J, Schmidt M, Earley TJ, Ranade S, Petrus MJ, Dubin AE, Patapoutian A (2010) Piezo1 and Piezo2 are essential components of distinct mechanically activated cation channels. Science 330:55–60.

Coste B, Murthy SE, Mathur J, Schmidt M, Mechioukhi Y, Delmas P, Patapoutian A (2015) Piezo1 ion channel pore properties are dictated by C-terminal region. Nat Commun 6:7223.

Cowan CW, Fariss RN, Sokal I, Palczewski K, Wensel TG (1998) High expression levels in cones of RGS9, the predominant GTPase accelerating protein of rods. Proc Natl Acad Sci U S A 95:5351–6.

Cox CD, Bavi N, Martinac B (2018) Bacterial mechanosensors. Annu Rev Physiol 80:71–93.

Crawford AC, Fettiplace R (1981) An electrical tuning mechanism in turtle cochlear hair cells. J Physiol 312:377–412.

Crawford AC, Fettiplace R (1985) The mechanical properties of ciliary bundles of turtle cochlear hair cells. J Physiol 364:359–79.

Crawford AC, Evans MG, Fettiplace R (1989) Activation and adaptation of transducer currents in turtle hair cells. J Physiol 419:405–34.

Crawford AC, Evans MG, Fettiplace R (1991) The actions of calcium on the mechano-electrical transducer current of turtle hair cells. J Physiol 434:369–98.

Cronin TW, Johnsen S, Marshall NJ, Warrant EJ (2014) Visual Ecology. Princeton, NJ: Princeton University Press.

Cueva JG, Mulholland A, Goodman MB (2007) Nanoscale organization of the MEC-4 DEG/ENaC sensory mechanotransduction channel in *Caenorhabditis elegans* touch receptor neurons. J Neurosci 27:14089–98.

Cygnar KD, Zhao H (2009) Phosphodiesterase 1C is dispensable for rapid response termination of olfactory sensory neurons. Nat Neurosci 12:454–62.

Czech-Damal NU, Dehnhardt G, Manger P, Hanke W (2013) Passive electroreception in aquatic mammals. J Comp Physiol A Neuroethol Sens Neural Behav Physiol 199:555–63.

D'Amours MR, Cote RH (1999) Regulation of photoreceptor phosphodiesterase catalysis by its non-catalytic cGMP-binding sites. Biochem J 340:863–9.

Dabdoub A, Payne R (1999) Protein kinase C activators inhibit the visual cascade in *Limulus* ventral photoreceptors at an early stage. J Neurosci 19:10262–9.

Dakin WJ (1928) The eyes of *Pecten, Spondylus, Amussium* and allied lamellibranchs, with a short discussion on their evolution. Proc R Soc London B 103:355–65.

Dallos P (2008) Cochlear amplification, outer hair cells and prestin. Curr Opin Neurobiol 18:370–6.

Dallos P, Evans BN, Hallworth R (1991) Nature of the motor element in electrokinetic shape changes of cochlear outer hair cells. Nature 350:155–7.

Dallos P, Wu X, Cheatham MA, Gao J, Zheng J, Anderson CT, Jia S, Wang X, Cheng WH, Sengupta S, He DZ, Zuo J (2008) Prestin-based outer hair cell motility is necessary for mammalian cochlear amplification. Neuron 58:333–9.

Dalton RP, Lomvardas S (2015) Chemosensory receptor specificity and regulation. Annu Rev Neurosci 38:331–49.

Damak S, Rong M, Yasumatsu K, Kokrashvili Z, Varadarajan V, Zou S, Jiang P, Ninomiya Y, Margolskee RF (2003) Detection of sweet and umami taste in the absence of taste receptor T1r3. Science 301:850–3.

Dang S, Feng S, Tien J, Peters CJ, Bulkley D, Lolicato M, Zhao J, Zuberbuhler K, Ye W, Qi L, Chen T, Craik CS, Jan YN, Minor DL, Jr., Cheng Y, Jan LY (2017) Cryo-EM structures of the TMEM16A calcium-activated chloride channel. Nature 552:426–9.

Darian-Smith I (1984a) The sense of touch: performance and peripheral neural processes. In: Handbook of Physiology, The Nervous System, Vol. III, Sensory Processes, Part 2 (Darian-Smith I, ed), pp 739–88. Bethesda, MD: American Physiological Society.

Darian-Smith I (1984b) Thermal sensibility. In: Handbook of Physiology, The Nervous System, Vol. III, Sensory Processes, Part 2 (Darian-Smith I, ed), pp 879–913. Bethesda, MD: American Physiological Society.

Darian-Smith I, Johnson KO (1977) Thermal sensibility and thermoreceptors. J Invest Dermatol 69:146–53.

Dawson TM, Arriza JL, Jaworsky DE, Borisy FF, Attramadal H, Lefkowitz RJ, Ronnett GV (1993) Beta-adrenergic receptor kinase-2 and beta-arrestin-2 as mediators of odorant-induced desensitization. Science 259:825–9.

de Bruyne M, Clyne PJ, Carlson JR (1999) Odor coding in a model olfactory organ: the *Drosophila* maxillary palp. J Neurosci 19:4520–32.

de Bruyne M, Foster K, Carlson JR (2001) Odor coding in the *Drosophila* antenna. Neuron 30:537–52.

Deckert A, Nagy K, Helrich CS, Stieve H (1992) Three components in the light-induced current of the *Limulus* ventral photoreceptor. J Physiol 453:69–96.

DeFazio RA, Dvoryanchikov G, Maruyama Y, Kim JW, Pereira E, Roper SD, Chaudhari N (2006) Separate populations of receptor cells and presynaptic cells in mouse taste buds. J Neurosci 26:3971–80.

del Pilar Gomez M, Nasi E (1995) Activation of light-dependent K$^+$ channels in ciliary invertebrate photoreceptors involves cGMP but not the IP3/Ca^{2+} cascade. Neuron 15:607–18.

Del Punta K, Puche A, Adams NC, Rodriguez I, Mombaerts P (2002) A divergent pattern of sensory axonal projections is rendered convergent by second-order neurons in the accessory olfactory bulb. Neuron 35:1057–66.

Delgado R, Munoz Y, Pena-Cortes H, Giavalisco P, Bacigalupo J (2014) Diacylglycerol activates the light-dependent channel TRP in the photosensitive microvilli of *Drosophila melanogaster* photoreceptors. J Neurosci 34:6679–86.

DeMaria S, Ngai J (2010) The cell biology of smell. J Cell Biol 191:443–52.

Denk W, Holt JR, Shepherd GM, Corey DP (1995) Calcium imaging of single stereocilia in hair cells: localization of transduction channels at both ends of tip links. Neuron 15:1311–21.

Desalvo JA, Hartline PH (1978) Spatial properties of primary infrared sensory neurons in Crotalidae. Brain Res 142:338–42.

Descartes R (1637/1987) Discours de la Methode plus La Dioptrique, Les Meteores, et la Geometrie. Paris: Librairie Artheme Fayard.

Dethier VG (1976) The Hungry Fly. Cambridge, MA: Harvard University Press.

Deupi X, Standfuss J, Schertler G (2012) Conserved activation pathways in G-protein-coupled receptors. Biochem Soc Trans 40:383–8.

Dewan A, Pacifico R, Zhan R, Rinberg D, Bozza T (2013) Non-redundant coding of aversive odours in the main olfactory pathway. Nature 497:486–9.

Dhaka A, Murray AN, Mathur J, Earley TJ, Petrus MJ, Patapoutian A (2007) TRPM8 is required for cold sensation in mice. Neuron 54:371–8.

Dibattista M, Amjad A, Maurya DK, Sagheddu C, Montani G, Tirindelli R, Menini A (2012) Calcium-activated chloride channels in the apical region of mouse vomeronasal sensory neurons. J Gen Physiol 140:3–15.

Dijkgraaf S (1967) Biological significance of the lateral line organs. In: Lateral Line Detectors (Cahn PH, ed), pp 83–95. Bloomington, ID: Indiana University Press.

Ding JD, Salinas RY, Arshavsky VY (2015) Discs of mammalian rod photoreceptors form through the membrane evagination mechanism. J Cell Biol 211:495–502.

Dizhoor, AM, Olshevskaya, EV, and Peshenko, IV (2010). Mg2+/Ca2+ cation binding cycle of guanylyl cyclase activating proteins (GCAPs): role in regulation of photoreceptor guanylyl cyclase. Mol Cell Biochem *334*, 117–124.

Do MT, Yau KW (2010) Intrinsically photosensitive retinal ganglion cells. Physiol Rev 90:1547–81.

Doolin RE, Gilbertson TA (1996) Distribution and characterization of functional amiloride-sensitive sodium channels in rat tongue. J Gen Physiol 107:545–54.

Dorlochter M, Stieve H (1997) The *Limulus* ventral photoreceptor: light response and the role of calcium in a classic preparation. Prog Neurobiol 53:451–515.

Doudna JA, Charpentier E (2014) Genome editing. The new frontier of genome engineering with CRISPR-Cas9. Science 346:1258096.

Doudna JA, Sternberg S (2017) A crack in creation: gene editing and the unthinkable power to control evolution. Boston: Houghton Mifflin Harcourt.

Dowling JE (1965) Foveal receptors of the monkey retina: fine structure. Science 147:57–9.

Dowling JE, Ripps H (1973) Effect of magnesium on horizontal cell activity in the skate retina. Nature 242:101–103.

Downes GB, Gautam N (1999) The G protein subunit gene families. Genomics 62:544–52.

Doyle DA, Morais Cabral J, Pfuetzner RA, Kuo A, Gulbis JM, Cohen SL, Chait BT, MacKinnon R (1998) The structure of the potassium channel: molecular basis of K+ conduction and selectivity. Science 280:69–77.

Duckworth WLH, Lyons MC, Towers B (2010) Galen on Anatomical Procedures, The Later Books. Cambridge, UK: Cambridge University Press.

Dulac C, Axel R (1995) A novel family of genes encoding putative pheromone receptors in mammals. Cell 83:195–206.

Dupre DJ, Robitaille M, Rebois RV, Hebert TE (2009) The role of Gbetagamma subunits in the organization, assembly, and function of GPCR signaling complexes. Annu Rev Pharmacol Toxicol 49:31–56.

Eakin RM (1965) Evolution of photoreceptors. Cold Spring Harb Symp Quant Biol 30:363–70.

Eastwood AL, Goodman MB (2012) Insight into DEG/ENaC channel gating from genetics and structure. Physiology (Bethesda) 27:282–90.

Eatock RA, Corey DP, Hudspeth AJ (1987) Adaptation of mechanoelectrical transduction in hair cells of the bullfrog's sacculus. J Neurosci 7:2821–36.

Ebrey T, Koutalos Y (2001) Vertebrate photoreceptors. Progr Retin Eye Res 20:49–94.

Ecker JL, Dumitrescu ON, Wong KY, Alam NM, Chen SK, LeGates T, Renna JM, Prusky GT, Berson DM, Hattar S (2010) Melanopsin-expressing retinal ganglion-cell photoreceptors: cellular diversity and role in pattern vision. Neuron 67:49–60.

Eckert R (1972) Bioelectric control of ciliary activity. Science 176:473–81.

Eckert R, Brehm P (1979) Ionic mechanisms of excitation in *Paramecium*. Annu Rev Biophys Bioeng 8:353–83.

Edelman NB, Fritz T, Nimpf S, Pichler P, Lauwers M, Hickman RW, Papadaki-Anastasopoulou A, Ushakova L, Heuser T, Resch GP, Saunders M, Shaw JA, Keays DA (2015) No evidence for intracellular magnetite in putative vertebrate magnetoreceptors identified by magnetic screening. Proc Natl Acad Sci U S A 112:262–7.

Edrington TCt, Bennett M, Albert AD (2008) Calorimetric studies of bovine rod outer segment disk membranes support a monomeric unit for both rhodopsin and opsin. Biophys J 95:2859–66.

Effertz T, Wiek R, Gopfert MC (2011) NompC TRP channel is essential for *Drosophila* sound receptor function. Curr Biol 21:592–7.

Effertz T, Nadrowski B, Piepenbrock D, Albert JT, Gopfert MC (2012) Direct gating and mechanical integrity of

Drosophila auditory transducers require TRPN1. Nat Neurosci 15:1198–200.

Emlen ST, Emlen JT (1966) A technique for recording migratory orientation of captive birds. The Auk 83:361–7.

Emlen ST, Wiltschko w, Demong NJ, Wiltschko R, Bergman S (1976) Magnetic direction finding: evidence for its use in migratory indigo buntings. Science 193:505–8.

Erlandson SC, McMahon C, Kruse AC (2018) Structural basis for G protein-coupled receptor signaling. Annu Rev Biophys Mar 2. doi:10.1146/annurev-biophys-070317-032931.

Erxleben C (1989) Stretch-activated current through single ion channels in the abdominal stretch receptor organ of the crayfish. J Gen Physiol 94:1071–83.

Estacion M, Sinkins WG, Schilling WP (2001) Regulation of *Drosophila* transient receptor potential-like (TrpL) channels by phospholipase C-dependent mechanisms. J Physiol 530:1–19.

Evans EF (1972) The frequency response and other properties of single fibres in the guinea-pig cochlear nerve. J Physiol 226:263–87.

Evans EF (1975) Cochlear nerve and cochlear nucleus. In: Handbook of Sensory Physiology (Keidel WD, Neff WD, eds). Berlin: Springer.

Ewing AW, Bennet-Clark HC (1968) The courtship songs of *Drosophila*. Behaviour 31:288–301.

Eyzaguirre C, Kuffler SW (1955) Processes of excitation in the dendrites and in the soma of single isolated sensory nerve cells of the lobster and crayfish. J Gen Physiol 39:87–119.

Fain GL (1975) Quantum sensitivity of rods in the toad retina. Science 187:838–41.

Fain GL (1986) Evidence for a role of messenger substances in phototransduction. In: The Molecular Mechanism of Photoreception (Stieve H, ed), pp 67–77. Berlin: Springer.

Fain GL (1999) Molecular and Cellular Physiology of Neurons. Cambridge, MA: Harvard University Press.

Fain GL (2006) Why photoreceptors die (and why they don't). Bioessays 28:344–54.

Fain GL (2011) Adaptation of mammalian photoreceptors to background light: putative role for direct modulation of phosphodiesterase. Molec Neurobiol 44:374–82.

Fain GL (2014) Molecular and Cellular Physiology of Neurons, Second Edition. Cambridge, MA: Harvard University Press.

Fain GL (2015) Phototransduction: making the chromophore to see through the murk. Curr Biol 25:R1126–7.

Fain GL, Granda AM, Maxwell JH (1977) Voltage signal of photoreceptors at visual threshold. Nature 265:181–3.

Fain GL, Quandt FN, Bastian BL, Gerschenfeld HM (1978) Contribution of a caesium-sensitive conductance increase to the rod photoresponse. Nature 272:466–9.

Fain GL, Lamb TD, Matthews HR, Murphy RL (1989) Cytoplasmic calcium as the messenger for light adaptation in salamander rods. J Physiol 416:215–43.

Fain GL, Matthews HR, Cornwall MC (1996) Dark adaptation in vertebrate photoreceptors. Trends Neurosci 19:502–7.

Fain GL, Matthews HR, Cornwall MC, Koutalos Y (2001) Adaptation in vertebrate photoreceptors. Physiol Rev 81:117–51.

Fain GL, Hardie R, Laughlin SB (2010) Phototransduction and the evolution of photoreceptors. Curr Biol 20:R114–24.

Fasick JI, Applebury ML, Oprian DD (2002) Spectral tuning in the mammalian short-wavelength sensitive cone pigments. Biochemistry 41:6860–5.

Fein A, Payne R, Corson DW, Berridge MJ, Irvine RF (1984) Photoreceptor excitation and adaptation by inositol 1,4,5-trisphosphate. Nature 311:157–60.

Feinstein P, Mombaerts P (2004) A contextual model for axonal sorting into glomeruli in the mouse olfactory system. Cell 117:817–31.

Fesenko EE, Kolesnikov SS, Lyubarsky AL (1985) Induction by cyclic GMP of cationic conductance in plasma membrane of retinal rod outer segment. Nature 313:310–13.

Fettiplace R (2016) Is TMC1 the hair cell mechanotransducer channel? Biophys J 111:3–9.

Fettiplace R (2017) Hair cell transduction, tuning, and synaptic transmission in the mammalian cochlea. Compr Physiol 7:1197–227.

Fettiplace R, Crawford AC (1978) The coding of sound pressure and frequency in cochlear hair cells of the terrapin. Proc R Soc B 203:209–18.

Fettiplace R, Fuchs PA (1999) Mechanisms of hair cell tuning. Annu Rev Physiol 61:809–34.

Fettiplace R, Kim KX (2014) The physiology of mechanoelectrical transduction channels in hearing. Physiol Rev 94:951–86.

Field GD, Rieke F (2002) Mechanisms regulating variability of the single photon responses of mammalian rod photoreceptors. Neuron 35:733–47.

Filipek S, Stenkamp RE, Teller DC, Palczewski K (2003) G protein-coupled receptor rhodopsin: a prospectus. Annu Rev Physiol 65:851–79.

Finger TE, Danilova V, Barrows J, Bartel DL, Vigers AJ, Stone L, Hellekant G, Kinnamon SC (2005) ATP signaling is crucial for communication from taste buds to gustatory nerves. Science 310:1495–9.

Finn JT, Grunwald ME, Yau K-W (1996) Cyclic nucleotide-gated ion channels: an extended family with diverse functions. Annual Review of Physiology 58:395–426.

Finn JT, Solessio EC, Yau KW (1997) A cGMP-gated cation channel in depolarizing photoreceptors of the lizard parietal eye. Nature 385:815–19.

Finn JT, Xiong WH, Solessio EC, Yau KW (1998) A cGMP-gated cation channel and phototransduction in depolarizing photoreceptors of the lizard parietal eye. Vision Res 38:1353–7.

Firestein S, Zufall F, Shepherd GM (1991) Single odor-sensitive channels in olfactory receptor neurons are also gated by cyclic nucleotides. J Neurosci 11:3565–72.

Firestein S, Picco C, Menini A (1993) The relation between stimulus and response in olfactory receptor cells of the tiger salamander. J Physiol 468:1–10.

Flock A (1965) Transducing mechanisms in the lateral line canal organ receptors. Cold Spring Harb Symp Quant Biol 30:133–45.

Flock A (1967) Ultrastructure and function in the lateral line organs. In: Lateral Line Detectors (Cahn PH, ed), pp 163–97. Bloomington, ID: Indiana University Press.

Foskett JK, White C, Cheung KH, Mak DO (2007) Inositol trisphosphate receptor Ca^{2+} release channels. Physiol Rev 87:593–658.

Frank M (1973) An analysis of hamster afferent taste nerve response functions. J Gen Physiol 61:588–618.

Frank M, Pfaffmann C (1969) Taste nerve fibers: a random distribution of sensitivities to four tastes. Science 164:1183–5.

Frank ME, Bieber SL, Smith DV (1988) The organization of taste sensibilities in hamster chorda tympani nerve fibers. J Gen Physiol 91:861–96.

Frank TM, Fein A (1991) The role of the inositol phosphate cascade in visual excitation of invertebrate microvillar photoreceptors. J Gen Physiol 97:697–723.

Frankel RB, Blakemore RP (1989) Magnetite and magneto-taxis in microorganisms. Bioelectromagnetics 10:223–37.

Frankel RB, Blakemore R, Wolfe RS (1979) Magnetite in freshwater magnetotactic bacteria. Science 203:1355–6.

Freeman EG, Dahanukar A (2015) Molecular neurobiology of *Drosophila* taste. Curr Opin Neurobiol 34:140–8.

Freitag J, Krieger J, Strotmann J, Breer H (1995) Two classes of olfactory receptors in *Xenopus laevis*. Neuron 15:1383–92.

Fuchs PA (2014) A 'calcium capacitor' shapes cholinergic inhibition of cochlear hair cells. J Physiol 592:3393–401.

Fujii S, Yavuz A, Slone J, Jagge C, Song X, Amrein H (2015) *Drosophila* sugar receptors in sweet taste perception, olfaction, and internal nutrient sensing. Curr Biol 25:621–7.

Fuortes MGF, Yeandle S (1964) Probability of occurrence of discrete potential waves in the eye of Limulus. J Gen Physiol 47:443–63.

Galizia CG, Sachse S, Rappert A, Menzel R (1999) The glomerular code for odor representation is species specific in the honeybee *Apis mellifera*. Nat Neurosci 2:473–8.

Gao Q, Yuan B, Chess A (2000) Convergent projections of *Drosophila* olfactory neurons to specific glomeruli in the antennal lobe. Nat Neurosci 3:780–5.

Garcia-Anoveros J, Samad TA, Zuvela-Jelaska L, Woolf CJ, Corey DP (2001) Transport and localization of the DEG/ENaC ion channel BNaC1alpha to peripheral mecha-nosensory terminals of dorsal root ganglia neurons. J Neurosci 21:2678–86.

Garger A, Richard EA, Lisman JE (2001) Inhibitors of gua-nylate cyclase inhibit phototransduction in limulus ventral photoreceptors. Vis Neurosci 18:625–32.

Ge J, Li W, Zhao Q, Li N, Chen M, Zhi P, Li R, Gao N, Xiao B, Yang M (2015) Architecture of the mammalian mech-anosensitive Piezo1 channel. Nature 527:64–9.

Geffeney SL, Cueva JG, Glauser DA, Doll JC, Lee TH, Montoya M, Karania S, Garakani AM, Pruitt BL, Goodman MB (2011) DEG/ENaC but not TRP channels are the major mechanoelectrical transduction channels in a *C. elegans* nociceptor. Neuron 71:845–57.

Geleoc GS, Lennan GW, Richardson GP, Kros CJ (1997) A quantitative comparison of mechanoelectrical transduc-tion in vestibular and auditory hair cells of neonatal mice. Proc R Soc Lond B Biol Sci 264:611–21.

Gesteland RC, Lettvin JY, Pitts WH (1965) Chemical trans-mission in the nose of the frog. J Physiol 181:525–59.

Getchell TV (1977) Analysis of intracellular recordings from salamander olfactory epithelium. Brain Res 123:275–86.

Ghitani N, Barik A, Szczot M, Thompson JH, Li C, Le Pichon CE, Krashes MJ, Chesler AT (2017) Specialized mechanosensory nociceptors mediating rapid responses to hair pull. Neuron 95:944–54.

Gibson F, Walsh J, Mburu P, Varela A, Brown KA, Antonio M, Beisel KW, Steel KP, Brown SD (1995) A type VII myosin encoded by the mouse deafness gene shaker-1. Nature 374:62–4.

Gilbertson TA, Roper SD, Kinnamon SC (1993) Proton cur-rents through amiloride-sensitive Na^+ channels in iso-lated hamster taste cells: enhancement by vasopressin and cAMP. Neuron 10:931–42.

Gogos JA, Osborne J, Nemes A, Mendelsohn M, Axel R (2000) Genetic ablation and restoration of the olfactory topographic map. Cell 103:609–20.

Goldstein EB (1970) Cone pigment regeneration in the iso-lated frog retina. Vision Res 10:1065–8.

Gomez MP, Nasi E (1994) The light-sensitive conductance of hyperpolarizing invertebrate photoreceptors: a patch-clamp study. J Gen Physiol 103:939–56.

Gomez MD, Nasi E (1996) Ion permeation through light-activated channels in rhabdomeric photoreceptors. Role of divalent cations. J Gen Physiol 107:715–30.

Gomez MP, Nasi E (1997a) Light adaptation in Pecten hyperpolarizing photoreceptors. Insensitivity to cal-cium manipulations. J Gen Physiol 109:371–84.

Gomez MP, Nasi E (1997b) Antagonists of the cGMP-gated conductance of vertebrate rods block the photocurrent in scallop ciliary photoreceptors. J Physiol 500:367–78.

Gomez MP, Nasi E (2000) Light transduction in inverte-brate hyperpolarizing photoreceptors: possible involvement of a Go-regulated guanylate cyclase. J Neurosci 20:5254–63.

Gomez Mdel P, Angueyra JM, Nasi E (2009) Light-transduction in melanopsin-expressing photoreceptors of *Amphioxus*. Proc Natl Acad Sci U S A 106:9081–6.

Gong Z, Son W, Chung YD, Kim J, Shin DW, McClung CA, Lee Y, Lee HW, Chang DJ, Kaang BK, Cho H, Oh U, Hirsh J, Kernan MJ, Kim C (2004) Two interdependent TRPV channel subunits, inactive and Nanchung, mediate hearing in *Drosophila*. J Neurosci 24:9059–66.

Goodman MB, Ernstrom GG, Chelur DS, O'Hagan R, Yao CA, Chalfie M (2002) MEC-2 regulates *C. elegans* DEG/ENaC channels needed for mechanosensation. Nature 415:1039–42.

Gopfert MC, Hennig RM (2016) Hearing in insects. Annu Rev Entomol 61:257–76.

Gopfert MC, Robert D (2001) Turning the key on *Drosophila* audition. Nature 411:908.

Gopfert MC, Robert D (2002) The mechanical basis of *Drosophila* audition. J Exp Biol 205:1199–208.

Gorman AL, McReynolds JS (1969) Hyperpolarizing and depolarizing receptor potentials in the scallop eye. Science 165:309–10.

Goulding EH, Tibbs GR, Siegelbaum SA (1994) Molecular mechanism of cyclic-nucleotide-gated channel activation. Nature 372:369–74.

Grabowski CT, Dethier VG (1954) The structure of the tarsal chemoreceptors of the blowfly, *Phormia regina* Meigen. J Morphol 94:1–20.

Gracheva EO, Ingolia NT, Kelly YM, Cordero-Morales JF, Hollopeter G, Chesler AT, Sanchez EE, Perez JC, Weissman JS, Julius D (2010) Molecular basis of infrared detection by snakes. Nature 464:1006–11.

Gracheva EO, Cordero-Morales JF, Gonzalez-Carcacia JA, Ingolia NT, Manno C, Aranguren CI, Weissman JS, Julius D (2011) Ganglion-specific splicing of TRPV1 underlies infrared sensation in vampire bats. Nature 476:88–91.

Graham DM, Wong KY, Shapiro P, Frederick C, Pattabiraman K, Berson DM (2008) Melanopsin ganglion cells use a membrane-associated rhabdomeric phototransduction cascade. J Neurophysiol 99:2522–32.

Grandl J, Kim SE, Uzzell V, Bursulaya B, Petrus M, Bandell M, Patapoutian A (2010) Temperature-induced opening of TRPV1 ion channel is stabilized by the pore domain. Nat Neurosci 13:708–14.

Grassé P-P (1970) Traité de Zoologie. Tome XIV Fasc. II: Reptiles: Caractères généraux et anatomie. Paris: Éditions Masson.

Grebe TW, Stock J (1998) Bacterial chemotaxis: the five sensors of a bacterium. Curr Biol 8:R154–7.

Grienberger C, Konnerth A (2012) Imaging calcium in neurons. Neuron 73:862–85.

Grosmaitre X, Vassalli A, Mombaerts P, Shepherd GM, Ma M (2006) Odorant responses of olfactory sensory neurons expressing the odorant receptor MOR23: a patch clamp analysis in gene-targeted mice. Proc Natl Acad Sci U S A 103:1970–5.

Grueber WB, Jan LY, Jan YN (2002) Tiling of the *Drosophila* epidermis by multidendritic sensory neurons. Development 129:2867–78.

Grynkiewicz G, Poenie M, Tsien RY (1985) A new generation of Ca^{2+} indicators with greatly improved fluorescence properties. J Biol Chem 260:3440–50.

Gu Y, Oberwinkler J, Postma M, Hardie RC (2005) Mechanisms of light adaptation in *Drosophila* photoreceptors. Curr Biol 15:1228–34.

Guharay F, Sachs F (1984) Stretch-activated single ion channel currents in tissue-cultured embryonic chick skeletal muscle. J Physiol 352:685–701.

Guler AD, Ecker JL, Lall GS, Haq S, Altimus CM, Liao HW, Barnard AR, Cahill H, Badea TC, Zhao H, Hankins MW, Berson DM, Lucas RJ, Yau KW, Hattar S (2008) Melanopsin cells are the principal conduits for rod-cone input to non-image-forming vision. Nature 453:102–5.

Gunther A, Einwich A, Sjulstok E, Feederle R, Bolte P, Koch KW, Solov'yov IA, Mouritsen H (2018) Double-cone localization and seasonal expression pattern suggest a role in magnetoreception for European robin cryptochrome 4. Curr Biol 28:211–23.

Guo YR, MacKinnon R (2017) Structure-based membrane dome mechanism for Piezo mechanosensitivity. eLife 6:e33660 doi:10.7554/eLife.33660.

Guth PS, Perin P, Norris CH, Valli P (1998) The vestibular hair cells: post-transductional signal processing. Prog Neurobiol 54:193–247.

Gutman GA, Chandy KG, Grissmer S, Lazdunski M, McKinnon D, Pardo LA, Robertson GA, Rudy B, Sanguinetti MC, Stuhmer W, Wang X (2005) International Union of Pharmacology. LIII. Nomenclature and molecular relationships of voltage-gated potassium channels. Pharmacol Rev 57:473–508.

Hackos DH, Korenbrot JI (1997) Calcium modulation of ligand affinity in the cyclic GMP-gated ion channels of cone photoreceptors. J Gen Physiol 110:515–28.

Haeberle H, Fujiwara M, Chuang J, Medina MM, Panditrao MV, Bechstedt S, Howard J, Lumpkin EA (2004) Molecular profiling reveals synaptic release machinery in Merkel cells. Proc Natl Acad Sci U S A 101:14503–8.

Hagiwara S, Szabo T, Enger PS (1965) Physiological properties of the electroreceptors in the electric eel, *Electrophorus electricus*. J Neurophysiol 28:775–83.

Halata Z (1975) The mechanoreceptors of the mammalian skin ultrastructure and morphological classification. Adv Anat Embryol Cell Biol 50:3–77.

Halpern BP (1998) Amiloride and vertebrate gustatory responses to NaCl. Neurosci Biobehav Rev 23:5–47.

Hamann W (1995) Mammalian cutaneous mechanoreceptors. Prog Biophys Mol Biol 64:81–104.

Hamdani el H, Doving KB (2007) The functional organization of the fish olfactory system. Prog Neurobiol 82:80–6.

Hanamori T, Miller IJ, Jr., Smith DV (1988) Gustatory responsiveness of fibers in the hamster glossopharyngeal nerve. J Neurophysiol 60:478–98.

Hanson MA, Stevens RC (2009) Discovery of new GPCR biology: one receptor structure at a time. Structure 17:8–14.

Hardie RC (1991) Voltage-sensitive potassium channels in *Drosophila* photoreceptors. J Neurosci 11:3079–95.

Hardie RC (1995) Photolysis of caged Ca^{2+} facilitates and inactivates but does not directly excite light-sensitive channels in *Drosophila* photoreceptors. J Neurosci 15:889–902.

Hardie RC (2014) Photosensitive TRPs. Handb Exp Pharmacol 223:795–826.

Hardie RC, Franze K (2012) Photomechanical responses in *Drosophila* photoreceptors. Science 338:260–3.

Hardie RC, Postma M (2008) Phototransduction in microvillar photoreceptors of *Drosophila* and other invertebrates. In: The Senses—A Comprehensive Reference. Vision. Volume 1 (Basbaum AI, Kaneko A, Shepherd GG, Westheimer G, eds), pp 77–130. Oxford: Academic Press.

Hardie RC, Voss D, Pongs O, Laughlin SB (1991) Novel potassium channels encoded by the Shaker locus in *Drosophila* photoreceptors. Neuron 6:477–86.

Hardie RC, Peretz A, Suss-Toby E, Rom-Glas A, Bishop SA, Selinger Z, Minke B (1993) Protein kinase C is required for light adaptation in *Drosophila* photoreceptors. Nature 363:634–7.

Hardie RC, Raghu P, Moore S, Juusola M, Baines RA, Sweeney ST (2001) Calcium influx via TRP channels is required to maintain PIP2 levels in *Drosophila* photoreceptors. Neuron 30:149–59.

Hardie RC, Martin F, Cochrane GW, Juusola MG, Georgiev P, Raghu P (2002) Molecular basis of amplification in *Drosophila* phototransduction: roles for G-protein, phospholipase C and diacylglycerol kinase. Neuron 36:689–701.

Hargrave PA (2001) Rhodopsin structure, function, and topography. The Friedenwald Lecture. Invest Ophthalmol Vis Sci 42:3–9.

Hargrave PA, McDowell JH, Curtis DR, Wang JK, Juszczak E, Fong SL, Rao JK, Argos P (1983) The structure of bovine rhodopsin. Biophys Struct Mech 9:235–44.

Harris GG, Frishkopf LS, Flock A (1970) Receptor potentials from hair cells of the lateral line. Science 167:76–9.

Hart NS (2001) The visual ecology of avian photoreceptors. Progr Retin Eye Res 20:675–703.

Hart S (1998) Beetle mania: an attraction to fire. Bioscience 48:3–5.

Hartline HK (1934) Intensity and duration in the excitation of single photoreceptor units. J Cellular Comparative Physiol 2:229–47.

Hartline HK (1938) The discharge of impulses in the optic nerve of *Pectin* in response to illumination of the eye. J Cellular Comparative Physiol 2:465–78.

Hartline HK, Graham CH (1932) Nerve impulses from single receptors in the eye. Journal of Cellular Comparative Physiol 1:227–95.

Hartline HK, Wagner HG, MacNichol EF, Jr (1952) The peripheral origin of nervous activity in the visual system. Cold Spr Harb Symp Quantitative Biol 17: 125–41.

Hattar S, Liao HW, Takao M, Berson DM, Yau KW (2002) Melanopsin-containing retinal ganglion cells: architecture, projections, and intrinsic photosensitivity. Science 295:1065–70.

Hattar S, Lucas RJ, Mrosovsky N, Thompson S, Douglas RH, Hankins MW, Lem J, Biel M, Hofmann F, Foster RG, Yau KW (2003) Melanopsin and rod-cone photoreceptive systems account for all major accessory visual functions in mice. Nature 424:76–81.

Hazelbauer GL (2012) Bacterial chemotaxis: the early years of molecular studies. Annu Rev Microbiol 66:285–303.

He W, Cowan CW, Wensel TG (1998) RGS9, a GTPase accelerator for phototransduction. Neuron 20:95–102.

He W, Lu L, Zhang X, El-Hodiri HM, Chen CK, Slep KC, Simon MI, Jamrich M, Wensel TG (2000) Modules in the photoreceptor RGS9-1.Gbeta 5L GTPase-accelerating protein complex control effector coupling, GTPase acceleration, protein folding, and stability. J Biol Chem 275:37093–100.

He W, Yasumatsu K, Varadarajan V, Yamada A, Lem J, Ninomiya Y, Margolskee RF, Damak S (2004) Umami taste responses are mediated by alpha-transducin and alpha-gustducin. J Neurosci 24:7674–80.

Heginbotham L, Abramson T, MacKinnon R (1992) A functional connection between the pores of distantly related ion channels as revealed by mutant K^+ channels. Science 258:1152–5.

Heiligenberg W (1973) Electrolocation of objects in the electric fish, *Eigenmannia* (Rhamphichthyidae, Gymnotoidei). J Comparative Physiol 87:137–64.

Heiligenberg W (1977) Principles of Electrolocation and Jamming Avoidance in Electric Fish: A Neuroethological Approach. Berlin: Springer.

Helmchen F, Denk W (2005) Deep tissue two-photon microscopy. Nat Method 2:932–40.

Helmholtz HLF (1877/1954) On the Sensations of Tone as a Physiological Basis for the Theory of Music, English translation of Die Lehre von den Tonempfindungen als physiologische Grundlage fur die Theorie der Musik, Fourth Edition. New York: Dover.

Hensel H (1973) Cutaneous thermoreceptors. In: Somatosensory System, Handbook of Sensory Physiology Vol. III (Iggo A, ed), pp 79–110. Berlin: Springer.

Hensel H, Andres KH, von During M (1974) Structure and function of cold receptors. Pflugers Arch 352:1–10.

Herman KG (1991) Light-stimulated rhabdom turnover in *Limulus* ventral photoreceptors maintained in vitro. J Comp Neurol 303:11–21.

Herrada G, Dulac C (1997) A novel family of putative pheromone receptors in mammals with a topographically organized and sexually dimorphic distribution. Cell 90:763–73.

Hestrin S (1987) The properties and function of inward rectification in rod photoreceptors of the tiger salamander. J Physiol 390:319–33.

Hestrin S, Korenbrot JI (1990) Activation kinetics of retinal cones and rods: response to intense flashes of light. J Neurosci 10:1967–73.

Hildebrand JG (1995) Analysis of chemical signals by the nervous system. Proc Natl Acad Sci U S A 92:67–74.

Hill KG (1983) The physiology of locust auditory receptors. II. Membrane potentials associated with the response of the receptor cell. J Comparative Physiol 152:483–93.

Hisatsune C, Yasumatsu K, Takahashi-Iwanaga H, Ogawa N, Kuroda Y, Yoshida R, Ninomiya Y, Mikoshiba K (2007) Abnormal taste perception in mice lacking the type 3 inositol 1,4,5-trisphosphate receptor. J Biol Chem 282:37225–31.

Hodgkin AL, McNaughton PA, Nunn BJ (1987) Measurement of sodium-calcium exchange in salamander rods. J Physiol 391:347–70.

Hodgkin NM, Bryant PJ (1978) Scanning electron microscopy of the adult of *Drosophila melanogaster*. In: The Genetics and Biology of *Drosophila*, Vol. 2c (Ashburner M, Wright TRF, eds), pp 337–58. London: Academic Press.

Hodgson ES, Lettvin JY, Roeder KD (1955) Physiology of a primary chemoreceptor unit. Science 122:417–18.

Hoff WD, Jung KH, Spudich JL (1997) Molecular mechanism of photosignaling by archaeal sensory rhodopsins. Annu Rev Biophys Biomol Struct 26:223–58.

Hoffman BU, Baba Y, Griffith TN, Mosharov EV, Woo SH, Roybal DD, Karsenty G, Patapoutian A, Sulzer D, Lumpkin EA (2018) Merkel cells activate sensory neural pathways through adrenergic synapses. Neuron 100:1401–1413.

Hofmann KP, Pulvermuller A, Buczylko J, Van Hooser P, Palczewski K (1992) The role of arrestin and retinoids in the regeneration pathway of rhodopsin. J Biol Chem 267:15701–6.

Holt JR, Corey DP, Eatock RA (1997) Mechanoelectrical transduction and adaptation in hair cells of the mouse utricle, a low-frequency vestibular organ. J Neurosci 17:8739–48.

Holt JR, Gillespie SK, Provance DW, Shah K, Shokat KM, Corey DP, Mercer JA, Gillespie PG (2002) A chemical-genetic strategy implicates myosin-1c in adaptation by hair cells. Cell 108:371–81.

Holton T, Hudspeth AJ (1986) The transduction channel of hair cells from the bull-frog characterized by noise analysis. J Physiol 375:195–227.

Hong K, Mano I, Driscoll M (2000) In vivo structure-function analyses of *Caenorhabditis elegans* MEC-4, a candidate mechanosensory ion channel subunit. J Neurosci 20:2575–88.

Hood DC, Hock PA (1973) Recovery of cone receptor activity in the frog's isolated retina. Vision Res 13:1943–51.

Hoon MA, Adler E, Lindemeier J, Battey JF, Ryba NJ, Zuker CS (1999) Putative mammalian taste receptors: a class of taste-specific GPCRs with distinct topographic selectivity. Cell 96:541–51.

Hore PJ, Mouritsen H (2016) The radical-pair mechanism of magnetoreception. Annu Rev Biophys 45:299–344.

Horio N, Yoshida R, Yasumatsu K, Yanagawa Y, Ishimaru Y, Matsunami H, Ninomiya Y (2011) Sour taste responses in mice lacking PKD channels. PLoS One 6:e20007.

Howard J, Ashmore JF (1986) Stiffness of sensory hair bundles in the sacculus of the frog. Hear Res 23:93–104.

Howard J, Hudspeth AJ (1987) Mechanical relaxation of the hair bundle mediates adaptation in mechanoelectrical transduction by the bullfrog's saccular hair cell. Proc Natl Acad Sci U S A 84:3064–8.

Howard J, Hudspeth AJ (1988) Compliance of the hair bundle associated with gating of mechanoelectrical transduction channels in the bullfrog's saccular hair cell. Neuron 1:189–99.

Hoy RR, Robert D (1996) Tympanal hearing in insects. Annu Rev Entomol 41:433–50.

Hsu YT, Molday RS (1993) Modulation of the cGMP-gated channel of rod photoreceptor cells by calmodulin. Nature 361:76–9.

Hu G, Wensel TG (2002) R9AP, a membrane anchor for the photoreceptor GTPase accelerating protein, RGS9-1. Proc Natl Acad Sci U S A 99:9755–60.

Hu J, Zhong C, Ding C, Chi Q, Walz A, Mombaerts P, Matsunami H, Luo M (2007) Detection of near-atmospheric concentrations of CO_2 by an olfactory subsystem in the mouse. Science 317:953–7.

Huang AL, Chen X, Hoon MA, Chandrashekar J, Guo W, Trankner D, Ryba NJ, Zuker CS (2006) The cells and logic for mammalian sour taste detection. Nature 442:934–8.

Huang J, Liu CH, Hughes SA, Postma M, Schwiening CJ, Hardie RC (2010) Activation of TRP channels by protons and phosphoinositide depletion in *Drosophila* photoreceptors. Curr Biol 20:189–97.

Huang L, Shanker YG, Dubauskaite J, Zheng JZ, Yan W, Rosenzweig S, Spielman AI, Max M, Margolskee RF (1999) Ggamma13 colocalizes with gustducin in taste receptor cells and mediates IP3 responses to bitter denatonium. Nat Neurosci 2:1055–62.

Huang YA, Maruyama Y, Stimac R, Roper SD (2008) Presynaptic (type III) cells in mouse taste buds sense sour (acid) taste. J Physiol 586:2903–12.

Huber A, Sander P, Gobert A, Bahner M, Hermann R, Paulsen R (1996) The transient receptor potential protein (Trp), a putative store-operated Ca^{2+} channel essential for phosphoinositide-mediated photoreception, forms a signaling complex with NorpA, InaC and InaD. Embo J 15:7036–45.

Hudspeth AJ (1982) Extracellular current flow and the site of transduction by vertebrate hair cells. J Neurosci 2:1–10.

Hudspeth AJ (1989) How the ear's works work. Nature 341:397–404.

Hudspeth AJ (2014) Integrating the active process of hair cells with cochlear function. Nat Rev Neurosci 15:600–14.

Hudspeth AJ, Corey DP (1977) Sensitivity, polarity, and conductance change in the response of vertebrate hair cells to controlled mechanical stimuli. Proc Natl Acad Sci U S A 74:2407–11.

Hughes HC (1999) Sensory Exotica. Cambridge, MA: MIT Press.

Hughes S, Hankins MW, Foster RG, Peirson SN (2012) Melanopsin phototransduction: slowly emerging from the dark. Prog Brain Res 199:19–40.

Iggo A, Muir AR (1969) The structure and function of a slowly adapting touch corpuscle in hairy skin. J Physiol 200:763–96.

Ikeda R, Cha M, Ling J, Jia Z, Coyle D, Gu JG (2014) Merkel cells transduce and encode tactile stimuli to drive Abeta-afferent impulses. Cell 157:664–75.

Ingram NT, Sampath AP, Fain GL (2016) Why are rods more sensitive than cones? J Physiol 594:5415–26.

Ishida AT, Fain GL (1981) D-aspartate potentiates the effects of L-glutamate on horizontal cells in goldfish retina. Proc Natl Acad Sci U S A 78:5890–4.

Ishimaru Y, Inada H, Kubota M, Zhuang H, Tominaga M, Matsunami H (2006) Transient receptor potential family members PKD1L3 and PKD2L1 form a candidate sour taste receptor. Proc Natl Acad Sci U S A 103:12569–74.

Isom LL, De Jongh KS, Catterall WA (1994) Auxiliary subunits of voltage-gated ion channels. Neuron 12:1183–94.

Jager S, Palczewski K, Hofmann KP (1996) Opsin/all-trans-retinal complex activates transducin by different mechanisms than photolyzed rhodopsin. Biochemistry 35:2901–8.

Jan LY, Jan YN (2012) Voltage-gated potassium channels and the diversity of electrical signalling. J Physiol 590:2591–9.

Jänig W (2018) Peripheral thermoreceptors in innocuous temperature detection. Handb Clin Neurol 156:47–56.

Jaramillo F, Hudspeth AJ (1991) Localization of the hair cell's transduction channels at the hair bundle's top by iontophoretic application of a channel blocker. Neuron 7:409–20.

Jennings HS (1906) Behavior of the Lower Organisms, reprinted in 1962 with an introduction by D. D. Jensen. Bloomington: Indiana University Press.

Jeon JH, Paik SS, Chun MH, Oh U, Kim IB (2013) Presynaptic localization and possible function of calcium-activated chloride channel anoctamin 1 in the mammalian retina. PLoS One 8:e67989.

Jia C, Halpern M (1996) Subclasses of vomeronasal receptor neurons: differential expression of G proteins (Gi alpha 2 and G(o alpha)) and segregated projections to the accessory olfactory bulb. Brain Res 719:117–28.

Jiang Y, Ruta V, Chen J, Lee A, MacKinnon R (2003b) The principle of gating charge movement in a voltage-dependent K^+ channel. Nature 423:42–8.

Jiang Z, Yue WWS, Chen L, Sheng Y, Yau KW (2018) Cyclic-nucleotide- and HCN-channel-mediated phototransduction in intrinsically photosensitive retinal ganglion cells. Cell 175:652–664.

Jin M, Li S, Moghrabi WN, Sun H, Travis GH (2005) Rpe65 is the retinoid isomerase in bovine retinal pigment epithelium. Cell 122:449–59.

Jin P, Bulkley D, Guo Y, Zhang W, Guo Z, Huynh W, Wu S, Meltzer S, Cheng T, Jan LY, Jan YN, Cheng Y (2017) Electron cryo-microscopy structure of the mechanotransduction channel NOMPC. Nature 547:118–22.

Johansen TK (1997) Aristotle on the Sense Organs. Cambridge: Cambridge University Press.

Johnson BA, Leon M (2007) Chemotopic odorant coding in a mammalian olfactory system. J Comp Neurol 503:1–34.

Johnson EC, Robinson PR, Lisman JE (1986) Cyclic GMP is involved in the excitation of invertebrate photoreceptors. Nature 324:468–70.

Johnson KO, Hsiao SS (1992) Neural mechanisms of tactual form and texture perception. Annu Rev Neurosci 15:227–50.

Johnson SL, Beurg M, Marcotti W, Fettiplace R (2011) Prestin-driven cochlear amplification is not limited by the outer hair cell membrane time constant. Neuron 70:1143–54.

Jones DT, Reed RR (1989) Golf: an olfactory neuron specific-G protein involved in odorant signal transduction. Science 244:790–5.

Jones GJ, Crouch RK, Wiggert B, Cornwall MC, Chader GJ (1989) Retinoid requirements for recovery of sensitivity after visual-pigment bleaching in isolated photoreceptors. Proc Natl Acad Sci U S A 86:9606–10.

Juilfs DM, Fulle HJ, Zhao AZ, Houslay MD, Garbers DL, Beavo JA (1997) A subset of olfactory neurons that selectively express cGMP-stimulated phosphodiesterase (PDE2) and guanylyl cyclase-D define a unique olfactory signal transduction pathway. Proc Natl Acad Sci U S A 94:3388–95.

Julius D (2013) TRP channels and pain. Annu Rev Cell Dev Biol 29:355–84.

Kachar B, Parakkal M, Kurc M, Zhao Y, Gillespie PG (2000) High-resolution structure of hair-cell tip links. Proc Natl Acad Sci U S A 97:13336–41.

Kadamur G, Ross EM (2012) Mammalian phospholipase C. Annu Rev Physiol 75:127-54. doi:10.1146/annurev-physiol-030212-183750.

Kaissling K-E (1971) Insect olfaction. In: Olfaction, Handbook of Sensory Physiology Vol. IV, Part I (Beidler LM, ed), pp 351–431. Berlin: Springer.

Kaissling K-E, Priesner E (1970) Die Riechschwelle des Seidenspinners. Naturwissenschaften 57:23–8.

Kajiya K, Inaki K, Tanaka M, Haga T, Kataoka H, Touhara K (2001) Molecular bases of odor discrimination: reconstitution of olfactory receptors that recognize overlapping sets of odorants. J Neurosci 21:6018–25.

Kalinec F, Holley MC, Iwasa KH, Lim DJ, Kachar B (1992) A membrane-based force generation mechanism in auditory sensory cells. Proc Natl Acad Sci U S A 89:8671–5.

Kalmijn AJ (1966) Electro-perception in sharks and rays. Nature 212:1232–3.

Kalmijn AJ (1971) The electric sense of sharks and rays. J Experiment Biol 55:371–83.

Kamikouchi A, Ishikawa Y (2016) Hearing in *Drosophila*. In: Insect Hearing (Pollack GS, Mason AC, Popper AN, Fay RR, eds), pp 239–61. Berlin: Springer.

Kamikouchi A, Shimada T, Ito K (2006) Comprehensive classification of the auditory sensory projections in the brain of the fruit fly *Drosophila melanogaster*. J Comp Neurol 499:317–56.

Kamikouchi A, Inagaki HK, Effertz T, Hendrich O, Fiala A, Gopfert MC, Ito K (2009) The neural basis of *Drosophila* gravity-sensing and hearing. Nature 458:165–71.

Kang L, Gao J, Schafer WR, Xie Z, Xu XZ (2010) *C. elegans* TRP family protein TRP-4 is a pore-forming subunit of a native mechanotransduction channel. Neuron 67: 381–91.

Karavitaki KD, Corey DP (2010) Sliding adhesion confers coherent motion to hair cell stereocilia and parallel gating to transduction channels. J Neurosci 30:9051–63.

Kashio M, Tominaga M (2017) The TRPM2 channel: a thermo-sensitive metabolic sensor. Channels (Austin) 11:426–33.

Kaupp UB (2010) Olfactory signalling in vertebrates and insects: differences and commonalities. Nat Rev Neurosci 11:188–200.

Kaupp UB, Seifert R (2002) Cyclic nucleotide-gated ion channels. Physiol Rev 82:769–824.

Kaupp UB, Niidome T, Tanabe T, Terada S, Bonigk W, Stuhmer W, Cook NJ, Kangawa K, Matsuo H, Hirose T, et al. (1989) Primary structure and functional expression from complementary DNA of the rod photoreceptor cyclic GMP-gated channel. Nature 342:762–6.

Kawai F, Horiguchi M, Suzuki H, Miyachi E (2001) Na(+) action potentials in human photoreceptors. Neuron 30:451–8.

Kawamura S (1993) Rhodopsin phosphorylation as a mechanism of cyclic GMP phosphodiesterase regulation by S-modulin. Nature 362:855–7.

Kawashima Y, Geleoc GS, Kurima K, Labay V, Lelli A, Asai Y, Makishima T, Wu DK, Della Santina CC, Holt JR, Griffith AJ (2011) Mechanotransduction in mouse inner ear hair cells requires transmembrane channel-like genes. J Clin Invest 121:4796–809.

Kazmierczak P, Sakaguchi H, Tokita J, Wilson-Kubalek EM, Milligan RA, Muller U, Kachar B (2007) Cadherin 23 and protocadherin 15 interact to form tip-link filaments in sensory hair cells. Nature 449:87–91.

Keil TA (1997) Functional morphology of insect mechanoreceptors. Microsc Res Tech 39:506–31.

Keller A, Zhuang H, Chi Q, Vosshall LB, Matsunami H (2007) Genetic variation in a human odorant receptor alters odour perception. Nature 449:468–72.

Kennedy HJ, Crawford AC, Fettiplace R (2005) Force generation by mammalian hair bundles supports a role in cochlear amplification. Nature 433:880–3.

Kennedy MJ, Lee KA, Niemi GA, Craven KB, Garwin GG, Saari JC, Hurley JB (2001) Multiple phosphorylation of rhodopsin and the in vivo chemistry underlying rod photoreceptor dark adaptation. Neuron 31:87–101.

Kernan M, Cowan D, Zuker C (1994) Genetic dissection of mechanosensory transduction: mechanoreception-defective mutations of *Drosophila*. Neuron 12:1195–206.

Kiang NY-S (1965) Discharge patterns of single fibers in the cat's auditory nerve. Cambridge, MA: MIT Press.

Kim J, Chung YD, Park DY, Choi S, Shin DW, Soh H, Lee HW, Son W, Yim J, Park CS, Kernan MJ, Kim C (2003) A TRPV family ion channel required for hearing in *Drosophila*. Nature 424:81–4.

Kim MS, Smith DP (2001) The invertebrate odorant-binding protein LUSH is required for normal olfactory behavior in *Drosophila*. Chem Senses 26:195–9.

Kim MS, Repp A, Smith DP (1998) LUSH odorant-binding protein mediates chemosensory responses to alcohols in *Drosophila melanogaster*. Genetics 150:711–21.

Kim S, Ma L, Yu CR (2011) Requirement of calcium-activated chloride channels in the activation of mouse vomeronasal neurons. Nat Commun 2:365.

Kim SE, Coste B, Chadha A, Cook B, Patapoutian A (2012) The role of *Drosophila* Piezo in mechanical nociception. Nature 483:209–12.

Kimchi T, Xu J, Dulac C (2007) A functional circuit underlying male sexual behaviour in the female mouse brain. Nature 448:1009–14.

Kimple AJ, Bosch DE, Giguere PM, Siderovski DP (2011) Regulators of G-protein signaling and their Galpha substrates: promises and challenges in their use as drug discovery targets. Pharmacol Rev 63:728–49.

Kindt KS, Finch G, Nicolson T (2012) Kinocilia mediate mechanosensitivity in developing zebrafish hair cells. Dev Cell 23:329–41.

King BL, Shi LF, Kao P, Clusin WT (2016) Calcium activated K(+) channels in the electroreceptor of the skate confirmed by cloning. Details of subunits and splicing. Gene 578:63–73.

Kirkwood A, Weiner D, Lisman JE (1989) An estimate of the number of G regulator proteins activated per excited rhodopsin in living *Limulus* ventral photoreceptors. Proc Natl Acad Sci U S A 86:3872–6.

Kiselev A, Socolich M, Vinos J, Hardy RW, Zuker CS, Ranganathan R (2000) A molecular pathway for light-dependent photorecptor apoptosis in *Drosophila*. Neuron 28:139–52.

Kitagawa M, Kusakabe Y, Miura H, Ninomiya Y, Hino A (2001) Molecular genetic identification of a candidate receptor gene for sweet taste. Biochem Biophys Res Commun 283:236–42.

Knowles A, Dartnall HJA (1977) The Eye, vol. 2B: The Photobiology of Vision. New York: Academic Press.

Koch KW, Stryer L (1988) Highly cooperative feedback control of retinal rod guanylate cyclase by calcium ions. Nature 334:64–6.

Kohn E, Katz B, Yasin B, Peters M, Rhodes E, Zaguri R, Weiss S, Minke B (2015) Functional cooperation between the IP3 receptor and phospholipase C secures the high sensitivity to light of *Drosophila* photoreceptors in vivo. J Neurosci 35:2530–46.

Kojima D, Terakita A, Ishikawa T, Tsukahara Y, Maeda A, Shichida Y (1997) A novel Go-mediated phototransduction cascade in scallop visual cells. J Biol Chem 272: 22979–82.

Korschen HG, Illing M, Seifert R, Sesti F, Williams A, Gotzes S, Colville C, Muller F, Dose A, Godde M, et al. (1995) A 240 kDa protein represents the complete beta subunit of the cyclic nucleotide-gated channel from rod photoreceptor. Neuron 15:627–36.

Korschen HG, Beyermann M, Muller F, Heck M, Vantler M, Koch KW, Kellner R, Wolfrum U, Bode C, Hofmann KP, Kaupp UB (1999) Interaction of glutamic-acid-rich proteins with the cGMP signalling pathway in rod photoreceptors. Nature 400:761–6.

Koutalos Y, Nakatani K, Yau KW (1995) The cGMP-phosphodiesterase and its contribution to sensitivity regulation in retinal rods. J Gen Physiol 106:891–921.

Krautwurst D, Yau KW, Reed RR (1998) Identification of ligands for olfactory receptors by functional expression of a receptor library. Cell 95:917–26.

Kretz O, Barbry P, Bock R, Lindemann B (1999) Differential expression of RNA and protein of the three pore-forming subunits of the amiloride-sensitive epithelial sodium channel in taste buds of the rat. J Histochem Cytochem 47:51–64.

Krieg M, Dunn AR, Goodman MB (2014) Mechanical control of the sense of touch by beta-spectrin. Nat Cell Biol 16:224–33.

Krispel CM, Chen D, Melling N, Chen YJ, Martemyanov KA, Quillinan N, Arshavsky VY, Wensel TG, Chen CK, Burns ME (2006) RGS expression rate-limits recovery of rod photoresponses. Neuron 51:409–16.

Krizaj D, Copenhagen DR (1998) Compartmentalization of calcium extrusion mechanisms in the outer and inner segments of photoreceptors. Neuron 21:249–56.

Kroese AB, van Netten SM (1989) Sensory transduction in lateral line hair cells. In: The Mechanosensory Lateral Line. Neurobiology and Evolution (Coombs S, Görner P, Münz H, eds), pp 265–84. Berlin: Springer.

Kros CJ, Marcotti W, van Netten SM, Self TJ, Libby RT, Brown SD, Richardson GP, Steel KP (2002) Reduced climbing and increased slipping adaptation in cochlear hair cells of mice with Myo7a mutations. Nat Neurosci 5:41–7.

Kuhlbrandt W (2004) Biology, structure and mechanism of P-type ATPases. Nat Rev Mol Cell Biol 5:282–95.

Kuhn H (1978) Light-regulated binding of rhodopsin kinase and other proteins to cattle photoreceptor membranes. Biochemistry 17:4389–95.

Kuhn H, Dreyer WJ (1972) Light dependent phosphorylation of rhodopsin by ATP. FEBS Lett 20:1–6.

Kuhn H, Cook JH, Dreyer WJ (1973) Phosphorylation of rhodopsin in bovine photoreceptor membranes. A dark reaction after illumination. Biochemistry 12:2495–502.

Kuhn H, Bennett N, Michel-Villaz M, Chabre M (1981) Interactions between photoexcited rhodopsin and GTP-binding protein: kinetic and stoichiometric analyses from light-scattering changes. Proc Natl Acad Sci U S A 78:6873–7.

Kung C, Martinac B, Sukharev S (2010) Mechanosensitive channels in microbes. Annu Rev Microbiol 64:313–29.

Kurahashi T, Menini A (1997) Mechanism of odorant adaptation in the olfactory receptor cell. Nature 385:725–9.

Kürten L, Schmidt U (1981) Thermoreception in the common vampirre bat. J Comparative Physiol 146A:223–8.

Kusuhara Y, Yoshida R, Ohkuri T, Yasumatsu K, Voigt A, Hubner S, Maeda K, Boehm U, Meyerhof W, Ninomiya Y (2013) Taste responses in mice lacking taste receptor subunit T1R1. J Physiol 591:1967–85.

Laberge F, Hara TJ (2001) Neurobiology of fish olfaction: a review. Brain Res Brain Res Rev 36:46–59.

Lacalli TC (2004) Sensory systems in amphioxus: a window on the ancestral chordate condition. Brain Behav Evol 64:148–62.

Lagnado L, McNaughton PA (1990) Electrogenic properties of the Na:Ca exchange. J Membr Biol 113:177–91.

Lalonde MR, Kelly ME, Barnes S (2008) Calcium-activated chloride channels in the retina. Channels (Austin) 2:252–60.

Lamb TD, Pugh EN, Jr. (2004) Dark adaptation and the retinoid cycle of vision. Progr Retin Eye Res 23:307–80.

Lazzerini Ospri L, Prusky G, Hattar S (2017) Mood, the circadian system, and melanopsin retinal ganglion cells. Annu Rev Neurosci 40:539–56.

Leal WS (2013) Odorant reception in insects: roles of receptors, binding proteins, and degrading enzymes. Annu Rev Entomol 58:373–91.

Lee JK, Strausfeld NJ (1990) Structure, distribution and number of surface sensilla and their receptor cells on the olfactory appendage of the male moth *Manduca sexta*. J Neurocytol 19:519–38.

Lehnert BP, Baker AE, Gaudry Q, Chiang AS, Wilson RI (2013) Distinct roles of TRP channels in auditory transduction and amplification in *Drosophila*. Neuron 77:115–28.

Leinders-Zufall T, Ma M, Zufall F (1999) Impaired odor adaptation in olfactory receptor neurons after inhibition of Ca^{2+}/calmodulin kinase II. J Neurosci 19:RC19.

Leinders-Zufall T, Lane AP, Puche AC, Ma W, Novotny MV, Shipley MT, Zufall F (2000) Ultrasensitive pheromone detection by mammalian vomeronasal neurons. Nature 405:792–6.

Leinders-Zufall T, Cockerham RE, Michalakis S, Biel M, Garbers DL, Reed RR, Zufall F, Munger SD (2007) Contribution of the receptor guanylyl cyclase GC-D to chemosensory function in the olfactory epithelium. Proc Natl Acad Sci U S A 104:14507–12.

Leskov IB, Klenchin VA, Handy JW, Whitlock GG, Govardovskii VI, Bownds MD, Lamb TD, Pugh EN, Jr., Arshavsky VY (2000) The gain of rod phototransduction: reconciliation of biochemical and electrophysiological measurements. Neuron 27:525–37.

Lesniak DR, Marshall KL, Wellnitz SA, Jenkins BA, Baba Y, Rasband MN, Gerling GJ, Lumpkin EA (2014) Computation identifies structural features that govern neuronal firing properties in slowly adapting touch receptors. eLife 3:e01488.

Levai O, Feistel T, Breer H, Strotmann J (2006) Cells in the vomeronasal organ express odorant receptors but project to the accessory olfactory bulb. J Comp Neurol 498:476–90.

Lewandowski BC, Sukumaran SK, Margolskee RF, Bachmanov AA (2016) Amiloride-insensitive salt taste is mediated by two populations of type III taste cells with distinct transduction mechanisms. J Neurosci 36:1942–53.

Leypold BG, Yu CR, Leinders-Zufall T, Kim MM, Zufall F, Axel R (2002) Altered sexual and social behaviors in trp2 mutant mice. Proc Natl Acad Sci U S A 99:6376–81.

Li HS, Montell C (2000) TRP and the PDZ protein, INAD, form the core complex required for retention of the signalplex in *Drosophila* photoreceptor cells. J Cell Biol 150:1411–22.

Li J, Mahajan A, Tsai MD (2006) Ankyrin repeat: a unique motif mediating protein-protein interactions. Biochemistry 45:15168–78.

Li L, Rutlin M, Abraira VE, Cassidy C, Kus L, Gong S, Jankowski MP, Luo W, Heintz N, Koerber HR, Woodbury CJ, Ginty DD (2011) The functional organization of cutaneous low-threshold mechanosensory neurons. Cell 147:1615–27.

Li M, Zhou X, Wang S, Michailidis I, Gong Y, Su D, Li H, Li X, Yang J (2017) Structure of a eukaryotic cyclic-nucleotide-gated channel. Nature 542:60–5.

Li RC, Ben-Chaim Y, Yau KW, Lin CC (2016) Cyclic-nucleotide-gated cation current and Ca^{2+}-activated Cl current elicited by odorant in vertebrate olfactory receptor neurons. Proc Natl Acad Sci U S A 113:11078–87.

Li, R.C., Lin, C.C., Ren, X., Wu, J.S., Molday, L.L., Molday, R.S., and Yau, K.W. (2018). Ca(2+)-activated Cl current predominates in threshold response of mouse olfactory receptor neurons. Proc Natl Acad Sci U S A 115: 5570-5575.

Liang X, Madrid J, Gartner R, Verbavatz JM, Schiklenk C, Wilsch-Brauninger M, Bogdanova A, Stenger F, Voigt A, Howard J (2013) A NOMPC-dependent membrane-microtubule connector is a candidate for the gating spring in fly mechanoreceptors. Curr Biol 23:755–63.

Liang Y, Fotiadis D, Filipek S, Saperstein DA, Palczewski K, Engel A (2003) Organization of the G protein-coupled receptors rhodopsin and opsin in native membranes. J Biol Chem 278:21655–62.

Liao M, Cao E, Julius D, Cheng Y (2013) Structure of the TRPV1 ion channel determined by electron cryo-microscopy. Nature 504:107–12.

Liberles SD (2014) Mammalian pheromones. Annu Rev Physiol 76:151–75.

Liberles SD, Buck LB (2006) A second class of chemosensory receptors in the olfactory epithelium. Nature 442:645–50.

Liberles SD, Horowitz LF, Kuang D, Contos JJ, Wilson KL, Siltberg-Liberles J, Liberles DA, Buck LB (2009) Formyl peptide receptors are candidate chemosensory receptors in the vomeronasal organ. Proc Natl Acad Sci U S A 106:9842–7.

Liberman MC, Gao J, He DZ, Wu X, Jia S, Zuo J (2002) Prestin is required for electromotility of the outer hair cell and for the cochlear amplifier. Nature 419:300–4.

Lieberman SJ, Hamasaki T, Satir P (1988) Ultrastructure and motion analysis of permeabilized *Paramecium* capable of motility and regulation of motility. Cell Motil Cytoskeleton 9:73–84.

Liman ER, Corey DP, Dulac C (1999) TRP2: a candidate transduction channel for mammalian pheromone sensory signaling. Proc Natl Acad Sci U S A 96:5791–6.

Liman ER, Zhang YV, Montell C (2014) Peripheral coding of taste. Neuron 81:984–1000.

Ling G, Gerard RW (1949) The membrane potential and metabolism of muscle fibers. J Cell Comp Physiol 34:413–38.

Lishko PV, Procko E, Jin X, Phelps CB, Gaudet R (2007) The ankyrin repeats of TRPV1 bind multiple ligands and modulate channel sensitivity. Neuron 54:905–18.

Lisman JE, Brown JE (1975) Effects of intracellular injection of calcium buffers on light adaptation in *Limulus* ventral photoreceptors. J Gen Physiol 66:489–506.

Lisman JE, Richard EA, Raghavachari S, Payne R (2002) Simultaneous roles for Ca²⁺ in excitation and adaptation in *Limulus* ventral photoreceptors. Adv Exp Med Biol 514:507–38.

Lissmann HW (1951) Continuous electrical signals from the tail of a fish, *Gymnarchus niloticus* Cuv. Nature 167:201–2.

Liu CH, Satoh AK, Postma M, Huang J, Ready DF, Hardie RC (2008) Ca²⁺-dependent metarhodopsin inactivation mediated by calmodulin and NINAC myosin III. Neuron 59:778–89.

Lloyd G (1975) Alcmaeon and the early history of dissection. Sudhoffs Archiv 59:113–47.

Loewenstein O, Wersäll J (1959) A functional interpretation of the electron-microscopic structure of the sensory hairs in the cristae of the elasmobranch *Raja clavata* in terms of directional sensitivity. Nature 184:1807–8.

Loewenstein WR (1971) Mechano-electric transduction in the Pacinian corpuscle. In: Principles of Receptor Physiology, Handbook of Sensory Physiology Vol. I (Loewenstein WR, ed), pp 269–90. Berlin: Springer.

Loewenstein WR, Altamirano-Orrego R (1958) The refractory state of the generator and propagated potentials in a pacinian corpuscle. J Gen Physiol 41:805–24.

Loewenstein WR, Mendelson M (1965) Components of receptor adaptation in a Pacinian corpuscle. J Physiol 177:377–97.

Loewenstein WR, Rathkamp R (1958) The sites for mechano-electric conversion in a Pacinian corpuscle. J Gen Physiol 41:1245–65.

Long SB, Campbell EB, Mackinnon R (2005a) Crystal structure of a mammalian voltage-dependent Shaker family K⁺ channel. Science 309:897–903.

Long SB, Campbell EB, Mackinnon R (2005b) Voltage sensor of Kv1.2: structural basis of electromechanical coupling. Science 309:903–8.

Long SB, Tao X, Campbell EB, MacKinnon R (2007) Atomic structure of a voltage-dependent K⁺ channel in a lipid membrane-like environment. Nature 450:376–82.

Longrigg J (1988) Anatomy in Alexandria in the third century BC. Br J Hist Sci 21:455–88.

Lowe G, Gold GH (1991) The spatial distributions of odorant sensitivity and odorant-induced currents in salamander olfactory receptor cells. J Physiol 442:147–68.

Lucas P, Ukhanov K, Leinders-Zufall T, Zufall F (2003) A diacylglycerol-gated cation channel in vomeronasal neuron dendrites is impaired in TRPC2 mutant mice: mechanism of pheromone transduction. Neuron 40:551–61.

Lucas RJ (2013) Mammalian inner retinal photoreception. Curr Biol 23:R125–33.

Ludwig J, Oliver D, Frank G, Klocker N, Gummer AW, Fakler B (2001) Reciprocal electromechanical properties of rat prestin: the motor molecule from rat outer hair cells. Proc Natl Acad Sci U S A 98:4178–83.

Luecke H, Schobert B, Lanyi JK, Spudich EN, Spudich JL (2001) Crystal structure of sensory rhodopsin II at 2.4 angstroms: insights into color tuning and transducer interaction. Science 293:1499–503.

Lumpkin EA, Hudspeth AJ (1995) Detection of Ca²⁺ entry through mechanosensitive channels localizes the site of mechanoelectric transduction in hair cells. Proc Natl Acad Sci U S A 92:10297–301.

Luo DG, Yue WW, Ala-Laurila P, Yau KW (2011) Activation of visual pigments by light and heat. Science 332:1307–12.

Lyall V, Alam RI, Phan DQ, Ereso GL, Phan TH, Malik SA, Montrose MH, Chu S, Heck GL, Feldman GM, DeSimone JA (2001) Decrease in rat taste receptor cell intracellular pH is the proximate stimulus in sour taste transduction. Am J Physiol Cell Physiol 281:C1005–13.

Madej MG, Ziegler CM (2018) Dawning of a new era in TRP channel structural biology by cryo-electron microscopy. Pflugers Arch 470:213–25.

Maison SF, Usubuchi H, Liberman MC (2013) Efferent feedback minimizes cochlear neuropathy from moderate noise exposure. J Neurosci 33:5542–52.

Makino ER, Handy JW, Li T, Arshavsky VY (1999) The GTPase activating factor for transducin in rod photoreceptors is the complex between RGS9 and type 5 G protein beta subunit. Proc Natl Acad Sci U S A 96:1947–52.

Maksimovic S, Nakatani M, Baba Y, Nelson AM, Marshall KL, Wellnitz SA, Firozi P, Woo SH, Ranade S, Patapoutian A, Lumpkin EA (2014) Epidermal Merkel cells are mechanosensory cells that tune mammalian touch receptors. Nature 509:617–21.

Malnic B (2007) Searching for the ligands of odorant receptors. Molec Neurobiol 35:175–81.

Malnic B, Hirono J, Sato T, Buck LB (1999) Combinatorial receptor codes for odors. Cell 96:713–23.

Malnic B, Gonzalez-Kristeller DC, Gutiyama LM (2010) Odorant receptors. In: The Neurobiology of Olfaction (Menini A, ed). Boca Raton, FL: CRC Press/Taylor Francis.

Mammano F, Ashmore JF (1996) Differential expression of outer hair cell potassium currents in the isolated cochlea of the guinea-pig. J Physiol 496:639–46.

Margolskee RF (2002) Molecular mechanisms of bitter and sweet taste transduction. J Biol Chem 277:1–4.

Maricich SM, Wellnitz SA, Nelson AM, Lesniak DR, Gerling GJ, Lumpkin EA, Zoghbi HY (2009) Merkel cells are essential for light-touch responses. Science 324:1580–2.

Maricich SM, Morrison KM, Mathes EL, Brewer BM (2012) Rodents rely on Merkel cells for texture discrimination tasks. J Neurosci 32:3296–300.

Martinac B, Buechner M, Delcour AH, Adler J, Kung C (1987) Pressure-sensitive ion channel in *Escherichia coli*. Proc Natl Acad Sci U S A 84:2297–301.

Martini S, Silvotti L, Shirazi A, Ryba NJ, Tirindelli R (2001) Co-expression of putative pheromone receptors in the sensory neurons of the vomeronasal organ. J Neurosci 21:843–8.

Mata NL, Ruiz A, Radu RA, Bui TV, Travis GH (2005) Chicken retinas contain a retinoid isomerase activity that catalyzes the direct conversion of all-trans-retinol to 11-cis-retinol. Biochemistry 44:11715–21.

Matsumoto H, Kurien BT, Takagi Y, Kahn ES, Kinumi T, Komori N, Yamada T, Hayashi F, Isono K, Pak WL, et al. (1994) Phosrestin I undergoes the earliest light-induced phosphorylation by a calcium/calmodulin-dependent protein kinase in *Drosophila* photoreceptors. Neuron 12:997–1010.

Matsunami H, Buck LB (1997) A multigene family encoding a diverse array of putative pheromone receptors in mammals. Cell 90:775–84.

Matsunami H, Montmayeur JP, Buck LB (2000) A family of candidate taste receptors in human and mouse. Nature 404:601–4.

Matsuyama T, Yamashita T, Imamoto Y, Shichida Y (2012) Photochemical properties of mammalian melanopsin. Biochemistry 51:5454–62.

Matthews HR, Fain GL (2001) A light-dependent increase in free Ca^{2+} concentration in the salamander rod outer segment. J Physiol 532:305–21.

Matthews HR, Murphy RL, Fain GL, Lamb TD (1988) Photoreceptor light adaptation is mediated by cytoplasmic calcium concentration. Nature 334:67–9.

Matthews HR, Fain GL, Murphy RL, Lamb TD (1990) Light adaptation in cone photoreceptors of the salamander: a role for cytoplasmic calcium. J Physiol 420:447–69.

Max M, Shanker YG, Huang L, Rong M, Liu Z, Campagne F, Weinstein H, Damak S, Margolskee RF (2001) Tas1r3, encoding a new candidate taste receptor, is allelic to the sweet responsiveness locus Sac. Nat Genet 28:58–63.

Mazzolini M, Marchesi A, Giorgetti A, Torre V (2010) Gating in CNGA1 channels. Pflugers Arch 459:547–55.

McCusker EC, Bagneris C, Naylor CE, Cole AR, D'Avanzo N, Nichols CG, Wallace BA (2012) Structure of a bacterial voltage-gated sodium channel pore reveals mechanisms of opening and closing. Nat Commun 3:1102.

McFadden J, Al-Khalili J (2014) Life on the Edge. New York: Broadway Books.

McIver SB (1985) Mechanoreception. In: Comprehensive Insect Physiology, Biochemistry and Pharmacology. Vol. 6, Nervous System: Sensory (Kerkut GA, Gilbert LI, eds), pp 71–132. Oxford: Pergamon.

McKemy DD (2013) The molecular and cellular basis of cold sensation. ACS Chem Neurosci 4:238–47.

McKemy DD, Neuhausser WM, Julius D (2002) Identification of a cold receptor reveals a general role for TRP channels in thermosensation. Nature 416:52–8.

McLaughlin SK, McKinnon PJ, Margolskee RF (1992) Gustducin is a taste-cell-specific G protein closely related to the transducins. Nature 357:563–9.

McReynolds JS, Gorman AL (1970a) Photoreceptor potentials of opposite polarity in the eye of the scallop, *Pecten irradians*. J Gen Physiol 56:376–91.

McReynolds JS, Gorman AL (1970b) Membrane conductances and spectral sensitivities of *Pecten* photoreceptors. J Gen Physiol 56:392–406.

McReynolds JS, Gorman AL (1974) Ionic basis of hyperpolarizing receptor potential in scallop eye: increase in permeability to potassium ions. Science 183:658–9.

Meech RW, Standen NB (1975) Potassium activation in *Helix aspersa* neurones under voltage clamp: a component mediated by calcium influx. J Physiol 249:211–39.

Meech RW, Strumwasser F (1970) Intracellular calcium injection activates potassium conductance in *Aplysia* nerve cells. Fed Proc 29:834a.

Mendez A, Burns ME, Roca A, Lem J, Wu LW, Simon MI, Baylor DA, Chen J (2000) Rapid and reproducible deactivation of rhodopsin requires multiple phosphorylation sites. Neuron 28:153–64.

Mendez A, Burns ME, Sokal I, Dizhoor AM, Baehr W, Palczewski K, Baylor DA, Chen J (2001) Role of guanylate cyclase-activating proteins (GCAPs) in setting the flash sensitivity of rod photoreceptors. Proc Natl Acad Sci U S A 98:9948–53.

Meyerhoff M (1928) The Book of the Ten Treatises on the Eye of Hunain Ibn Is-Haq. Cairo: Government Press.

Michael AK, Fribourgh JL, Van Gelder RN, Partch CL (2017) Animal cryptochromes: divergent roles in light perception, circadian timekeeping and beyond. Photochem Photobiol 93:128–40.

Michalakis S, Zong X, Becirovic E, Hammelmann V, Wein T, Wanner KT, Biel M (2011) The glutamic acid-rich protein is a gating inhibitor of cyclic nucleotide-gated channels. J Neurosci 31:133–41.

Millecchia R, Mauro A (1969a) The ventral photoreceptor cells of *Limulus*. 3. A voltage-clamp study. J Gen Physiol 54:331–51.

Millecchia R, Mauro A (1969b) Ventral photoreceptor cells of *Limulus* II. J Gen Physiol 54:310–30.

Miller AN, Long SB (2012) Crystal structure of the human two-pore domain potassium channel K2P1. Science 335:432–6.

Miller MR (1978) Scanning electron microscope studies of the papilla basilaris of some turtles and snakes. Am J Anat 151:409–35.

Ming D, Ruiz-Avila L, Margolskee RF (1998) Characterization and solubilization of bitter-responsive receptors that couple to gustducin. Proc Natl Acad Sci U S A 95:8933–8.

Minke B, Hardie RC (2000) Genetic dissection of *Drosophila* phototransduction. In: Handbook of Biological Physics, vol. 3 (Stavenga DG, DeGrip WJ, Pugh EN, Jr, eds), pp 449–525. Amsterdam: Elsevier Science.

Miyawaki A (2016) Roger Y. Tsien (1952–2016). Cell 167:298–300.

Molday RS (1998) Photoreceptor membrane proteins, phototransduction, and retinal degenerative diseases. The Friedenwald Lecture. Invest Ophthalmol Vis Sci 39:2491–513.

Molday RS, Kaupp UB (2000) Ion channels of vertebrate photoreceptors. In: Handbook of Biological Physics, Vol. 3 (Stavenga DG, DeGrip WJ, Pugh EN, Jr, eds), pp 143–81. Amsterdam: Elsevier Science.

Molday RS, Zhong M, Quazi F (2009) The role of the photoreceptor ABC transporter ABCA4 in lipid transport and Stargardt macular degeneration. Biochim Biophys Acta 1791:573–83.

Moller P (1995) Electric Fishes. New York: Chapman and Hall.

Mombaerts P (1999a) Seven-transmembrane proteins as odorant and chemosensory receptors. Science 286:707–11.

Mombaerts P (1999b) Molecular biology of odorant receptors in vertebrates. Annu Rev Neurosci 22:487–509.

Mombaerts P (2004) Genes and ligands for odorant, vomeronasal and taste receptors. Nat Rev Neurosci 5:263–78.

Mombaerts P, Wang F, Dulac C, Chao SK, Nemes A, Mendelsohn M, Edmondson J, Axel R (1996) Visualizing an olfactory sensory map. Cell 87:675–86.

Montealegre ZF, Jonsson T, Robson-Brown KA, Postles M, Robert D (2012) Convergent evolution between insect and mammalian audition. Science 338:968–71.

Montell C (1998) TRP trapped in fly signaling web. Curr Opin Neurobiol 8:389–97.

Montell C (1999) Visual transduction in *Drosophila*. Annu Rev Cell Dev Biol 15:231–68.

Montmayeur JP, Liberles SD, Matsunami H, Buck LB (2001) A candidate taste receptor gene near a sweet taste locus. Nat Neurosci 4:492–8.

Morgans CW, El Far O, Berntson A, Wassle H, Taylor WR (1998) Calcium extrusion from mammalian photoreceptor terminals. J Neurosci 18:2467–74.

Mori K, Takahashi YK, Igarashi KM, Yamaguchi M (2006) Maps of odorant molecular features in the mammalian olfactory bulb. Physiol Rev 86:409–33.

Morita H (1959) Initiation of spike potentials in contact chemosensory hairs of insects. III. D.C. stimulation and generator potential of labellar chemoreceptor of *Calliphora*. J Cell Comp Physiol 54:177–87.

Morrison EE, Costanzo RM (1990) Morphology of the human olfactory epithelium. J Comp Neurol 297:1–13.

Morshedian A, Fain GL (2015) Single-photon sensitivity of lamprey rods with cone-like outer segments. Curr Biol 25:484–7.

Morshedian A, Fain GL (2017) Light adaptation and the evolution of vertebrate photoreceptors. J Physiol 595:4947–60.

Morshedian, A, Kaylor, JJ, Ng, SY, Tsan, A, Frederiksen, R, Xu, T, Yuan, L, Sampath, AP, Radu, RA, Fain, GL, Travis, GH. (2019). Light-Driven Regeneration of Cone Visual Pigments through a Mechanism Involving RGR Opsin in Muller Glial Cells. Neuron *102*, 1172-1183 e1175. doi: 10.1016/j.neuron.2019.04.004.

Morshedian A, Woodruff ML, Fain GL (2018) Role of recoverin in rod photoreceptor light adaptation. J Physiol 596:1513–26.

Morth JP, Pedersen BP, Buch-Pedersen MJ, Andersen JP, Vilsen B, Palmgren MG, Nissen P (2011) A structural overview of the plasma membrane Na$^+$,K$^+$-ATPase and H+-ATPase ion pumps. Nat Rev Mol Cell Biol 12:60–70.

Mouritsen H (2018) Long-distance navigation and magnetoreception in migratory animals. Nature 558:50–9.

Mowbray SL, Sandgren MO (1998) Chemotaxis receptors: a progress report on structure and function. J Struct Biol 124:257–75.

Mueller KL, Hoon MA, Erlenbach I, Chandrashekar J, Zuker CS, Ryba NJ (2005) The receptors and coding logic for bitter taste. Nature 434:225–9.

Munger SD, Leinders-Zufall T, Zufall F (2009) Subsystem organization of the mammalian sense of smell. Annu Rev Physiol 71:115–40.

Murakami M, Kijima H (2000) Transduction ion channels directly gated by sugars on the insect taste cell. J Gen Physiol 115:455–66.

Murray RG, Murray A, Fujimoto S (1969) Fine structure of gustatory cells in rabbit taste buds. J Ultrastruct Res 27:444–61.

Murray RW (1965) Electroreceptor mechanisms: the relation of impulse frequency to stimulus strength and responses to pulsed stimuli in the ampullae of Lorenzini of elasmobranchs. J Physiol 180:592–606.

Murray RW (1974) The ampullae of Lorenzini. In: Handbook of Sensory Physiology, Vol. III/3, Electroreceptors and Other Specialized Receptors in Lower Vertebrates (Fessard A, ed), pp 125–46. Berlin: Springer.

Murthy SE, Dubin AE, Patapoutian A (2017) Piezos thrive under pressure: mechanically activated ion channels in health and disease. Nat Rev Mol Cell Biol 18:771–83.

Murthy VN (2011) Olfactory maps in the brain. Annu Rev Neurosci 34:233–58.

Naitoh Y, Eckert R (1969) Ionic mechanisms controlling behavioral responses of *Paramecium* to mechanical stimulation. Science 164:963–5.

Naitoh Y, Eckert R (1973) Sensory mechanisms in *Paramecium*. II. Ionic basis of the hyperpolarizing mechanoreceptor potential. J Exp Biol 59:53–65.

Naitoh Y, Kaneko H (1972) Reactivated triton-extracted models of paramecium: modification of ciliary movement by calcium ions. Science 176:523–4.

Nakajima S, Onodera K (1969a) Membrane properties of the stretch receptor neurones of crayfish with particular reference to mechanisms of sensory adaptation. J Physiol 200:161–85.

Nakajima S, Onodera K (1969b) Adaptation of the generator potential in the crayfish stretch receptors under constant length and constant tension. J Physiol 200:187–204.

Nakamura T, Gold GH (1987a) A cyclic nucleotide-gated conductance in olfactory receptor cilia. Nature 325:442–4.

Nakamura T, Gold GH (1987b) A cyclic nucleotide-gated conductance in olfactory receptor cilia. Nature 325:442–4.

Nakatani K, Yau KW (1988a) Calcium and magnesium fluxes across the plasma membrane of the toad rod outer segment. J Physiol 395:695–729.

Nakatani K, Yau KW (1988b) Calcium and light adaptation in retinal rods and cones. Nature 334:69–71.

Nara K, Saraiva LR, Ye X, Buck LB (2011) A large-scale analysis of odor coding in the olfactory epithelium. J Neurosci 31:9179–91.

Nasi E, Gomez MD (1991) Light-activated channels in scallop photoreceptors: recordings from cell-attached and perfused excised patches. Biophys J 59:540a.

Nasi E, Gomez MD, Payne R (2000) Phototransduction mechanisms in microvillar and ciliary photoreceptors of invertebrates. In: Handbook of Biological Physics, Vol. 3 (Stavenga DG, DeGrip WJ, Pugh EN, Jr, eds), pp 389–448. Amsterdam: Elsevier Science.

Nathans J, Thomas D, Hogness DS (1986) Molecular genetics of human color vision: the genes encoding blue, green, and red pigments. Science 232:193–202.

Navaratnam DS, Bell TJ, Tu TD, Cohen EL, Oberholtzer JC (1997) Differential distribution of Ca^{2+} activated K^+ channel splice variants among hair cells along the tonotopic axis of the chick cochlea. Neuron 19:1077–85.

Neher E, Sakmann B (1976) Single-channel currents recorded from membrane of denervated frog muscle fibers. Nature 260:799–802.

Nelson G, Hoon MA, Chandrashekar J, Zhang Y, Ryba NJ, Zuker CS (2001) Mammalian sweet taste receptors. Cell 106:381–90.

Nelson G, Chandrashekar J, Hoon MA, Feng L, Zhao G, Ryba NJ, Zuker CS (2002) An amino-acid taste receptor. Nature 416:199–202.

Nelson TM, Lopezjimenez ND, Tessarollo L, Inoue M, Bachmanov AA, Sullivan SL (2010) Taste function in mice with a targeted mutation of the pkd1l3 gene. Chem Senses 35:565–77.

Neuhaus EM, Gisselmann G, Zhang W, Dooley R, Stortkuhl K, Hatt H (2005) Odorant receptor heterodimerization in the olfactory system of *Drosophila melanogaster*. Nat Neurosci 8:15–17.

Newman EA, Hartline HK (1982) The infrared "vision" of snakes. Sci Am 246:116–27.

Ngai J, Dowling MM, Buck L, Axel R, Chess A (1993) The family of genes encoding odorant receptors in the channel catfish. Cell 72:657–66.

Nickle B, Robinson PR (2007) The opsins of the vertebrate retina: insights from structural, biochemical, and evolutionary studies. Cell Mol Life Sci 64:2917–32.

Nie Y, Vigues S, Hobbs JR, Conn GL, Munger SD (2005) Distinct contributions of T1R2 and T1R3 taste receptor subunits to the detection of sweet stimuli. Curr Biol 15:1948–52.

Niemeyer BA, Suzuki E, Scott K, Jalink K, Zuker CS (1996) The *Drosophila* light-activated conductance is composed of the two channels TRP and TRPL. Cell 85:651–9.

Nießner C, Denzau S, Gross JC, Peichl L, Bischof HJ, Fleissner G, Wiltschko W, Wiltschko R (2011) Avian ultraviolet/violet cones identified as probable magnetoreceptors. PLoS One 6:e20091.

Niimura Y, Nei M (2005) Evolutionary dynamics of olfactory receptor genes in fishes and tetrapods. Proc Natl Acad Sci U S A 102:6039–44.

Nin F, Yoshida T, Sawamura S, Ogata G, Ota T, Higuchi T, Murakami S, Doi K, Kurachi Y, Hibino H (2016) The unique electrical properties in an extracellular fluid of the mammalian cochlea; their functional roles, homeostatic processes, and pathological significance. Pflugers Arch 468:1637–49.

Nobili R, Mammano F, Ashmore J (1998) How well do we understand the cochlea? Trends Neurosci 21:159–67.

Noble GK, Schmidt A (1937) The structure and function of the facial and labial pits of snakes. Proc Am Philos Soc 77:263–88.

Nolte J, Brown JE (1969) The spectral sensitivities of single cells in the median ocellus of *Limulus*. J Gen Physiol 54:636–49.

Nunn BJ, Schnapf JL, Baylor DA (1984) Spectral sensitivity of single cones in the retina of *Macaca fascicularis*. Nature 309:264–6.

Obara S, Bennett MVL (1972) Mode of operation of ampullae of Lorenzini of skate, *Raja*. J Gen Physiol 60:534–57.

O'Hagan R, Chalfie M, Goodman MB (2005) The MEC-4 DEG/ENaC channel of *Caenorhabditis elegans* touch receptor neurons transduces mechanical signals. Nat Neurosci 8:43–50.

Ohkuma M, Kawai F, Horiguchi M, Miyachi E (2007) Patch-clamp recording of human retinal photoreceptors and bipolar cells. Photochem Photobiol 83:317–22.

Ohmori H (1985) Mechano-electrical transduction currents in isolated vestibular hair cells of the chick. J Physiol 359:189–217.

Ohmori H (1988) Mechanical stimulation and Fura-2 fluorescence in the hair bundle of dissociated hair cells of the chick. J Physiol 399:115–37.

Ohshima K, Hirai S, Nishida A, Hiramatsu K (1999) Ultrastructure and serotonin immunocytochemistry of the parietal-pineal complex in the Japanese grass lizard, *Takydromus tachydromoides*. Tissue Cell 31:126–37.

Ohyama T, Hackos DH, Frings S, Hagen V, Kaupp UB, Korenbrot JI (2000) Fraction of the dark current carried by Ca(2+) through cGMP-gated ion channels of intact rod and cone photoreceptors. J Gen Physiol 116:735–54.

Oka Y, Butnaru M, von Buchholtz L, Ryba NJ, Zuker CS (2013) High salt recruits aversive taste pathways. Nature 494:472–5.

Okada T, Le Trong I, Fox BA, Behnke CA, Stenkamp RE, Palczewski K (2000) X-ray diffraction analysis of three-dimensional crystals of bovine rhodopsin obtained from mixed micelles. J Struct Biol 130:73–80.

Okada T, Ernst OP, Palczewski K, Hofmann KP (2001a) Activation of rhodopsin: new insights from structural and biochemical studies. Trends Biochem Sci 26:318–24.

Okada Y, Fujiyama R, Miyamoto T, Sato T (2001b) Saccharin activates cation conductance via inositol 1,4,5-trisphosphate production in a subset of isolated rod taste cells in the frog. Eur J Neurosci 13:308–14.

Okazawa M, Takao K, Hori A, Shiraki T, Matsumura K, Kobayashi S (2002) Ionic basis of cold receptors acting as thermostats. J Neurosci 22:3994–4001.

Oliver D, He DZ, Klocker N, Ludwig J, Schulte U, Waldegger S, Ruppersberg JP, Dallos P, Fakler B (2001) Intracellular anions as the voltage sensor of prestin, the outer hair cell motor protein. Science 292:2340–3.

Olt J, Allen CE, Marcotti W (2016) In vivo physiological recording from the lateral line of juvenile zebrafish. J Physiol 594:5427–38.

Ortega A, Zhulin IB, Krell T (2017) Sensory repertoire of bacterial chemoreceptors. Microbiol Mol Biol Rev 81:e00033-17.

Owens DM, Lumpkin EA (2014) Diversification and specialization of touch receptors in skin. Cold Spring Harb Perspect Med 4(6). pii:a013656.

Pace U, Hanski E, Salomon Y, Lancet D (1985) Odorant-sensitive adenylate cyclase may mediate olfactory reception. Nature 316:255–8.

Pak WL (1995) *Drosophila* in vision research. The Friedenwald Lecture. Invest Ophthalmol Vis Sci 36:2340–57.

Palczewski K, Buczylko J, Ohguro H, Annan RS, Carr SA, Crabb JW, Kaplan MW, Johnson RS, Walsh KA (1994) Characterization of a truncated form of arrestin isolated from bovine rod outer segments. Protein Sci 3:314–24.

Palczewski K, Kumasaka T, Hori T, Behnke CA, Motoshima H, Fox BA, Le Trong I, Teller DC, Okada T, Stenkamp RE, Yamamoto M, Miyano M (2000) Crystal structure of rhodopsin: a G protein-coupled receptor. Science 289:739–45.

Palkar R, Lippoldt EK, McKemy DD (2015) The molecular and cellular basis of thermosensation in mammals. Curr Opin Neurobiol 34:14–19.

Pan B, Geleoc GS, Asai Y, Horwitz GC, Kurima K, Ishikawa K, Kawashima Y, Griffith AJ, Holt JR (2013) TMC1 and TMC2 are components of the mechanotransduction channel in hair cells of the mammalian inner ear. Neuron 79:504–15.

Pan B, Akyuz N, Liu XP, Asai Y, Nist-Lund C, Kurima K, Derfler BH, Gyorgy B, Limapichat W, Walujkar S, Wimalasena LN, Sotomayor M, Corey DP, Holt JR (2018) TMC1 forms the pore of mechanosensory transduction channels in vertebrate inner ear hair cells. Neuron 99:736–53.

Pantages E, Dulac C (2000) A novel family of candidate pheromone receptors in mammals. Neuron 28:835–45.

Papazian DM, Schwarz TL, Tempel BL, Jan YN, Jan LY (1987) Cloning of genomic and complementary DNA from Shaker, a putative potassium channel gene from *Drosophila*. Science 237:749–53.

Parkinson JS, Hazelbauer GL, Falke JJ (2015) Signaling and sensory adaptation in *Escherichia coli* chemoreceptors: 2015 update. Trends Microbiol 23:257–66.

Payandeh J, Scheuer T, Zheng N, Catterall WA (2011) The crystal structure of a voltage-gated sodium channel. Nature 475:353–8.

Payne R, Demas J (2000) Timing of Ca(2+) release from intracellular stores and the electrical response of *Limulus* ventral photoreceptors to dim flashes. J Gen Physiol 115:735–48.

Payne R, Walz B, Levy S, Fein A (1988) The localization of calcium release by inositol trisphosphate in *Limulus* photoreceptors and its control by negative feedback. Philos Trans R Soc Lond B Biol Sci 320:359–79.

Payne R, Flores TM, Fein A (1990) Feedback inhibition by calcium limits the release of calcium by inositol trisphosphate in *Limulus* ventral photoreceptors. Neuron 4:547–55.

Pedemonte N, Galietta LJ (2014) Structure and function of TMEM16 proteins (anoctamins). Physiol Rev 94:419–59.

Peier AM, Moqrich A, Hergarden AC, Reeve AJ, Andersson DA, Story GM, Earley TJ, Dragoni I, McIntyre P, Bevan S, Patapoutian A (2002) A TRP channel that senses cold stimuli and menthol. Cell 108:705–15.

Peinado G, Osorno T, Gomez Mdel P, Nasi E (2015) Calcium activates the light-dependent conductance in melanopsin-expressing photoreceptors of amphioxus. Proc Natl Acad Sci U S A 112:7845–50.

Peng AW, Effertz T, Ricci AJ (2013) Adaptation of mammalian auditory hair cell mechanotransduction is independent of calcium entry. Neuron 80:960–72.

Penn RD, Hagins WA (1969) Signal transmission along retinal rods and the origin of the electroretinographic a-wave. Nature 223:201–4.

Perry RJ, McNaughton PA (1991) Response properties of cones from the retina of the tiger salamander. J Physiol 433:561–87.

Pfaff DW, editor (1985) Taste, Olfaction, and the Central Nervous System. A Festschrift in Honor of Carl Pfaffmann. New York: Rockefeller University Press.

Pfaffmann C (1939) Specific gustatory impulses. J Physiol 96:41P–42P.

Phillips AM, Bull A, Kelly LE (1992) Identification of a *Drosophila* gene encoding a calmodulin-binding protein with homology to the trp phototransduction gene. Neuron 8:631–42.

Pickles JO (1988) An Introduction to the Physiology of Hearing, Second Edition. London: Academic Press.

Pickles JO, Comis SD, Osborne MP (1984) Cross-links between stereocilia in the guinea pig organ of Corti, and their possible relation to sensory transduction. Hear Res 15:103–12.

Polans A, Baehr W, Palczewski K (1996) Turned on by Ca^{2+}! The physiology and pathology of Ca(2+)-binding proteins in the retina. Trends Neurosci 19:547–54.

Pollack GS, Balakrishnan R (1997) Taste sensilla of flies: function, central neuronal projections, and development. Microsc Res Tech 39:532–46.

Polyak SL (1941) The Retina. Chicago: University of Chicago Press.

Poo M, Cone RA (1974) Lateral diffusion of rhodopsin in the photoreceptor membrane. Nature 247:438–41.

Potter SM, Zheng C, Koos DS, Feinstein P, Fraser SE, Mombaerts P (2001) Structure and emergence of specific olfactory glomeruli in the mouse. J Neurosci 21:9713–23.

Prakriya M, Lewis RS (2015) Store-operated calcium channels. Physiol Rev 95:1383–436.

Probst FJ, Fridell RA, Raphael Y, Saunders TL, Wang A, Liang Y, Morell RJ, Touchman JW, Lyons RH, Noben-Trauth K, Friedman TB, Camper SA (1998) Correction of deafness in shaker-2 mice by an unconventional myosin in a BAC transgene. Science 280:1444–7.

Pugh EN, Jr., Lamb TD (1993) Amplification and kinetics of the activation steps in phototransduction. Biochim Biophys Acta 1141:111–49.

Qiu X, Muller U (2018) Mechanically gated ion channels in mammalian hair cells. Front Cell Neurosci 12:100.

Quattrochi LE, Stabio ME, Kim I, Ilardi MC, Michelle Fogerson P, Leyrer ML, Berson DM (2018) The M6 cell: a small-field bistratified photosensitive retinal ganglion cell. J Comp Neurol 527(1):297–311.

Raghu P, Usher K, Jonas S, Chyb S, Polyanovsky A, Hardie RC (2000a) Constitutive activity of the light-sensitive channels TRP and TRPL in the *Drosophila* diacylglycerol kinase mutant, rdgA. Neuron 26:169–79.

Raghu P, Colley NJ, Webel R, James T, Hasan G, Danin M, Selinger Z, Hardie RC (2000b) Normal phototransduction in *Drosophila* photoreceptors lacking an InsP(3) receptor gene. Mol Cell Neurosci 15:429–45.

Ranade SS, Woo SH, Dubin AE, Moshourab RA, Wetzel C, Petrus M, Mathur J, Begay V, Coste B, Mainquist J, Wilson AJ, Francisco AG, Reddy K, Qiu Z, Wood JN, Lewin GR, Patapoutian A (2014) Piezo2 is the major transducer of mechanical forces for touch sensation in mice. Nature 516:121–5.

Ranganathan R, Harris GL, Stevens CF, Zuker CS (1991) A *Drosophila* mutant defective in extracellular calcium-dependent photoreceptor deactivation and rapid desensitization. Nature 354:230–2.

Rao VR, Oprian DD (1996) Activating mutations of rhodopsin and other G protein-coupled receptors. Annu Rev Biophys Biomol Struct 25:287–314.

Rasmussen SG, Choi HJ, Rosenbaum DM, Kobilka TS, Thian FS, Edwards PC, Burghammer M, Ratnala VR, Sanishvili R, Fischetti RF, Schertler GF, Weis WI, Kobilka BK (2007) Crystal structure of the human beta2 adrenergic G-protein-coupled receptor. Nature 450:383–7.

Rasmussen SG, DeVree BT, Zou Y, Kruse AC, Chung KY, Kobilka TS, Thian FS, Chae PS, Pardon E, Calinski D, Mathiesen JM, Shah ST, Lyons JA, Caffrey M, Gellman SH, Steyaert J, Skiniotis G, Weis WI, Sunahara RK, Kobilka BK (2011) Crystal structure of the beta2 adrenergic receptor-Gs protein complex. Nature 477:549–55.

Rebrik TI, Korenbrot JI (1998) In intact cone photoreceptors, a Ca^{2+}-dependent, diffusible factor modulates the cGMP-gated ion channels differently than in rods. J Gen Physiol 112:537–48.

Rebrik TI, Kotelnikova EA, Korenbrot JI (2000) Time course and Ca(2+) dependence of sensitivity modulation in cyclic GMP-gated currents of intact cone photoreceptors. J Gen Physiol 116:521–34.

Reed DR, Nanthakumar E, North M, Bell C, Bartoshuk LM, Price RA (1999) Localization of a gene for bitter-taste perception to human chromosome 5p15. Am J Hum Genet 64:1478–80.

Reid G, Flonta ML (2001) Physiology. Cold current in thermoreceptive neurons. Nature 413:480.

Reingruber J, Pahlberg J, Woodruff ML, Sampath AP, Fain GL, Holcman D (2013) Detection of single photons by toad and mouse rods. Proc Natl Acad Sci U S A 110:19378–83.

Reingruber J, Holcman D, Fain GL (2015) How rods respond to single photons: key adaptations of a G-protein cascade that enable vision at the physical limit of perception. Bioessays 37:1243–52.

Reisert J, Matthews HR (1998) Na+-dependent Ca^{2+} extrusion governs response recovery in frog olfactory receptor cells. J Gen Physiol 112:529–35.

Reisert J, Matthews HR (2001) Simultaneous recording of receptor current and intraciliary Ca^{2+} concentration in salamander olfactory receptor cells. J Physiol 535:637–45.

Reisert J, Reingruber J (2018) Ca^{2+}-activated Cl^- current ensures robust and reliable signal amplification in vertebrate olfactory receptor neurons. Proc Natl Acad Sci U S A 116(3):1053–8.

Reisert J, Zhao H (2011) Perspectives on: information and coding in mammalian sensory physiology: response kinetics of olfactory receptor neurons and the implications in olfactory coding. J Gen Physiol 138:303–10.

Reisert J, Lai J, Yau KW, Bradley J (2005) Mechanism of the excitatory Cl^- response in mouse olfactory receptor neurons. Neuron 45:553–61.

Reiter E, Ahn S, Shukla AK, Lefkowitz RJ (2012) Molecular mechanism of beta-arrestin-biased agonism at seven-transmembrane receptors. Annu Rev Pharmacol Toxicol 52:179–97.

Ressler KJ, Sullivan SL, Buck LB (1994) Information coding in the olfactory system: evidence for a stereotyped and highly organized epitope map in the olfactory bulb. Cell 79:1245–55.

Reuss H, Mojet MH, Chyb S, Hardie RC (1997) In vivo analysis of the *Drosophila* light-sensitive channels, TRP and TRPL. Neuron 19:1249–59.

Ricci AJ, Gray-Keller M, Fettiplace R (2000) Tonotopic variations of calcium signalling in turtle auditory hair cells. J Physiol 524 Pt 2:423–36.

Ricci AJ, Crawford AC, Fettiplace R (2002) Mechanisms of active hair bundle motion in auditory hair cells. J Neurosci 22:44–52.

Ricci AJ, Crawford AC, Fettiplace R (2003) Tonotopic variation in the conductance of the hair cell mechanotransducer channel. Neuron 40:983–90.

Richard EA, Lisman JE (1992) Rhodopsin inactivation is a modulated process in *Limulus* photoreceptors. Nature 356:336–8.

Richardson GP, de Monvel JB, Petit C (2011) How the genetics of deafness illuminates auditory physiology. Annu Rev Physiol 73:311–34.

Riesgo-Escovar JR, Piekos WB, Carlson JR (1997) The *Drosophila* antenna: ultrastructural and physiological studies in wild-type and lozenge mutants. J Comp Physiol 180:151–60.

Riviere S, Challet L, Fluegge D, Spehr M, Rodriguez I (2009) Formyl peptide receptor-like proteins are a novel family of vomeronasal chemosensors. Nature 459:574–7.

Roberts WM, Jacobs RA, Hudspeth AJ (1990) Colocalization of ion channels involved in frequency selectivity and synaptic transmission at presynaptic active zones of hair cells. J Neurosci 10:3664–84.

Robinson PR, Cohen GB, Zhukovsky EA, Oprian DD (1992) Constitutively active mutants of rhodopsin. Neuron 9:719–25.

Rodriguez I, Feinstein P, Mombaerts P (1999) Variable patterns of axonal projections of sensory neurons in the mouse vomeronasal system. Cell 97:199–208.

Rodriguez I, Punta KD, Rothman A, Ishii T, Mombaerts P (2002) Multiple new and isolated families within the mouse superfamily of V1r vomeronasal receptors. Nat Neurosci 5:134–40.

Roeder KD (1967) Nerve Cells and Insect Behavior. Cambridge MA: Harvard University Press.

Ronnett GV, Moon C (2002) G proteins and olfactory signal transduction. Annu Rev Physiol 64:189–222.

Roper SD (2007) Signal transduction and information processing in mammalian taste buds. Pflugers Arch 454:759–76.

Roper SD, Chaudhari N (2017) Taste buds: cells, signals and synapses. Nat Rev Neurosci 18:485–97.

Rosenbaum DM, Rasmussen SG, Kobilka BK (2009) The structure and function of G-protein-coupled receptors. Nature 459:356–63.

Rosenblatt KP, Sun ZP, Heller S, Hudspeth AJ (1997) Distribution of Ca^{2+}-activated K^+ channel isoforms along the tonotopic gradient of the chicken's cochlea. Neuron 19:1061–75.

Ross EM, Wilkie TM (2000) GTPase-activating proteins for heterotrimeric G proteins: regulators of G protein signaling (RGS) and RGS-like proteins. Annu Rev Biochem 69:795–827.

Rubin BD, Katz LC (1999) Optical imaging of odorant representations in the mammalian olfactory bulb. Neuron 23:499–511.

Ruiz M, Karpen JW (1999) Opening mechanism of a cyclic nucleotide-gated channel based on analysis of single channels locked in each liganded state. J Gen Physiol 113:873–95.

Ruiz ML, Karpen JW (1997) Single cyclic nucleotide-gated channels locked in different ligand-bound states. Nature 389:389–92.

Ruiz-Avila L, Wong GT, Damak S, Margolskee RF (2001) Dominant loss of responsiveness to sweet and bitter compounds caused by a single mutation in alpha-gustducin. Proc Natl Acad Sci U S A 98:8868–73.

Russell IJ, Sellick PM (1983) Low-frequency characteristics of intracellularly recorded receptor potentials in guinea-pig cochlear hair cells. J Physiol 338:179–206.

Russell IJ, Kossl M, Richardson GP (1992) Nonlinear mechanical responses of mouse cochlear hair bundles. Proc R Soc Lond B Biol Sci 250:217–27.

Ryan AF, Dallos P (1984) Physiology of the cochlea. In: Hearing Disorders, Second Edition (Northern JL, ed), pp 253–66. Boston: Little Brown and Company.

Ryba NJ, Tirindelli R (1997) A new multigene family of putative pheromone receptors. Neuron 19:371–9.

Rydqvist B, Purali N (1993) Transducer properties of the rapidly adapting stretch receptor neurone in the crayfish (*Pacifastacus leniusculus*). J Physiol 469:193–211.

Rydqvist B, Purali N, Lannergren J (1994) Visco-elastic properties of the rapidly adapting stretch receptor muscle of the crayfish. Acta Physiol Scand 150:151–9.

Rydqvist B, Lin JH, Sand P, Swerup C (2007) Mechanotransduction and the crayfish stretch receptor. Physiol Behav 92:21–8.

Rytz R, Croset V, Benton R (2013) Ionotropic receptors (IRs): chemosensory ionotropic glutamate receptors in *Drosophila* and beyond. Insect Biochem Mol Biol 43:888–97.

Sagoo MS, Lagnado L (1997) G-protein deactivation is rate-limiting for shut-off of the phototransduction cascade. Nature 389:392–5.

Sainz E, Korley JN, Battey JF, Sullivan SL (2001) Identification of a novel member of the T1R family of putative taste receptors. J Neurochem 77:896–903.

Saito H, Kubota M, Roberts RW, Chi Q, Matsunami H (2004) RTP family members induce functional expression of mammalian odorant receptors. Cell 119:679–91.

Saito H, Chi Q, Zhuang H, Matsunami H, Mainland JD (2009) Odor coding by a mammalian receptor repertoire. Sci Signal 2:ra9.

Sakai K, Imamoto Y, Su CY, Tsukamoto H, Yamashita T, Terakita A, Yau KW, Shichida Y (2012) Photochemical nature of parietopsin. Biochemistry 51:1933–41.

Sakmann B, Neher E, eds (1995) Single-Channel Recording. New York: Plenum.

Sakmar TP (1998) Rhodopsin: a prototypical G protein-coupled receptor. Prog Nucleic Acid Res Mol Biol 59:1–34.

Sakmar TP, Menon ST, Marin EP, Awad ES (2002) RHODOPSIN: insights from recent structural studies. Annu Rev Biophys Biomol Struct 31:443–84.

Sampath AP, Matthews HR, Cornwall MC, Fain GL (1998) Bleached pigment produces a maintained decrease in outer segment Ca^{2+} in salamander rods. J Gen Physiol 111:53–64.

Sampath AP, Matthews HR, Cornwall MC, Bandarchi J, Fain GL (1999) Light-dependent changes in outer segment free-Ca^{2+} concentration in salamander cone photoreceptors. J Gen Physiol 113:267–77.

Sand O (1975) Effects of different ionic environments on the mechano-sensitivity of lateral line organs in the mudpuppy. J Comp Physiol 102:27–42.

Santos-Sacchi J (1991) Reversible inhibition of voltage-dependent outer hair cell motility and capacitance. J Neurosci 11:3096–110.

Santos-Sacchi J, Tan W (2018) The frequency response of outer hair cell voltage-dependent motility is limited by kinetics of prestin. J Neurosci 38:5495–506.

Saotome K, Murthy SE, Kefauver JM, Whitwam T, Patapoutian A, Ward AB (2018) Structure of the mechanically activated ion channel Piezo1. Nature 554:481–6.

Sato K, Pellegrino M, Nakagawa T, Nakagawa T, Vosshall LB, Touhara K (2008) Insect olfactory receptors are heteromeric ligand-gated ion channels. Nature 452:1002–6.

Sato K, Tanaka K, Touhara K (2011) Sugar-regulated cation channel formed by an insect gustatory receptor. Proc Natl Acad Sci U S A 108:11680–5.

Satoh AK, Ready DF (2005) Arrestin1 mediates light-dependent rhodopsin endocytosis and cell survival. Curr Biol 15:1722–33.

Schmidt TM, Chen SK, Hattar S (2011) Intrinsically photosensitive retinal ganglion cells: many subtypes, diverse functions. Trends Neurosci 34:572–80.

Schmidt TM, Alam NM, Chen S, Kofuji P, Li W, Prusky GT, Hattar S (2014) A role for melanopsin in alpha retinal ganglion cells and contrast detection. Neuron 82:781–8.

Schmitz H, Bleckmann H (1998) The photomechanic infrared receptor for the detection of forest fires in the beetle *Melanophila acuminata* (Coleoptera: Buprestidae). J Comp Physiol 182A:647–57.

Schmitz H, Murtz M, Bleckmann H (2000) Responses of the infrared sensilla of *Melanophila acuminata* (Coleoptera: Buprestidae) to monochromatic infrared stimulation. J Comp Physiol 186:543–9.

Schneeweis DM, Schnapf JL (1995) Photovoltage of rods and cones in the macaque retina. Science 268:1053–6.

Schwander M, Kachar B, Muller U (2010) Review series: the cell biology of hearing. J Cell Biol 190:9–20.

Schwartz IR, Pappas GD, Bennett MVL (1975) The fine structure of electrocytes in weakly electric teleosts. J Neurocytol 4:87–114.

Schwarzer A, Schauf H, Bauer PJ (2000) Binding of the cGMP-gated channel to the Na/Ca-K exchanger in rod photoreceptors. J Biol Chem 275:13448–54.

Schwenk K (1994) Why snakes have forked tongues. Science 263:1573–7.

Scott K (2018) Gustatory processing in *Drosophila melanogaster*. Annu Rev Entomol 63:15–30.

Scott K, Zuker CS (1998) Assembly of the *Drosophila* phototransduction cascade into a signalling complex shapes elementary responses. Nature 395:805–8.

Scott K, Becker A, Sun Y, Hardy R, Zuker C (1995) Gq alpha protein function in vivo: genetic dissection of its role in photoreceptor cell physiology. Neuron 15:919–27.

Scott K, Sun Y, Beckingham K, Zuker CS (1997) Calmodulin regulation of *Drosophila* light-activated channels and receptor function mediates termination of the light response in vivo. Cell 91:375–83.

Self T, Mahony M, Fleming J, Walsh J, Brown SD, Steel KP (1998) Shaker-1 mutations reveal roles for myosin VIIA in both development and function of cochlear hair cells. Development 125:557–66.

Senthilan PR, Piepenbrock D, Ovezmyradov G, Nadrowski B, Bechstedt S, Pauls S, Winkler M, Mobius W, Howard J, Gopfert MC (2012) *Drosophila* auditory organ genes and genetic hearing defects. Cell 150:1042–54.

Shapiro MS, Zagotta WN (2000) Structural basis for ligand selectivity of heteromeric olfactory cyclic nucleotide-gated channels. Biophys J 78:2307–20.

Shaw J, Boyd A, House M, Woodward R, Mathes F, Cowin G, Saunders M, Baer B (2015) Magnetic particle-mediated magnetoreception. J R Soc Interface 12:0499.

Shi Y, Radlwimmer FB, Yokoyama S (2001) Molecular genetics and the evolution of ultraviolet vision in vertebrates. Proc Natl Acad Sci U S A 98:11731–6.

Shichida Y, Matsuyama T (2009) Evolution of opsins and phototransduction. Philos Trans R Soc Lond B Biol Sci 364:2881–95.

Shields VD, Hildebrand JG (2001) Recent advances in insect olfaction, specifically regarding the morphology and sensory physiology of antennal sensilla of the female sphinx moth *Manduca sexta*. Microsc Res Tech 55:307–29.

Sigworth FJ, Neher E (1980) Single Na+ channel currents observed in cultured rat muscle cells. Nature 287:447–9.

Silbering AF, Rytz R, Grosjean Y, Abuin L, Ramdya P, Jefferis GS, Benton R (2011) Complementary function and integrated wiring of the evolutionarily distinct *Drosophila* olfactory subsystems. J Neurosci 31:13357–75.

Silbering AF, Bell R, Galizia CG, Benton R (2012) Calcium imaging of odor-evoked responses in the *Drosophila* antennal lobe. J Vis Exp 14(61). pii:2976.

Silva L, Antunes A (2017) Vomeronasal receptors in vertebrates and the evolution of pheromone detection. Annu Rev Anim Biosci 5:353–70.

Sklar PB, Anholt RR, Snyder SH (1986) The odorant-sensitive adenylate cyclase of olfactory receptor cells. Differential stimulation by distinct classes of odorants. J Biol Chem 261:15538–43.

Solessio E, Engbretson GA (1993) Antagonistic chromatic mechanisms in photoreceptors of the parietal eye of lizards. Nature 364:442–5.

Solmsen F (1961) Greek philosophy and the discovery of nerves. Museum Helveticum 18:150–97.

Song K, Wang H, Kamm GB, Pohle J, Reis FC, Heppenstall P, Wende H, Siemens J (2016) The TRPM2 channel is a hypothalamic heat sensor that limits fever and can drive hypothermia. Science 353:1393–8.

Song Y, Cygnar KD, Sagdullaev B, Valley M, Hirsh S, Stephan A, Reisert J, Zhao H (2008) Olfactory CNG channel desensitization by Ca^{2+}/CaM via the B1b subunit affects response termination but not sensitivity to recurring stimulation. Neuron 58:374–86.

Sonoda T, Lee SK, Birnbaumer L, Schmidt TM (2018) Melanopsin phototransduction is repurposed by ipRGC subtypes to shape the function of distinct visual circuits. Neuron 99:754–67.

Soucy ER, Albeanu DF, Fantana AL, Murthy VN, Meister M (2009) Precision and diversity in an odor map on the olfactory bulb. Nat Neurosci 12:210–20.

Spudich EN, Zhang W, Alam M, Spudich JL (1997) Constitutive signaling by the phototaxis receptor sensory rhodopsin II from disruption of its protonated Schiff base-Asp-73 interhelical salt bridge. Proc Natl Acad Sci U S A 94:4960–5.

Spudich JL, Yang CS, Jung KH, Spudich EN (2000) Retinylidene proteins: structures and functions from archaea to humans. Annu Rev Cell Dev Biol 16:365–92.

Stabio ME, Sabbah S, Quattrochi LE, Ilardi MC, Fogerson PM, Leyrer ML, Kim MT, Kim I, Schiel M, Renna JM, Briggman KL, Berson DM (2018) The M5 cell: a color-opponent intrinsically photosensitive retinal ganglion cell. Neuron 97:251.

Staden Hv (1989) Herophilus, the Art of Medicine in Early Alexandria. Cambridge: Cambridge University Press.

Steinbach AB (1974) Transmission from receptor cells to afferent nerve fibers. In: Synaptic Transmission and Neuronal Interaction (Bennett MVL, ed), pp 105–40. New York: Raven.

Steinberg RH, Fisher SK, Anderson DH (1980) Disc morphogenesis in vertebrate photoreceptors. J Comp Neurol 190:501–18.

Stengl M (2010) Pheromone transduction in moths. Front Cell Neurosci 4:133.

Stenkamp RE, Filipek S, Driessen CA, Teller DC, Palczewski K (2002) Crystal structure of rhodopsin: a template for cone visual pigments and other G protein-coupled receptors. Biochim Biophys Acta 1565:168–82.

Stephan AB, Shum EY, Hirsh S, Cygnar KD, Reisert J, Zhao H (2009) ANO2 is the cilial calcium-activated chloride channel that may mediate olfactory amplification. Proc Natl Acad Sci U S A 106:11776–81.

Stephan AB, Tobochnik S, Dibattista M, Wall CM, Reisert J, Zhao H (2012) The Na(+)/Ca(2+) exchanger NCKX4 governs termination and adaptation of the mammalian olfactory response. Nat Neurosci 15:131–7.

Stern J, Chinn K, Bacigalupo J, Lisman J (1982) Distinct lobes of *Limulus* ventral photoreceptors. I. Functional and anatomical properties of lobes revealed by removal of glial cells. J Gen Physiol 80:825–37.

Stiles WS, Crawford BH (1932) Equivalent adaptation levels in localized retinal areas. In: Report of a Joint Discussion on Vision, June 3, 1932, Imperial College of Science, pp 194–211. London: Physical Society.

Stirling CE, Lee A (1980) [3H]ouabain autoradiography of frog retina. J Cell Biol 85:313–24.

Stocker RF (1994) The organization of the chemosensory system in *Drosophila melanogaster*: a review. Cell Tissue Res 275:3–26.

Stocker RF (2001) *Drosophila* as a focus in olfactory research: mapping of olfactory sensilla by fine structure, odor specificity, odorant receptor expression, and central connectivity. Microsc Res Tech 55:284–96.

Stowers L, Kuo TH (2015) Mammalian pheromones: emerging properties and mechanisms of detection. Curr Opin Neurobiol 34:103–9.

Stowers L, Holy TE, Meister M, Dulac C, Koentges G (2002) Loss of sex discrimination and male–male aggression in mice deficient for TRP2. Science 295:1493–500.

Stryer L (1986) Cyclic GMP cascade of vision. Annu Rev Neurosci 9:87–119.

Su CY, Luo DG, Terakita A, Shichida Y, Liao HW, Kazmi MA, Sakmar TP, Yau KW (2006) Parietal-eye phototransduction components and their potential evolutionary implications. Science 311:1617–21.

Sueur J, Windmill JF, Robert D (2006) Tuning the drum: the mechanical basis for frequency discrimination in a Mediterranean cicada. J Exp Biol 209:4115–28.

Suga H, Schmid V, Gehring WJ (2008) Evolution and functional diversity of jellyfish opsins. Curr Biol 18:51–5.

Sutherland EW (1972) Studies on the mechanism of hormone action. Science 177:401–8.

Svoboda K, Yasuda R (2006) Principles of two-photon excitation microscopy and its applications to neuroscience. Neuron 50:823–39.

Syeda R, Florendo MN, Cox CD, Kefauver JM, Santos JS, Martinac B, Patapoutian A (2016) Piezo1 channels are inherently mechanosensitive. Cell Rep 17:1739–46.

Szabo T (1974) Anatomy of the specialized lateral line organs of electroreception. In: Handbook of Sensory Physiology, Vol. III/3, Electroreceptors and Other Specialized Receptors in Lower Vertebrates (Fessard A, ed), pp 13–58. Berlin: Springer.

Szabo T, Fessard A (1974) Physiology of electroreceptors. In: Handbook of Sensory Physiology, Vol. III/3, Electroreceptors and Other Specialized Receptors in Lower Vertebrates (Fessard A, ed), pp 59–124. Berlin: Springer.

Szamier RB, Wachtel AW (1970) Special cutaneous receptor organs of fish. VI. The tuberous and ampullary organs of *Hypopomus*. J Ultrastruct Res 30:450–71.

Takumida M, Yajin K (1996) Scanning electron microscopic observation of the statocyst in the crayfish *Procambarus clarkii* Girard. Auris Nasus Larynx 23:133–9.

Tan CH, McNaughton PA (2016) The TRPM2 ion channel is required for sensitivity to warmth. Nature 536:460–3.

Tao Cheng JH, Hirosawa K, Nakajima Y (1981) Ultrastructure of the crayfish stretch receptor in relation to its function. J Comp Neurol 200:1–21.

Tempel BL, Papazian DM, Schwarz TL, Jan YN, Jan LY (1987) Sequence of a probable potassium channel component encoded at Shaker locus of *Drosophila*. Science 237:770–5.

Ter Hofstede HM, Ratcliffe JM (2016) Evolutionary escalation: the bat-moth arms race. J Exp Biol 219:1589–602.

Terashima SI, Goris RC (1975) Electrophysiology of snake infrared receptors. Prog Neurobiol 2:311–32.

Terashima S, Liang YF (1991) Temperature neurons in the crotaline trigeminal ganglia. J Neurophysiol 66:623–34.

Terashima SI, Goris RC, Katsuki Y (1968) Generator potential of crotaline snake infrared receptor. J Neurophysiol 31:682–8.

Thurm U (1964) Mechanoreceptors in the cuticle of the honey bee: fine structure and stimulus mechanism. Science 145:1063–5.

Thurm U (1965a) An insect mechanoreceptor, part II: receptor potentials. Cold Spring Harb Symp Quant Biol 30:83–94.

Thurm U (1965b) An insect mechanoreceptor, part I: fine structure and adequate stimulus. Cold Spring Harb Symp Quant Biol 30:75–82.

Thurm U (1983) Mechano-electric transduction. In: Biophysics (Hoppe W, Lohmann W, Markl H, Ziegler H, eds), pp 666–71. Berlin: Springer.

Thurm U, Küppers J (1980) Epithelial physiology of insect sensilla. In: Insect Biology in the Future (Locke M, Smith DS, eds), pp 735–63. New York: Academic Press.

Tizzano M, Dvoryanchikov G, Barrows JK, Kim S, Chaudhari N, Finger TE (2008) Expression of Galpha14 in sweet-transducing taste cells of the posterior tongue. BMC Neurosci 9:110.

Tomita T (1965) Electrophysiological study of the mechanisms subserving color coding in the fish retina. Cold Spring Harbor Symposia on Quantitative Biology 30.

Toomey MB, Collins AM, Frederiksen R, Cornwall MC, Timlin JA, Corbo JC (2015) A complex carotenoid palette tunes avian colour vision. J R Soc Interface 12:20150563.

Torre V, Matthews HR, Lamb TD (1986) Role of calcium in regulating the cyclic GMP cascade of phototransduction in retinal rods. Proc Natl Acad Sci U S A 83:7109–13.

Touhara K, Vosshall LB (2009) Sensing odorants and pheromones with chemosensory receptors. Annu Rev Physiol 71:307–32.

Touhara K, Sengoku S, Inaki K, Tsuboi A, Hirono J, Sato T, Sakano H, Haga T (1999) Functional identification and reconstitution of an odorant receptor in single olfactory neurons. Proc Natl Acad Sci U S A 96:4040–5.

Toyoshima C, Kanai R, Cornelius F (2011) First crystal structures of Na$^+$,K$^+$-ATPase: new light on the oldest ion pump. Structure 19:1732–8.

Trautmann A (1982) Curare can open and block ionic channels associated with cholinergic receptors. Nature 298:272–5.

Treiber CD, Salzer MC, Riegler J, Edelman N, Sugar C, Breuss M, Pichler P, Cadiou H, Saunders M, Lythgoe M, Shaw J, Keays DA (2012) Clusters of iron-rich cells in the upper beak of pigeons are macrophages not magneto-sensitive neurons. Nature 484:367–70.

Treiber CD, Salzer M, Breuss M, Ushakova L, Lauwers M, Edelman N, Keays DA (2013) High resolution anatomical mapping confirms the absence of a magnetic sense system in the rostral upper beak of pigeons. Commun Integr Biol 6:e24859.

Troemel ER (1999) Chemosensory signaling in *C. elegans*. Bioessays 21:1011–20.

Trujillo-Cenoz O (1965) Some aspects of the structural organization of the intermediate retina of dipterans. J Ultrastruct Res 13:1–33.

Tsang SH, Burns ME, Calvert PD, Gouras P, Baylor DA, Goff SP, Arshavsky VY (1998) Role for the target enzyme in deactivation of photoreceptor G protein in vivo. Science 282:117–21.

Tsang SH, Woodruff ML, Chen CK, Yamashita CY, Cilluffo MC, Rao AL, Farber DB, Fain GL (2006) GAP-independent termination of photoreceptor light response by excess gamma subunit of the c-GMP-phosphodiesterase. J Neurosci 26:4472–80.

Tskhovrebova LA, Popov VI, Pavlenko VK, Lednev VV (1991) The spatial organization of the cytoskeleton in crayfish stretch receptor. Eur J Cell Biol 56:132–8.

Tsunoda S, Zuker CS (1999) The organization of INAD-signaling complexes by a multivalent PDZ domain protein in *Drosophila* photoreceptor cells ensures sensitivity and speed of signaling. Cell Calcium 26:165–71.

Tu YH, Cooper AJ, Teng B, Chang RB, Artiga DJ, Turner HN, Mulhall EM, Ye W, Smith AD, Liman ER (2018) An evolutionarily conserved gene family encodes proton-selective ion channels. Science 359:1047–50.

Uchida N, Takahashi YK, Tanifuji M, Mori K (2000) Odor maps in the mammalian olfactory bulb: domain organization and odorant structural features. Nat Neurosci 3:1035–43.

Ueno K, Kohatsu S, Clay C, Forte M, Isono K, Kidokoro Y (2006) Gsalpha is involved in sugar perception in *Drosophila melanogaster*. J Neurosci 26:6143–52.

Ukhanov K, Payne R (1997) Rapid coupling of calcium release to depolarization in *Limulus polyphemus* ventral photoreceptors as revealed by microphotolysis and confocal microscopy. J Neurosci 17:1701–9.

Ukhanov K, Ukhanova M, Taylor CW, Payne R (1998) Putative inositol 1,4,5-trisphosphate receptor localized to endoplasmic reticulum in *Limulus* photoreceptors. Neuroscience 86:23–8.

van Netten SM, Khanna SM (1994) Stiffness changes of the cupula associated with the mechanics of hair cells in the fish lateral line. Proc Natl Acad Sci U S A 91:1549–53.

Van Petegem F (2012) Ryanodine receptors: structure and function. J Biol Chem 287:31624–32.

Vandenbeuch A, Clapp TR, Kinnamon SC (2008) Amiloride-sensitive channels in type I fungiform taste cells in mouse. BMC Neurosci 9:1.

Vassar R, Chao SK, Sitcheran R, Nunez JM, Vosshall LB, Axel R (1994) Topographic organization of sensory projections to the olfactory bulb. Cell 79:981–91.

Viana F, de la Pena E, Belmonte C (2002) Specificity of cold thermotransduction is determined by differential ionic channel expression. Nat Neurosci 5:254–60.

Vinos J, Jalink K, Hardy RW, Britt SG, Zuker CS (1997) A G protein-coupled receptor phosphatase required for rhodopsin function. Science 277:687–90.

Vishnivetskiy SA, Raman D, Wei J, Kennedy MJ, Hurley JB, Gurevich VV (2007) Regulation of arrestin binding by rhodopsin phosphorylation level. J Biol Chem 282:32075–83.

Vogt RG, Riddiford LM (1981) Pheromone binding and inactivation by moth antennae. Nature 293:161–3.

Volland S, Hughes LC, Kong C, Burgess BL, Linberg KA, Luna G, Zhou ZH, Fisher SK, Williams DS (2015) Three-dimensional organization of nascent rod outer segment disk membranes. Proc Natl Acad Sci U S A 112:14870–5.

von Békésy G (1960) Experiments in Hearing. New York: McGraw-Hill.

von der Emde G (2001) Electric fields and electroreception: how electrosensory fish perceive their environment. In: Ecology of Sensing (Barth FG, Schmid A, eds). Berlin: Springer.

von der Emde G, Schwarz S, Gomez L, Budelli R, Grant K (1998) Electric fish measure distance in the dark. Nature 395:890–4.

von Frisch K (1967) The Dance Lanaguage and Orientation of Bees. Cambridge, MA: Harvard University Press.

Vondran T, Apel K-H, Schmitz H (1995) The infrared receptor of *Melanophila acuminata* DeGree (Coleoptera: Buprestidae): ultrastructural study of a unique insect thermoreceptor and its possible descent from a hair mechanoreceptor. Tissue Cell 27:645–58.

Vosshall LB (2001) The molecular logic of olfaction in *Drosophila*. Chem Senses 26:207–13.

Vosshall LB, Amrein H, Morozov PS, Rzhetsky A, Axel R (1999) A spatial map of olfactory receptor expression in the *Drosophila* antenna. Cell 96:725–36.

Vosshall LB, Wong AM, Axel R (2000) An olfactory sensory map in the fly brain. Cell 102:147–59.

Vriens J, Nilius B, Voets T (2014) Peripheral thermosensation in mammals. Nat Rev Neurosci 15:573–89.

Wade NJ (1998) A Natural History of Vision. Cambridge, MA: MIT Press.

Wald G (1959) Life and light. Sci Am 201(4):92–108.

Wald G (1968) The molecular basis of visual excitation. Nature 219:800–7.

Walker MT, Brown RL, Cronin TW, Robinson PR (2008) Photochemistry of retinal chromophore in mouse melanopsin. Proc Natl Acad Sci U S A 105:8861–5.

Walker RG, Willingham AT, Zuker CS (2000) A *Drosophila* mechanosensory transduction channel. Science 287:2229–34.

Walsh T, Walsh V, Vreugde S, Hertzano R, Shahin H, Haika S, Lee MK, Kanaan M, King MC, Avraham KB (2002) From flies' eyes to our ears: mutations in a human class III myosin cause progressive nonsyndromic hearing loss DFNB30. Proc Natl Acad Sci U S A 99:7518–23.

Waltman B (1966) Electrical properties and fine structure of the ampulllary canals of Lorenzini. Acta Physiol Scand 66(Suppl 264):1–60.

Wang JS, Kefalov VJ (2009) An alternative pathway mediates the mouse and human cone visual cycle. Curr Biol 19:1665–9.

Wang JS, Kefalov VJ (2011) The cone-specific visual cycle. Prog Retin Eye Res 30:115–28.

Wang JS, Estevez ME, Cornwall MC, Kefalov VJ (2009) Intra-retinal visual cycle required for rapid and complete cone dark adaptation. Nat Neurosci 12:295–302.

Wang X, Wang T, Jiao Y, von Lintig J, Montell C (2010) Requirement for an enzymatic visual cycle in *Drosophila*. Curr Biol 20:93–102.

Wang Y, Wright NJ, Guo H, Xie Z, Svoboda K, Malinow R, Smith DP, Zhong Y (2001) Genetic manipulation of the odor-evoked distributed neural activity in the *Drosophila* mushroom body. Neuron 29:267–76.

Wang Y, Deshpande M, Payne R (2002) 2-aminoethoxydiphenyl borate inhibits phototransduction and blocks voltage-gated potassium channels in *Limulus* ventral photoreceptors. Cell Calcium 32(4):209–16.

Warren JW, Proske U (1968) Infrared receptors in the facial pits of the Australian python *Morelia spilotes*. Science 159:439–41.

Wei J, Zhao AZ, Chan GC, Baker LP, Impey S, Beavo JA, Storm DR (1998) Phosphorylation and inhibition of olfactory adenylyl cyclase by CaM kinase II in neurons: a mechanism for attenuation of olfactory signals. Neuron 21:495–504.

Weiss LA, Dahanukar A, Kwon JY, Banerjee D, Carlson JR (2011) The molecular and cellular basis of bitter taste in *Drosophila*. Neuron 69:258–72.

Weitz D, Ficek N, Kremmer E, Bauer PJ, Kaupp UB (2002) Subunit stoichiometry of the CNG channel of rod photoreceptors. Neuron 36:881–9.

Wersäll J, Bagger-Sjöbäck D (1974) Morphology of the vestibular sense organ. In: Handbook of Sensory Physiology, Vol. VI/1, Vestibular System Part 1: Basic Mechanisms (Kornhuber HH, ed), pp 124–70. Berlin: Springer.

Wettschureck N, Offermanns S (2005) Mammalian G proteins and their cell type specific functions. Physiol Rev 85:1159–204.

White RH, Lord E (1975) Diminution and enlargement of the mosquito rhabdom in light and darkness. J Gen Physiol 65:583–98.

White RH, Sundeen CD (1967) The effect of light and light deprivation upon the ultrastructure of the larval mosquito eye. I. Polyribosomes and endoplasmic reticulum. J Exp Zool 164:461–77.

Wicher D, Schafer R, Bauernfeind R, Stensmyr MC, Heller R, Heinemann SH, Hansson BS (2008) *Drosophila* odorant receptors are both ligand-gated and cyclic-nucleotide-activated cation channels. Nature 452:1007–11.

Wiersma CA (1967) Invertebrate Nervous Systems. Chicago: University of Chicago Press.

Wilden U, Kuhn H (1982) Light-dependent phosphorylation of rhodopsin: number of phosphorylation sites. Biochemistry 21:3014–22.

Wilden U, Hall SW, Kuhn H (1986) Phosphodiesterase activation by photoexcited rhodopsin is quenched when rhodopsin is phosphorylated and binds the intrinsic 48-kDa protein of rod outer segments. Proc Natl Acad Sci U S A 83:1174–8.

Willis WD, Jr., Coggeshall RE (1991) Sensory Mechanisms in the Spinal Cord. New York: Plenum.

Wilson RI (2013) Early olfactory processing in *Drosophila*: mechanisms and principles. Annu Rev Neurosci 36:217–41.

Wilson VJ, Melvill Jones G (1979) Mammalian Vestibular Physiology. New York: Plenum.

Wiltschko R, Wiltschko W (1995) Magnetic Orientation in Animals. Berlin: Springer.

Wiltschko W, Wiltschko R (1972) Magnetic compass of European robins. Science 176:62–4.

Wiltschko W, Wiltschko R (2001) The geomagnetic field and its role in directional orientation. In: Ecology of Sensing (Barth FG, Schmid A, eds), pp 289–312. Berlin: Springer.

Wiltschko W, Wiltschko R (2005) Magnetic orientation and magnetoreception in birds and other animals. J Comp Physiol A Neuroethol Sens Neural Behav Physiol 191:675–93.

Wiltschko W, Munro U, Ford H, Wiltschko R (1993) Red light disrupts magnetic orientation of migratory birds. Nature 364:525–7.

Wollmuth LP, Hille B (1992) Ionic selectivity of I_h channels of rod photoreceptors in tiger salamanders. J Gen Physiol 100:749–65.

Wong GT, Gannon KS, Margolskee RF (1996) Transduction of bitter and sweet taste by gustducin. Nature 381:796–800.

Wong ST, Trinh K, Hacker B, Chan GC, Lowe G, Gaggar A, Xia Z, Gold GH, Storm DR (2000) Disruption of the type III adenylyl cyclase gene leads to peripheral and behavioral anosmia in transgenic mice. Neuron 27:487–97.

Woo SH, Ranade S, Weyer AD, Dubin AE, Baba Y, Qiu Z, Petrus M, Miyamoto T, Reddy K, Lumpkin EA, Stucky CL, Patapoutian A (2014) Piezo2 is required for Merkel-cell mechanotransduction. Nature 509:622–6.

Woo SH, Lukacs V, de Nooij JC, Zaytseva D, Criddle CR, Francisco A, Jessell TM, Wilkinson KA, Patapoutian A (2015) Piezo2 is the principal mechanotransduction channel for proprioception. Nat Neurosci 18:1756–62.

Woodruff ML, Sampath AP, Matthews HR, Krasnoperova NV, Lem J, Fain GL (2002) Measurement of cytoplasmic calcium concentration in the rods of wild- type and transducin knock-out mice. J Physiol 542:843–54.

Wu A, Dvoryanchikov G, Pereira E, Chaudhari N, Roper SD (2015) Breadth of tuning in taste afferent neurons varies with stimulus strength. Nat Commun 6:8171.

Wu Z, Grillet N, Zhao B, Cunningham C, Harkins-Perry S, Coste B, Ranade S, Zebarjadi N, Beurg M, Fettiplace R, Patapoutian A, Mueller U (2017) Mechanosensory hair cells express two molecularly distinct mechanotransduction channels. Nat Neurosci 20:24–33.

Xiong WH, Solessio EC, Yau KW (1998) An unusual cGMP pathway underlying depolarizing light response of the vertebrate parietal-eye photoreceptor. Nat Neurosci 1:359–65.

Xu J, Dodd RL, Makino CL, Simon MI, Baylor DA, Chen J (1997) Prolonged photoresponses in transgenic mouse rods lacking arrestin. Nature 389:505–9.

Xu P, Atkinson R, Jones DN, Smith DP (2005) *Drosophila* OBP LUSH is required for activity of pheromone-sensitive neurons. Neuron 45:193–200.

Xu W, Zhang HJ, Anderson A (2012) A sugar gustatory receptor identified from the foregut of cotton bollworm *Helicoverpa armigera*. J Chem Ecol 38:1513–20.

Xu XZ, Chien F, Butler A, Salkoff L, Montell C (2000) TRPgamma, a *Drosophila* TRP-related subunit, forms a regulated cation channel with TRPL. Neuron 26:647–57.

Xue T, Do MT, Riccio A, Jiang Z, Hsieh J, Wang HC, Merbs SL, Welsbie DS, Yoshioka T, Weissgerber P, Stolz S, Flockerzi V, Freichel M, Simon MI, Clapham DE, Yau KW (2011) Melanopsin signalling in mammalian iris and retina. Nature 479:67–73.

Yack JE (2004) The structure and function of auditory chordotonal organs in insects. Microsc Res Tech 63:315–37.

Yan C, Zhao AZ, Bentley JK, Loughney K, Ferguson K, Beavo JA (1995) Molecular cloning and characterization of a calmodulin-dependent phosphodiesterase enriched in olfactory sensory neurons. Proc Natl Acad Sci U S A 92:9677–81.

Yan Z, Zhang W, He Y, Gorczyca D, Xiang Y, Cheng LE, Meltzer S, Jan LY, Jan YN (2013) *Drosophila* NOMPC is a mechanotransduction channel subunit for gentle-touch sensation. Nature 493:221–5.

Yang C, Delay RJ (2010) Calcium-activated chloride current amplifies the response to urine in mouse vomeronasal sensory neurons. J Gen Physiol 135:3–13.

Yang F, Cui Y, Wang K, Zheng J (2010) Thermosensitive TRP channel pore turret is part of the temperature activation pathway. Proc Natl Acad Sci U S A 107:7083–8.

Yang RB, Garbers DL (1997) Two eye guanylyl cyclases are expressed in the same photoreceptor cells and form homomers in preference to heteromers. J Biol Chem 272:13738–42.

Yang RB, Foster DC, Garbers DL, Fulle HJ (1995) Two membrane forms of guanylyl cyclase found in the eye. Proc Natl Acad Sci U S A 92:602–6.

Yao CA, Carlson JR (2010) Role of G-proteins in odor-sensing and CO_2-sensing neurons in *Drosophila*. J Neurosci 30:4562–72.

Yao CA, Ignell R, Carlson JR (2005) Chemosensory coding by neurons in the coeloconic sensilla of the *Drosophila* antenna. J Neurosci 25:8359–67.

Yao J, Liu B, Qin F (2011) Modular thermal sensors in temperature-gated transient receptor potential (TRP) channels. Proc Natl Acad Sci U S A 108:11109–14.

Yarmolinsky DA, Zuker CS, Ryba NJ (2009) Common sense about taste: from mammals to insects. Cell 139:234–44.

Yau KW, Hardie RC (2009) Phototransduction motifs and variations. Cell 139:246–64.

Yau KW, Nakatani K (1984) Electrogenic Na–Ca exchange in retinal rod outer segment. Nature 311:661–3.

Yau KW, Matthews G, Baylor DA (1979) Thermal activation of the visual transduction mechanism in retinal rods. Nature 279:806–7.

Ye W, Chang RB, Bushman JD, Tu YH, Mulhall EM, Wilson CE, Cooper AJ, Chick WS, Hill-Eubanks DC, Nelson MT, Kinnamon SC, Liman ER (2016) The K^+ channel KIR2.1 functions in tandem with proton influx to mediate sour taste transduction. Proc Natl Acad Sci U S A 113:E229–38.

Yeandle S (1958) Evidence of quantized slow potentials in the eye of *Limulus*. Am J Ophthalmol 46:82–7.

Yorozu S, Wong A, Fischer BJ, Dankert H, Kernan MJ, Kamikouchi A, Ito K, Anderson DJ (2009) Distinct sensory representations of wind and near-field sound in the *Drosophila* brain. Nature 458:201–5.

Yoshida R, Shigemura N, Sanematsu K, Yasumatsu K, Ishizuka S, Ninomiya Y (2006) Taste responsiveness of fungiform taste cells with action potentials. J Neurophysiol 96:3088–95.

Young RW (1976) Visual cells and the concept of renewal. Invest Ophthalmol Vis Sci 15:700–25.

Yue, WWS, Silverman, D, Ren, X, Frederiksen, R, Sakai, K, Yamashita, T, Shichida, Y, Cornwall, MC, Chen, J, and

Yau, KW (2019). Elementary response triggered by transducin in retinal rods. Proc Natl Acad Sci U S A 116: 5144-5153.

Zagotta WN, Siegelbaum SA (1996) Structure and function of cyclic nucleotide-gated channels. Annu Rev Neurosci 19:235–63.

Zakon HH (1984) The electroreceptive periphery. In: Electroreception (Bullock TH, Heiligenberg W, eds), pp 103–56. New York: John Wiley and Sons.

Zanazzi G, Matthews G (2009) The molecular architecture of ribbon presynaptic terminals. Molec Neurobiol 39:130–48.

Zhang C, Yan J, Chen Y, Chen C, Zhang K, Huang X (2014) The olfactory signal transduction for attractive odorants in *Caenorhabditis elegans*. Biotechnol Adv 32:290–5.

Zhang HJ, Anderson AR, Trowell SC, Luo AR, Xiang ZH, Xia QY (2011) Topological and functional characterization of an insect gustatory receptor. PLoS One 6:e24111.

Zhang T, Cao LH, Kumar S, Enemchukwu NO, Zhang N, Lambert A, Zhao X, Jones A, Wang S, Dennis EM, Fnu A, Ham S, Rainier J, Yau KW, Fu Y (2016) Dimerization of visual pigments in vivo. Proc Natl Acad Sci U S A 113:9093–8.

Zhang W, Cheng LE, Kittelmann M, Li J, Petkovic M, Cheng T, Jin P, Guo Z, Gopfert MC, Jan LY, Jan YN (2015) Ankyrin repeats convey force to gate the NOMPC mechanotransduction channel. Cell 162:1391–403.

Zhang X, Wensel TG, Kraft TW (2003a) GTPase regulators and photoresponses in cones of the eastern chipmunk. J Neurosci 23:1287–97.

Zhang Y, Hoon MA, Chandrashekar J, Mueller KL, Cook B, Wu D, Zuker CS, Ryba NJ (2003b) Coding of sweet, bitter, and umami tastes: different receptor cells sharing similar signaling pathways. Cell 112:293–301.

Zhang YV, Ni J, Montell C (2013) The molecular basis for attractive salt-taste coding in *Drosophila*. Science 340:1334–8.

Zhang Z, Zhao Z, Margolskee R, Liman E (2007) The transduction channel TRPM5 is gated by intracellular calcium in taste cells. J Neurosci 27:5777–86.

Zhao GQ, Zhang Y, Hoon MA, Chandrashekar J, Erlenbach I, Ryba NJ, Zuker CS (2003) The receptors for mammalian sweet and umami taste. Cell 115:255–66.

Zhao H, Ivic L, Otaki JM, Hashimoto M, Mikoshiba K, Firestein S (1998) Functional expression of a mammalian odorant receptor. Science 279:237–42.

Zhao Q, Zhou H, Chi S, Wang Y, Wang J, Geng J, Wu K, Liu W, Zhang T, Dong MQ, Wang J, Li X, Xiao B (2018) Structure and mechanogating mechanism of the Piezo1 channel. Nature 554:487–92.

Zhao X, Pack W, Khan NW, Wong KY (2016) Prolonged inner retinal photoreception depends on the visual retinoid cycle. J Neurosci 36:4209–17.

Zhao Y, Yamoah EN, Gillespie PG (1996) Regeneration of broken tip links and restoration of mechanical transduction in hair cells. Proc Natl Acad Sci U S A 93:15469–74.

Zheng J, Ma L (2014) Structure and function of the thermo-TRP channel pore. Curr Top Membr 74:233–57.

Zheng J, Shen W, He DZ, Long KB, Madison LD, Dallos P (2000) Prestin is the motor protein of cochlear outer hair cells. Nature 405:149–55.

Zheng J, Trudeau MC, Zagotta WN (2002) Rod cyclic nucleotide-gated channels have a stoichiometry of three CNGA1 subunits and one CNGB1 subunit. Neuron 36:891–6.

Zhong H, Molday LL, Molday RS, Yau KW (2002) The heteromeric cyclic nucleotide-gated channel adopts a 3A:1B stoichiometry. Nature 420:193–8.

Zhong L, Hwang RY, Tracey WD (2010) Pickpocket is a DEG/ENaC protein required for mechanical nociception in *Drosophila* larvae. Curr Biol 20:429–34.

Zimmerman A, Bai L, Ginty DD (2014) The gentle touch receptors of mammalian skin. Science 346:950–4.

Zipser B, Bennett MVL (1976) Interaction of electrosensory and electromotor signals in lateral line lobe of a Mormyrid fish. J Neurophysiol 39(4):713–21.

Zou Z, Horowitz LF, Montmayeur JP, Snapper S, Buck LB (2001) Genetic tracing reveals a stereotyped sensory map in the olfactory cortex. Nature 414:173–9.

Index

ABCR/Rim protein, 195
Accessory olfactory bulbs, 155–156
Accessory olfactory epithelium, 154–156
Actin, 19–20, 23, 105–106, 109
Adaptation, 17, 31, 114–116, 133–136, 145, 148–149, 181, 193–195, 209–213, 218, 220
Adenylyl cyclase, 57–59, 62–64, 75, 144–145
Adrian, Edgar Douglas, Lord, 4, 174
Aequorin, 67, 190
Amiloride, 79–81, 170–172
Amphioxus photoreceptors, 68–69
Ampullary receptors, 224, 226–230
  anatomy, 224–226
  physiology, 227–230
Ankyrin repeats, 43, 87–91, 190
Anoctamin 2, 146–7. *See also* TMEM16
Arachidonic acid, 57, 64–65, 187
Aristotle, 1–2, 144
Arrestin, 58, 60, 62, 145, 148, 187, 193, 204–205, 208
Atomic force microscope, 188–189
Avoiding reaction, 76–77

β-Adrenergic receptor, 60–61, 96–98, 177, 179
β-Adrenergic receptor kinase (βARK), 148
Bacterorhodopsin (sensory rhodopsin of archaeobacteria), 178–179, 181
Basilar membrane, 1, 4, 121–127
Basilar papilla, 127–130
Békésy, Georg von, 4, 121, 124
Benign positional vertigo, 120–121
Bitter detection (taste)
  insects, 163
  mammals, 165–169, 174–6
Bombykol, 34–36

Ca^{2+}-activated chloride channel, 110, 146–148, 156, 199–200, 221. *See also* TMEM16

Ca^{2+}-activated potassium channel (K$_{Ca}$), 128–130, 228–229
Ca^{2+} ATPase, 48, 65, 200
Ca^{2+} channels (voltage-gated), 40, 67, 128–130, 141, 175, 197, 199–200, 227–229
Ca^{2+} indicator dyes. *See* Fluorescent Ca^{2+} indicator dyes
Cadherin, 108–109, 111
Caged Ca^{2+}. *See* DM-nitrophen
Cajal. *See* Ramón y Cajal
Calmodulin, 23, 41, 43, 59, 68, 148–149, 190, 193, 211
CaM kinase, 59, 145, 148
Campaniform sensillum, 85–86, 90–91
Capsaicin receptor, 42–44. *See also* TRPV1
Capsule (electroreception), 224–226
Channels. *See* Ion channels and under specific channel types
Che proteins, 133–136, 181
Chemoreceptor (cell), 6, 27, 139, 160–161, 217, 221
Chemotaxis, 132–136
  adaptation, 133–136
  Che proteins, 133–136, 181
  flagellum and flagellar motor, 132–134
  methyl-accepting chemotaxis proteins (MCPs), 133–136
  phosphorylation, 133–136, 181
Chorda tympani nerve, 164, 171, 175–176
Chordotonal organ, 85–87, 99–104
Chromophore. *See* Retinal
Cilium and Cilia, 20–22, 27, 30, 36, 50–55, 73, 76–77, 85–87, 102, 104, 106, 116–117, 119, 144–147, 181, 195, 225–226
Circadian clock, 73, 213, 215
Circumferential endings, 94, 97
Cochlea, 118, 121–127
  See also Organ of Corti
Cochlear amplifier, 124–127

Coding
  audition, 5, 121–122
  olfaction, 141–142, 149–151
  taste, 174–176
  touch, 92–94
Cold spots, 217–218
Columella, 127–128
Command potential, 53–54
Cone photoreceptors
  adaptation, 209–211
  anatomy, 195
  cyclic-nucleotide-gated channels, 198–200
  inner-segment channels, 200
  lamellae, 21, 195, 197
  photopigments, 180
  PDE6, 197
  pigment regeneration, 211–212
  recovery (response turnoff), 205–208
  responses, 6, 203–204
  sensitivity, 203–204
  synapse and synaptic transmitter release, 196–197
  transduction, 197–208
Crayfish stretch receptor, 81–85
  adaptation, 83–85
  anatomy, 81–82
  single channels, 84–85
  tension clamp, 84
  voltage clamp (current response), 84
  voltage response, 82–83
CRISPR/Cas9, 13
Crista ampullaris, 117–120
Cryogenic electron microscopy (cryo-EM), 10–11, 14–16, 43–44, 68–71, 90–91, 138–140
Cryptochromes, 237–238
Cupula, 116–119
Cyclic-nucleotide-gated channels
  discovery, 68–70
  gating, 70–73
  ion permeability, 68–73, 182, 200–202

Cyclic-nucleotide-gated channels (*Cont.*)
  olfactory (vertebrate), 144–145, 149, 154
  photoreceptor (vertebrate), 198–202, 204, 211
  structure, 68, 70–71

Dark current, 200
Decartes, René, 2
DEG/ENaC channels, 79–81, 91, 163–164, 170–172
De Rerum Natura (of Lucretius), 1
Diacylglycerol (DAG), 23, 57–59, 64, 185, 187–189
Diacylglycerol kinase (DAG kinase), 187–188
Disks (of rod photoreceptors)
  anatomy, 195–196
  formation, 21–22
  renewal, 25–26
  role in transduction, 197, 203–204
DM-nitrophen (caged Ca²⁺), 193–194
Dorsal root ganglion (DRG), 91, 94, 218, 220
Driving force, 44, 48–49, 52–55, 111, 115, 123–124, 201

Effector enzymes, 57–58, 62–65
  adenylyl cyclase, 62–63, 144–145
  cyclic-nucleotide phosphodiesterase, 22, 57, 59, 62–64, 73–75, 145, 148–149, 181, 197–198, 204–205, 208–209, 211, 213
  phospholipase A₂ (PLA₂), 63–65
  phospholipase C (PLC), 23, 58–59, 62–65, 68, 164–169, 181, 185, 187–189, 193, 195, 213
Electric organs, 230–233
Electrical resonance, 127–130
Electrocytes, 230–231
Electrolocation, 230–233
Electrophorus (electric eel), 233
Electroreception, 224–233
  ampullary organs, 224–230
  tuberous receptors, 230–233
Ellipsoid body, 195
Endocochlear potential (endolymphatic potential), 123–124
Endolymph, 122–124
Enteroreceptors, 217
Epley and Semont maneuver, 121
Equivalent background light, 213
Erasistratus of Ceos, 2

FAD and FADH, 237–238
Feedback amplifier, 53
Flagella, 132–134, 181, 234

Fluorescent Ca²⁺ indicator dyes, 67–69, 107, 142, 147, 207. *See also* GCaMP
Free radicals, 237–238
Funnel cage (Emlen), 235–236

G protein (heterotrimeric), 57, 61–62. *See also* Gα₀, Gα_olf, Gα_q/11, Gustducin, Transducin
G-protein receptor kinase (GRK), 58, 60, 145.
  See also β-Adrenergic receptor kinase, Rhodopsin kinase
G-protein-coupled receptor, 60–61.
  activation, 60
  inactivation, 58, 60–61
  structure, 60
  See also under individual receptor types, e.g. β-Adrenergic receptor, ORs (vertebrate), Rhodopsin, *T1R*, *T2R*.
Gα₀, 154–155, 181–182
Gα_olf, 144–145
Gα_q/11, 58, 62, 64, 181, 185, 213. *See also* NORPA
Galen, 2
Gating compliance (of hair cell), 112–113
Gβ5L, 205, 208–209
GCaMP, 101, 103, 107, 174–176
Gene cloning, 9–10
Gene expression, 10
Geniculate ganglion, 174–176
GHK (Goldman-Hodgkin-Katz) equation. *See* Goldman voltage equation
Glomerulus (in olfaction)
  insects, 141–142
  vertebrates, 151–153
Glossopharyngeal nerve, 164, 171
Glutamic-acid-rich protein (GARP), 199
Goldman voltage equation, 48–49, 55–56, 168–169, 181
GR proteins, 161–163
Green fluorescent protein (GFP), 96–97, 143, 150–152, 157
GTPase-activating proteins (GAPs), 62, 205, 208–209
Guanylyl cyclase
  membrane bound, 58–60, 154
  vertebrate photoreceptor, 205–208, 209–210
Guanylyl-cyclase-activating proteins (GCAPs), 206–211
Gustation
  C. elegans, 160
  fish, 160

insects, 160–164
mammals, 164–175
Gustducin (Gα_gus), 62, 73, 75, 166–167, 169

Hair cells, 49–56, 104–130
  adaptation, 114–116
  anatomy, 50, 104–106, 108, 117–120
  Ca²⁺ and hair cell function, 56, 107, 109–111, 115–116, 128–130
  channels, 110–111
  electrical resonance, 127–130
  gating, 51–55, 104
  inner hair cells, 121–3
  ion selectivity of channels, 56, 107
  myosin, 109, 115–116
  outer hair cells, 124–127
  prestin, 125–126
  synapses, 105–106, 121, 127
  tip links, 106–109
  voltage clamping, 51–55
  voltage response, 49–51, 104
Hair plate sensillum, 85–88
Harmonin (hair-cell protein), 109
Hartline, Haldan Keffer, 4, 6, 28, 181, 183
HCN channels (I_h), 199–202
Helmholtz, Hermann von, 2–4
Herophilus of Chalcedon, 2
Holding potential. *See* Command potential
Hooke's law, 112–113
Humain Ibn Is-Hâq, 2
Hydropathy analysis, 10, 12–15, 43
5-Hydroxytryptamine (5-HT, serotonin), 144, 164

I_h channels. *See* HCN channels
Inactive (TRP channel gene), 103–104
INAD (inactivation-no-after-potential D protein), 22–23, 187, 190, 193
Inclination compass, 234–235, 237–238
Infrared (IR) detection, 221–225
Inner hair cells, 121–3
Inositol 1,4,5-trisphosphate.
  See IP₃
Insect mechanoreceptors, 85–91
  mutations, 88
  NOMPC, 88–91
  response, 87–88
  structure, 85–87
  type I, 85–87, 91
  type II, 87, 91
Intracellular recording, 6, 50, 82–83, 99–100, 181, 202
Inter-photoreceptor retinol binding protein (IRBP), 211–212

Intrinsically photosensitive retinal ganglion cells (ipRGCs), 213–215
Inward rectifier potassium channel. *See* $K_{IR}2.1$
Ion channels
gating, 15, 18–19, 40–42, 43–44, 51–55, 88–91, 106–109, 138–140, 166–169, 221
structure, 13–16, 42–44, 68–71, 90–91, 138–140.
See also under individual channel types
Ion homeostasis, 47–48, 205–208
Ion selectivity (of channels), 55–56.
See also under individual channel types
Ionotropic sensory transduction, 18–19
IP_3 (inositol 1,4,5-trisphosphate) and IP_3 receptors, 57–59, 65–68, 166–169, 183, 185, 188, 190, 192–195
IR proteins (ionotropic receptors), 139–140, 163

Jacobson's organ. *See* Vomeronasal organ
Jamming avoidance, 233
Johnston's organ, 101–104
anatomy, 101–102
channels proteins, 104
sensory neurons, 101–103

$K_{CS}A$ (potassium channel of *Streptomyces lividens*), 38–40
Kinocilium (of hair cell), 50–55, 104–106, 108, 112–117, 119, 225–226
$K_{IR}2.1$ (potassium channel), 172–174
$K_V2.1$ (potassium channel), 38–40, 44–45, 68

Labeled lines, 5–6, 174–175
Laminar cells, 31–32
Lamprey, 203–204
Lanceolate endings, 93–94, 97
Lateral line organs, 99, 104, 116–117
Limulus (horseshoe crab), 33, 183–184, 188–195
Linolenic acid, 187
Lisman's Law, 2
Lucretius, 1–2
LUSH, 138

Magnetic field (of Earth), 234–235
Magnetite, 234–235, 237

Magnetoreception, 233–239
bacteria, 233–235
migrating birds, 235–237
mechanism, 237–238
Magnetotaxis, 233–235
Mammalian touch receptors, 91–97
anatomy, 91–92, 93–94
responses, 92–93, 94–97
See also Meissner corpuscle, Lanceolate ending, Merkel cell, Ruffini organ, Pacinian corpuscle
MEC (mechanosensory) proteins, 81
Mechanoreceptor, 4, 6, 15, 18, 20, 26–27, 29, 31, 42, 79–97, 101, 104, 110, 113, 160, 190, 217, 221
Meissner corpuscle (RA fibers), 91–93
Melanophila (beetles), 221–222
Melanopsin, 213–214
Membrane potentials, 44–49
Merkel cells (SAI fibers), 91–97
anatomy, 94–95
patch-clamp recordings, 94–96
Piezo2, 96
transmitter, 96–97
Metabotropic sensory transduction, 18–19, 57–60.
See also G proteins, G-protein-coupled receptors, Effector enzymes, Second messengers
Metarhodopsin, 181, 211–213
Metal microelectrodes, 5–6
Methyl-accepting chemotaxis proteins (MCPs), 133–136
Microtubule cells, 78–79
Microtubules, 20–21, 78–82, 85–86, 88–91, 105–106
Microvilli, 19–21, 23, 25–27, 30, 36, 49–50, 56, 69, 85, 90–91, 104–106, 130, 143, 154, 156,165, 181, 183, 185, 188, 215, 226, 230
MRO_1 and MRO_2. *See* Crayfish stretch receptor
$M_{SC}L$, 78
$M_{SC}S$, 78
Müller cells, 212
Myosin, 20, 23, 109, 115–116, 187, 193

Na^+-K^+ ATPase, 44–49, 123–124, 195, 199–200
Na^+/K^+-Ca^{2+} transport protein, 48, 146, 148, 200, 205–208
Nanchung (TRP channel gene), 104
NCKX. *See* Na^+/K^+-Ca^{2+} transport protein
Necklace glomeruli, 154
Neher, Erwin, 7
Nernst equation, 45–49, 55

Noise (in sensory transduction), 34, 180, 185
NompA, 91, 104
NOMPC (TRPN1), 87–91, 104
ankyrin repeats, 87–91
mutations, 87–88
structure, 90–91
NORPA (*Drosophila* $G\alpha_q$), 181, 185, 187–188, 213

Odorant binding proteins
insects, 137–138
vertebrates, 143
Ohm's law, 52, 68, 227
Oil droplets, 238
Olfaction
insects, 135–142
vertebrates, 142–156
silk moth, 34–36
Olfactory bulbs, 151–153. *See also* Accessory olfactory bulbs
Olfactory epithelium (primary)
adaptation, 148–149
anatomy, 142–143
coding, 149–151
receptor proteins, 143–144
transduction, 144–148
See also Accessory olfactory epithelium
Ommatidium, 185–186, 188–189
Opn4. *See* Melanopsin
Opsin. *See* Rhodopsin
Orai channels, 67, 221
Organ of Corti, 121–127
anatomy, 121–123
basilar membrane, 121–123
tectorial membrane, 121, 123
ORs (odorant receptors)
insects, 138–140
vertebrates, 143–144, 151, 154–156
Otoconia, 120
Otolith organs, 50–51, 117, 119–120
Otolithic membrane, 50, 120
Otop1 (gene for proton channel), 172–174
Outer hair cells, 105, 108, 121–127

P loop (P region), 37–40, 43, 45, 68–70, 88, 193
Pacinian corpuscle, 26–28, 92–93, 97
Paramecium, 76–78
Parietal eye, 73–75
cyclic-nucleotide-gated channels, 73, 75
photoreceptors, 73–74
responses, 73–74
structure, 73–74
transduction, 73, 75

Patch-clamp recording
  discovery, 7
  inside-out (excised), 7–8, 42–43, 68–70, 71–72, 73, 75, 144, 167–168, 173, 188–191, 220
  on-cell (cell-attached), 7–8, 84–85, 144–146, 188–191
  outside-out, 8–9, 89, 138–139, 162–163
  whole-cell, 8, 13–14, 54–55, 80, 90, 96, 103, 107, 110–111, 114–115, 148–150, 162–163, 169–170, 173, 189–194, 218–220
Peripherin, 195
Pfaffmann, Carl, 174–175
Pheromones and pheromone detection, 34–36, 136–137, 139–140, 154–156
Phosphodiesterase (PDE), 22, 57, 59, 62–64, 73–75, 145, 148–149, 181, 197–198, 204–205, 208–209, 211, 213
Phospholipase C (PLC and PLCβ), 23, 58–59, 62–65, 164–169, 185, 187–189, 193, 195, 213
Photoisomerization, 178–180, 211–213
Photoreceptors
  Amphioxus, 68–69
  Deinopis subrufa (spider), 24–25
  Drosophila, 33–34, 183, 185–195
  lamprey, 203–204
  Limulus (horseshoe crab), 33, 183–184, 188–195
  lizard parietal eye, 73–75
  primate (monkey), 203
  scallop (*Pecten*), 27–30, 181–182
  vertebrates (rods and cones). *See* Rod photoreceptors, Cone photoreceptors
Pickpocket genes (*ppks*), 91, 163–164
Piezo proteins, 13–16, 18, 42–44, 91, 96–98, 110
  discovery, 13
  gating, 15–16
  hydropathy analysis, 13–15
  Piezo 1, 13–14, 15–16
  Piezo 2, 13–14, 96–98, 110
  structure, 14–15
Pigment regeneration (photoreceptors), 211–213
Pit organs, 222–225
  anatomy, 222–223
  optics, 224–225
  sensitivity, 223–224
Pit vipers, 222–225
PKC. *See* Protein kinase C
Plato, 1

Polycystic-kidney-disease-like channels (PDK1L2 and PDK2L1), 172–173
Potassium channel
  permeability, 39–40
  structure, 37–40
  See also Ca^{2+}-activated potassium channel, $K_{CS}A$, $K_{IR}2.1$, $K_V2.1$
Prestin, 125–126
Primary olfactory epithelium. *See* Olfactory epithelium
Primary receptor cell, 31, 140. *See also* Crayfish stretch receptor, Olfaction, Taste receptor cell (insect), Thermoreception, Touch reception
Proprioception, 217
Protein kinase C (PKC), 23, 59, 65, 187, 189, 193
Protocadherin, 108–109
Proton channel, 172–174
Purinergic receptors (receptors for ATP), 164, 175
Pythons (IR detection), 222

Quantum bumps, 33–34, 187–188

R9AP, 205, 208–209
Ramón y Cajal, 2
Rattlesnakes, 222–225
Receptor expression enhancing protein (REEP1), 151
Receptor-transporting proteins (RTP1 and RTP2), 151
Recoverin, 211
Renewal of sensory membrane, 22–26
  Deinopis subrufa (spider), 24–25
  vertebrate rod, 25–26
Resonance. *See* Electrical resonance
Retinal, 4, 60, 178–180, 195, 211–213
Retinal G-protein-coupled receptor opsin (RGR opsin), 212
RGS9, 205, 208–209, 213
Rhabdomere, 23, 183, 185–186, 188–190
Rhodopsin, 4, 9–10, 22–23, 25, 32–34, 60–61, 73, 151, 179–181, 193–195, 197–198, 204–205, 211–213
Rhodopsin kinase, 204–205, 211
Ribeye, 197
Rod photoreceptor
  adaptation, 209–213
  anatomy, 25–26, 195–197
  Ca^{2+} concentration, 205–211
  cyclic-nucleotide-gated channels, 68–73, 198–202, 204, 211
  disks, 25–26, 195–196, 203–204
  guanylyl cyclase and GCAPs, 206–211

inner segment, 25, 195–197, 199–200
inner segment channels, 199–200
outer segment, 21–22, 25–26, 29, 68–70, 195–197
PDE6, 62–64, 181, 197–198, 204–205, 208–209
pigment regeneration, 211–213
recovery (response turnoff), 204–209
renewal of sensory membrane, 25–26
synaptic terminal (spherule), 196–197
transducin, 22, 62, 73, 166, 197–198, 204–205
transduction, 197–208
Rom-1, 195
RPE65, 211
Ruffini organs (SAII fibers?), 92–3
Ryanodine receptors, 66–67

Sacculus, 50–51, 104–106, 112–114, 117, 120
Sakmann, Bert, 7
Salicylate, 125
Salt bridge, 179–180
Salt detection (taste)
  insects, 161, 163
  mammals, 168–172, 174–6
Sans (hair-cell protein), 109
Sarco/endoplasmic reticulum Ca^{2+} ATPase (SERCA), 65–67
Scaffolding proteins, 21–23
Scallop (*Pecten*), 27–30, 181–182
Schiff base, 179–180
Scolopale and scolopidium, 85–87, 99–104
Second messengers, 57–59
  Ca^{2+}, 57–59, 65–69, 145–148, 166–169, 185, 188, 190, 205–211, 216, 220
  cAMP, 57–59, 62–65, 142, 144–146, 149, 154
  cGMP, 40–41, 57–60, 63–65, 68–75, 154, 181–182, 188–191, 197–200, 205–206, 209
  DAG, 23, 57–59, 64, 185, 187–189
  IP$_3$, 57–59, 65–68, 166–169, 183, 185, 188, 190, 192–195
  nitric oxide (NO), 59
Secondary receptor cell, 31, 105, 196. *See also* Electroreception, Hair cells, Photoreceptors (except amphioxus and scallop), Taste receptor cell (vertebrate), Synaptic ribbon

Semicircular canals, 116–120

Sensillum, 26–27, 29, 34–36, 85–88, 90–91, 99–101, 136–137, 140–141, 160–163

Sensillum liquor, 137

Sensory membrane, 19–21

Sex pheromone detection by male moth, 34–36

Shaker (K+ channel gene and protein), 37–38, 40

Signalplex, 22, 187, 190

Silk moth (*Bombyx mori*), 34–36

Single-photon responses
  Drosophila photoreceptor, 33–34, 197–198.
  vertebrate rod, 32–33, 197, 203–204
  See also Quantum bumps

Skate, 224–226, 228–230

Sodium-potassium ATPase. *See* Na+-K+ ATPase

Sour detection, 172–176

Statocysts, 27–29

STIM proteins, 66–67

Streptomyces lividens K+ channel. *See* K$_{CS}$A

Stretch-sensitive channels, 13–16, 42. *See also* MEC proteins, M$_{SC}$L, M$_{SC}$S, NOMPC, Piezo proteins, TMC1 and TMC2

Stria vascularis, 122–124

Suction-electrode recording, 33, 147–148, 201–203, 205–211

Superchiasmatic nucleus, 213

Superior olivary complex, 105, 127

Sweet detection (taste)
  insects, 161–163
  mammals, 165–169, 174–6

Synaptic ribbon, 31–32, 105, 196–197, 225–226, 230

T body, 31–32

T1R genes (*Tas1R*), 60, 165–169, 174

T2R genes (*Tas2R*), 165–166, 167, 169, 174

TAARs (trace amine-associated receptors), 144, 154

Taste. *See* Gustation

Taste bud, 164–176
  anatomy, 164–165

Taste papilla, 164–165

Taste receptor cell
  fly, 26–27, 160–163
  vertebrate, 164–174

Taste receptor proteins. *See* DEG/ENaC, GR proteins, IR proteins, K$_{IR}$2.1, *T1R*, *T2R*

Tectorial membrane, 121, 123, 126, 127

Temperature-sensitive channels, 42–44, 218–222

Thermoreception and thermoreceptors, 42–44, 91, 217–221

Tip links, 106–112, 115–116

TMC1 and TMC2 (hair-cell channel proteins), 110–111

TMEM16, 110, 146–147

Touch domes, 94, 96–97

Touch reception
  bacteria, 78
  C. elegans, 78–81
  insects, 85–91
  mammals, 91–97
  Paramecium, 76–78

Transducin, 22, 62, 73, 166, 197–198, 204–205

Transducisome, 22, 187, 190

Transfection, 10–11

TRP (*Drosophila* light-activated channel), 187, 190–193, 195

TRP family of ion channels, 42–45, 54, 87–91, 97, 103–104, 130, 156, 164, 166–169, 172, 179, 187, 190–193, 195, 213, 218–222, 224, 239

TRPC2, 156

TRPL (*Drosophila* light-activated channel), 187, 190–192

TRPM5, 164–169

TRPM8, 218–221

TRPN1. *See* NOMPC

TRPV1, 42–44, 221, 222

Tsien, Roger, 67

Tuberous receptors, 230–233
  anatomy, 230
  physiology, 232–233

Tumble, 132–136, 181

Turtle ear, 127–130

Tympanal organs, 99–101

Umami, 159, 164–169, 172, 174–6

Usher proteins and syndrome, 109

Utriculus, 117–118, 120, 128

V1Rs (vomeronasal receptors), 154–155

V2Rs (vomeronasal receptors), 154–155

Vampire bats, 222

Vestibular system, 50–55, 104, 110, 114, 116–121, 124
  otolith organs (sacculus and utriculus), 50–51, 104–106, 112–114, 120
  semicircular canals, 116–120

Voltage clamp, 8, 10, 23, 51–55. *See also* Patch-clamp recording, whole-cell

Vomeronasal organ, 19, 154–156, 166
  receptors proteins, 154–156
  transduction, 154

Wald, George, 4

Warm spots, 217–218

Water detection (insect taste), 163–164

X-ray crystallography, 10, 38–40, 60–61

Zugunruhe, 236